STREATHAM NOW & THEN AGAIN

by

John W Brown

With best wishes
John
XMAS 2011

LOCAL HISTORY PUBLICATIONS

316 GREEN LANE, STREATHAM, LONDON SW16 3AS

Published by
Local History Publications
316 Green Lane
Streatham
London SW16 3AS
Copyright 2011 John W Brown
ISBN 978 1 873520 82 6

All rights reserved. No part of this publication may be reproduced, stored in a retrieval system, or transmitted, in any form, or by any means, electronic, mechanical, photocopying, recording or otherwise, without the prior permission of the publisher and the copyright holders.

ACKNOWLEDGEMENTS

My thanks and appreciation to the following for the use of their photographs and illustrations: Brian Bloice, David Clark, Colin Crocker, Tony Fletcher, Graham and Marion Gower, Andrew Hadden, Sarah Hawes, Kevin Kelly, John Kinsman, Local History Publications, London South Bank University, St. Leonard's Church Archives and The Streatham Society.

DEDICATION

This publication is dedicated to Keith Holdaway, a member of the Streatham Society Local History Group and archivist and leader of the Norwood Society Local History Group. Keith was an early pioneer in promoting interest in local history through his talks, walks and the many exhibitions and displays he has organised and mounted. He is always generous in sharing his knowledge and his extensive local history collection of old photographs, postcards and ephemera. He has also compiled a comprehensive contemporary photographic record of Norwood which will be a valuable resource to future generations of local history researchers.

CONTENTS

Introduction 3	142-144 Wellfield Road
SOUTH STREATHAM 5	St. Paul's Chapel & St. Peter's Hall 27
The Lahore Kebab House 6	Sunnyhill School 28
Lidl Supermarket 7	**CENTRAL STREATHAM** 29
The Pied Bull 8	Streatham Green 30
Beehive Coffee Tavern 9	Bedford House 31
Bank Parade 10	Pratts and Payne 32
Hambly Mansions 11	75-89 Streatham High Road 33
394 Streatham High Road 12	Carpet Right 34
GREYHOUND LANE AND STREATHAM VALE 13	Norfolk House Road 35
Sanders House 14	Belle Vue Terrace 36
Pathfield Road 15	Tesco Express 37
The Railway Hotel 16	Norwich House 38
Streatham Common Station 17	**STREATHAM HILL** 39
Holy Redeemer Church 18	Daily Fresh Foods 40
Crumb's Bakery, 132 Streatham Vale 19	45 Streatham Hill 41
Monti's Day Nursery, 6 Lilian Road 20	Telford Court 42
STREATHAM VILLAGE 21	Telford Parade Mansion 43
48-50 Sunnyhill Road 22	Wavertree Court 44
Leigham Arms 23	Brixton Hill Bus Depot 45
Wellfield Centre 24	Streatham Hill Reservoir 46
The Old Smithy 25	Christchurch House 47
54 Wellfield Road 26	Brixton Hill United Reform Church 48

STREATHAM NOW & THEN AGAIN

INTRODUCTION

One of the most difficult tasks in compiling the first volume of *Streatham Now and Then* was in selecting the old photographs to be featured in the book. Many wonderful old views had to be omitted for want of space.

With the production of this second volume I have been able to include many of these pictures, some of which have never been published before.

This book follows the same route through Streatham as its predecessor, journeying along Streatham High Road from the southern boundary of the town to the northern end of Streatham Hill.

Along the way we explore the history of some of the buildings we encounter, many of which can trace their origins back to the 19th century, and some, like the Pied Bull public house by Streatham Common (see page 9), have a lifespan that now crosses four centuries.

Even the apparently mundane and uninteresting of buildings, such as the Carpet Right store at 130-132 Streatham High Road (see page 34), have a fascinating tale to tell as it originated as one of Streatham's earliest cinemas - the Streatham Picture Theatre commonly known as *The Golden Domes*.

Some of the town's historic sites remain hidden from public view, as with the Streatham Hill Reservoir (see page 46). Originally constructed in 1832, when it covered an area of over 3 acres, it was subsequently doubled in size and now stores 4.3 million gallons of water.

Other locations have national significance, such as the Daily Fresh Foods store at 113-117 Streatham Hill (see page 40), as it was here that the first self-service supermarket in Britain opened on 26th November 1951.

Two detours along the route enable us to explore other areas of our town. The first takes us along Greyhound Lane and Streatham Vale. This is an ancient trackway which led to an isolated and deserted part of Streatham parish known as Lonesome. Streatham Common Railway Station and bridge marks the change of name of this street with Greyhound Lane being mainly developed in the late 19th century and the Streatham Vale Estate being laid out after the First World War.

We also discover the fascinating history of Sunnyhill and Wellfield Roads, an area known today as Streatham Village. Terraces of old Victorian cottages line these streets where are also to be found shops, chapels, a school, an old smithy and a pub - all contributing to the village atmosphere of the area.

Over the many years I have been researching the history of Streatham I have discovered that almost every part of our town has its own intriguing story to tell.

For those wishing to learn more about Streatham's fascinating past a warm welcome awaits them at the monthly meetings of the Local History Group of the Streatham Society. These are held at 8pm on the first Monday of the month at the Woodlawns Centre at 16 Leigham Court Road, Streatham, SW16.

I hope you will find this second exploration of Streatham Now & Then of interest and I would be pleased to hear from readers of their recollections of the area in days gone by.

John W Brown
316 Green Lane,
Streatham,
London SW16 3AS.

SOUTH STREATHAM

The hamlets of South Streatham, centred at the top of Greyhound Lane opposite Streatham Common, and Lower Streatham, located on the High Road between Green Lane and the River Graveney, were among the last parts of Streatham to be fully developed.

These areas remained semi-rural, with fields, pasture and market gardens, up until the 1920s when post-First World War development claimed the last vestiges of Streatham's countryside for house building.

At this time the hamlets still retained a number of ancient buildings dating back to the 17th, 18th and early 19th centuries. However, with redevelopment and Wandsworth council's inter-war slum clearance programme, many of the older properties were demolished to make way for new, modern buildings.

Two ancient cottages which suffered this fate were Waterloo House and Woplin's Cottage, in Lower Streatham. Had they survived they would have given the area a glimpse of its rural past and would have provided a picturesque scene far superior to that now offered by the buildings which replaced them.

Be that as it may, we are fortunate that one of the area's oldest buildings, the Pied Bull public house, dating from at least 1736, still survives. Although much altered over the centuries, local residents have supped ale in this old inn for almost 300 years.

Some of the buildings erected in the Victorian era also remain to show the transformation of Streatham from a small Surrey town into a bustling suburb of South London. Such is the case with Bank Parade and Hambly Mansions opposite Streatham Common. One only has to view the upper floors of these buildings to see that they are beautifully designed to provide an impressive backdrop to the Common.

STREATHAM NOW & THEN AGAIN

Bill Watts and the White Aces perform at the Sussex in the late 1930s

LAHORE KEBAB HOUSE

The Lahore Kebab House stands on the northern bank of the River Graveney at 668 Streatham High Road.

The building was erected in 1937 as the Sussex public house. This originated as on old beer shop known as the Sussex Arms which stood on the southern corner of Colmer Road and Streatham High Road.

In the early 1870s it was run by George Staples and here local residents could buy a range of bottled ales or bring their own jugs which would be filled from the pumps with a variety of draft beers.

Between the wars it was a common sight to see young children carrying jugs full of beer from the shop back home to their thirsty parents.

When the beer shop was closed a door at the side of the building, in Colmer Road, gained access to the rear of the premises where a knock on the back window would enable trade to be undertaken out of hours beyond the knowledge of the local constabulary.

The old beer shop closed in 1937 when the new pub known simply as *The Sussex* opened.

In the years leading up to the Second World War the pub was a popular Saturday night dance venue. Bill Watts and the White Aces often played their syncopated rhythm here. Bill lived at 127 Ellison Road and was a talented trumpet player who was ably assisted by his friends Wallie and Archie on their saxophones.

In the 1950s and 60s the pub was well known for its restaurant which was a popular lunch-time dining venue.

Among the past patrons of the Sussex are Steve Brookstein, the 2004 winner of the ITV talent show *The X Factor*, and the famous author Kingsley Amis, who, as a youth, lived in nearby Norbury.

In 1993 the pub was refurbished at a cost of £140,000 and reopened on 28th March as the Brass Farthing.

When this venture failed the old name was restored but this failed to guarantee its future and the last pint was pulled here in February 2002.

Steve Brookstein

By January 2005 the upper part of the building had been converted into a complex of 12 one and two bed apartments known as Sussex Court.

The former pub premises, on the ground floor, was transformed into a large, open-plan restaurant, originally trading as The Feeding Tree and now as the Original Lahore Kebab House, a branch of the very first Pakistani restaurant to be established in Britain, at Whitechapel, in the early 1970s.

The Sussex Tavern 1973

LIDL SUPERMARKET

Lidl Supermarket occupies a site where once stood the oldest cottage in Streatham, known as Woplin's Cottage, after the name of the family who once lived there.

Daniel Woplin came to Streatham from Essex in 1825 and originally lived in one of four small cottages that occupied the triangle of land at the junction of the High Road and Green Lane. There his son Charles was born in 1826 and his grandson, Charles (Jnr), was born in 1864.

Charles (Snr) later moved to the ancient cottage at the junction of Hermitage Lane and the High Road, and when he died his son, Charles Jnr, continued to live there.

My grandmother use to make Charlie his Sunday dinner and, as a young boy, my father would take this from his home in Danbrook Road to Woplin's cottage. He remembered the building as being very dark with tiny rooms.

The small, single storey, wooden cottage was painted with black tar to help keep the weather out, on top of which sat an old rusty-red tiled roof.

Woplin's Cottage in the mid 1930s.

Either side of its bright green door were two small windows with wooden shutters.

On entering the house you descended down a step as the floor was below street level, the earthen surface having been worn away over the centuries until it was as hard as concrete.

The four small rooms were barely 6 feet 6 inches high and all had whitewashed ceilings to help make the rooms brighter.

The kitchen floor was covered with old flagstones and a closed-in bake-oven wound its way up behind one of the two chimneys in the room.

Over the open hearth was a bar on which hooks hung on which bacon would be smoked over the open fire.

Water for the family came from a well in the small back yard and the tiny cottage was lit by oil lamps until gas was eventually connected to the property.

The origins of this ancient dwelling are lost in the antiquity of time but a large old beam in the roof had a date of 16?? carved on it, the last two digits being concealed and therefore unreadable.

This surviving remnant of 17th century Streatham was demolished in 1936 when a development of single-storey lock-up shops, called Commodore Parade, were erected on the site.

These in their turn were torn down in 2006 to make way for a Lidl supermarket which opened here on 25th February 2010.

Commondore Parade in the 1970s

THE PIED BULL

The Pied Bull has been serving ale for at least the past 275 years.

The earliest known reference to the inn occurs in 1736 when the parish was fined 2/6d (12½p) for not repairing the highway between "the Pyed Bull and the Smith's shop" opposite St. Leonard's Church.

The pub we see today probably dates from the 1750s and was possibly rebuilt to accommodate the large number of visitors then coming to Streatham to take the waters at the mineral wells at the top of Streatham Common.

In August 1773 a strange suicide took place here when, after drinking some wine, a gentleman ordered a bed for the night insisting that he have clean sheets.

In the morning the landlord, thinking the man had overslept, knocked on his door at nine, and again at eleven, but received no answer. At twelve he got a constable and they broke into the room where they found the man with a pistol in his hand with which he had shot himself. By his side was a second gun charged and ready to fire.

In the 1790s the inn was described as being "a genteel public house" and was one of the coaching inns on the London to Brighton Road.

Young and Bainbridge first leased the premises in 1832 and it has remained in their ownership ever since.

In the 19th century inquests were regularly held here, particularly those for dead bodies discovered in the undergrowth on Streatham Common or fished out of the ponds there.

In 1881 considerable excitement occurred when a balloon flying from Crystal Palace went out of control whilst trying to land on Streatham Common and was blown towards the pub, almost crashing into the roof before landing safely in a field off nearby Kempshott Road.

The Pied Bull was the original headquarters of the Herne Hill Harriers from their founding in 1889 to 1899, and the runners would change here before setting off on their cross country runs across the local countryside.

The building was extensively renovated in the 1930s when the original ground-floor bow window was removed and the tiled frontage we see today was installed.

The Pied Bull won the Evening Standard Pub of the Year competition in 1973 and comedian Jimmy Edwards unveiled the plaque which commemorates this achievement which can be seen above the entrance to the old Billiard Hall.

A troop of horsemen, believed to be a recruiting party from the Hampshire Carabiniers Yeomanry Cavalry, visit the Pied Bull in the spring of 1901. The troop had it headquarters in Cherry Orchard Road, Croydon, where they had a drill hall, Morris Tube range, saddlery stores and club rooms. Riders from the troop toured the area seeking recruits to replace those members of the company serving in South Africa during the Boer War.

STREATHAM NOW & THEN AGAIN

BEEHIVE COFFEE TAVERN

At the Streatham High Road entrance to Sainsbury's supermarket stands the offices of Henry Hughes & Hughes, Solicitors.

This impressive red brick building started life as the Beehive Coffee Tavern.

Situated strategically next to the Pied Bull Public House, the building was designed by George and Peto for the Immanuel Church Temperance Association. It was built in 1878 by R & E Smith of Balham, at a cost of £3,838, and was opened by Lord Cairns, the Lord Chancellor.

The Rev. Stenton Eardley, the vicar of Immanuel, was concerned that workers from the adjacent P B Cows rubber factory spent much of their time (and money!) in the Pied Bull.

He hoped that by providing an alternative place for the men to spend their time he could reduce the amount of drunkenness among the workers.

The Beehive contained a temperance bar, a club and games room, a reading room, a large hall and 8 small bedrooms.

It provided a teetotal environment for the consumption of meals and acted as a venue for local concerts and social events.

The main hall could accommodate 500 people and was used for lantern lectures. The first recorded public screening of a film is reported to have taken place at the Beehive in 1897.

The main hall was also the venue for Lady Key's annual New Year's Day dinner for the poor of South Streatham which she hosted each year from the 1850s until her death in 1901.

Her husband, Sir Kingsmill Key, would carve 40lb joints of the finest beef into thick slices for up to 100 paupers to enjoy and Lady Key, assisted by her family and servants, would pile the pensioners plates with generous portions of boiled mutton, turnips, potatoes and savoys, followed by plum pudding; all washed down with large mugs of hot, sweet tea or coffee

After this sumptuous feast brief addresses would be given by the vicars of Immanuel and St. Andrew's churches, and then each pauper would receive a gift of blankets or cloth.

The Beehive was last used as a coffee tavern around 1909, since when it has served as a school gym, a builders merchants, a church bookshop and as offices for P.B.Cows rubber factory. After a complete renovation in 1987-9 it has been occupied by Henry Hughes & Hughes, solicitors, and a nursery.

Rev. Stenton Eardley

The Beehive Coffee Tavern in the late 1890s with the manager, Henry Hawkins, standing in the doorway and his niece, Deborah Clara Lovell, who helped in the tavern standing on the right.

BANK PARADE

Bank Parade, 426-450 Streatham High Road, was built in 1890 as the date in the brickwork over the door at the junction of the High Road and Greyhound Lane testifies.

Standing four stories high, this was the largest building in South Streatham. Shops occupied the ground floor with flats on the three floors above, each of which enjoyed magnificent views over Streatham Common.

Bank Parade was built on the site of two large houses which faced the Common called *High Elms* and *Westwell*.

High Elms was a long, low, white house and was occupied by a number of notable residents. John Creswicke lived here from 1739-61 and Warren Hastings (1732-1818), the first Governor General of India, stayed here when young. It was also the home of Sir John Milton, Assistant Accountant General to the War Office in 1871-8, and Sir Henry Muggeridge, Sheriff of London.

Westwell was a large, red-brick house and was so called for being the house to the west of the original Streatham Mineral Wells located at the top of Streatham Common on the site now occupied by the Rookery gardens.

For many years Westwell was the home of Andrew Hamilton (1793-1853) who donated part of his garden for the building of Immanuel Church in 1854. The Hamiltons also donated £1,000 towards the cost of constructing the church and in return for their generosity the family were allowed to appoint the first three vicars of Immanuel after which the patronage was transferred to the Rector of Streatham.

Another notable resident of Westwell was Major General Charles Frederick Parkinson, Staff Officer of Pensioners, who lived here in 1880-81.

Bank Parade was built by William Marriage of Croydon, to the designs of Tooly & Son of 66 Canon Street.

For many years Bank Parade was regarded as South Streatham's premier shopping centre with many of the areas leading traders operating from the building.

Notable among these was Thomas Edwin Park, a local butcher, who was a keen collector of landscape pictures, amassing a collection of more than 150 works, most of them by the local artist Holland Tringham.

The building originally had a glass canopy, supported on ornate cast iron pillars, which provided a covered walkway in front of the shops which was removed in 1939. Bank Parade remains today one of the most impressive buildings in South Streatham and is a fine example of late Victorian architecture.

HAMBLY MANSIONS

Hambly Mansions stands on the western side of the High Road, opposite Streatham Common, on a site once occupied by Hambly House, a large 18th century mansion.

It was named after William Hambly, a native of Cornwall, who came to Streatham in the late 17th century. The building is thought to have been erected around 1710 and provided a comfortable country residence for William, his wife Sarah and their son Peter. William's descendants continued to reside at Hambly House until at least 1784 when they moved to Ashstead in Surrey.

Hambly House Academy in 1820

The property was subsequently acquired by John Henry Lord of Camberwell who, together with his wife Elizabeth, established the Hambly House Academy here. Following his death in January 1787 John Wilson became Master of the School and under his headship the Academy flourished by specialising in preparing young boys for careers in banking and commerce.

On 14th March 1831 John Wilson died, aged 79, and he was interred in his family's large tomb in St. Leonard's graveyard. His son, Samuel Mitchell Wilson, took over the running of the school but his period as head was short lived as he died in December 1835 aged 34. His widow, Mary, was left to run the school which could not have been an easy task for a woman to undertake at that time. No doubt she was ably assisted by one of the school masters, Joseph Cox Dear, as on July 15th 1837 she married him and he became the new Master of the Academy.

The school continued in existence until the 1870s when it was demolished and Hambly Mansions were erected on the site in 1877. The mansions originally comprised three houses which were built for P B Cow Senior, the owner of the local rubber factory which occupied the site now covered by the Sainsbury Supermarket at Streatham Common.

Hambly Mansions were designed by the local architect, Sir Ernest George RA (1839-1922), who was President of the Royal Institute of British Architects from 1908 to 1910 and lived in Ryecroft Road.

In the early 1900s the ground floor of the building was converted into shops and Potter Perrin have sold bathroom and kitchen fittings here since the early 1970s. Few passers-by who gaze at the wide variety of sanitary fittings now displayed in the shop windows here can be aware that this was once the site of one of Streatham's leading public schools.

Architect's drawing of Hambly Mansions in 1877

STREATHAM NOW & THEN AGAIN

394 STREATHAM HIGH ROAD

Standing on the northern corner of the High Road junction with Natal Road is No. 394 Streatham High Road. It was built around 1859 for John Powell whose family resided here until the late 1870s.

In the mid 1880s the house was converted into a girls' school run by Miss Maria Cater in 1887 at which time the house was called Netherwood. From 1891 Miss Mary Withiel was in charge of the school and the house was then known as Airedale.

In 1900 St. Helen's Girls' School moved here. This School was founded in the early 1890s by Miss Edith Salmon and her younger sister, Christine, and was originally located at 14 Bournevale Road before moving to 12 Stanthorpe Road c1893.

The junior, or kindergarten, branch of the School was situated close by at No. 145 Hopton Road.

In January 1908, after a long illness, Edith Salmon died. Her obituary described her as "one of the most erudite schoolmistresses in the district ... No one ever left the school without carrying with them fond memories of the happy days spent within, and without its walls, and the kind, thoughtful and generous principal in whom love was a first virtue."

St. Helen's School in the early 1900s

Following Edith's death, her sister, Christine, known affectionately as Chrissie, took over the running of the establishment, and under her guidance the school continued to flourish.

The school played an active part in raising funds for comforts for the troops in the First World War and it frequently performed in local entertainments.

In 1916 the school entertained sixty wounded soldiers in the Baptist Church Hall in Natal Road providing songs and musical turns as well as serving refreshments.

Ida Lupino

After the First World War St. Helen's continued to prosper and once again the need arose for larger accommodation. This led to the acquisition of the former vicarage of Immanuel Church, at No. 22 Streatham Common Northside, where the school moved in 1929 and remained until it closed in 1965.

One of the school's most famous former pupils was the actress Ida Lupino, who lived in nearby Leigham Court Road.

After St. Helen's School vacated 394 Streatham High Road in 1929 the front of the premises was rebuilt to accommodate shops at ground floor level with the upper floors converted into flats.

This building is scheduled for demolition to make way for the Streatham Hub development and a new ice rink and swimming pool will be erected on the site.

St. Helen's School Junior Class (left) and a lesson in the school garden in the early 1900s (right).

STREATHAM NOW & THEN AGAIN

GREYHOUND LANE AND STREATHAM VALE

Greyhound Lane is one of the ancient trackways of Streatham. It led to an area of deserted wasteland on the parish border with Mitcham called Lonesome - a name that indicated its remote and isolated position.

Before it was developed, Greyhound Lane was no more than a muddy track through the fields.

By the early 1700s a small cluster of cottages had been erected at the eastern end of Greyhound Lane, opposite Streatham Common. These were mainly occupied by farm labourers who worked in the surrounding fields.

At the centre of this hamlet stood the Greyhound Inn, where those living at Lonesome used to call for their letters as postmen would not journey down the lane.

Up to the time of the development of the Streatham Vale Estate in the 1920s, the main inhabitants of Lonesome were occasional nomadic groups of gypsies and tramps.

The area was treated with much caution by the locals and few would journey alone at night beyond Streatham Common Station.

The station opened in 1862 and was originally known as Greyhound Lane Station, its name being changed to Streatham Common Station in 1870.

By the 1880s the impact of the railway began to be felt when roads of large terraced houses were built on the surrounding farmland at the eastern end of the lane.

Greyhound Lane Farm was one of the last of Streatham's ancient rural holdings to succumb to house building and by the 1930s the surburbanisation of the area was complete with the development of the Streatham Vale Estate at the western end of the Lane.

STREATHAM NOW & THEN AGAIN

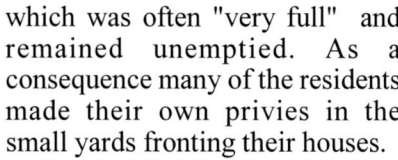

Above and below: Greyhound Square in the early 1930s

SANDERS HOUSE

Sanders House was erected at the western end of Greyhound Lane by Wandsworth Council in 1935/6. It was named after Alderman Sidney Sanders (1871-1942), a prominent Streatham councillor and local businessman.

In 1902 he, together with William Metherell, opened a jewellery shop at 187 Streatham High Road. This business flourished and through acquisition and merger grew to a chain of 127 jewellery shops which traded under the name of James Walker.

Although Sanders House is associated with the glamour of the jewellery industry, the buildings which formerly occupied this site had far less prestigious connections.

Here once stood 20, small two-storeyed terraced properties, arranged to form three sides of a square which opened onto Greyhound Lane. At the centre of the square stood the tiny yards belonging to each house, which were often strewn with rubbish and with washing dangling from a jungle of clothes lines.

Known as Greyhound Square its origins dated back to the 18th century when a few small cottages were built here to accommodate local farm labourers.

Around 1819 the square of brick-built buildings was formed. These were probably erected to accommodate some of the workers employed at the nearby Streatham Silk Mill which was built by Stephen Wilson around 1820. The mill survives today as the coffee shop and offices at Sainsbury's supermarket, situated opposite Streatham Common.

Greyhound Square was so badly built and maintained that the buildings quickly deteriorated into a slum and by 1831 the Vestry Clerk reported that it was "almost impossible to describe the filth" there.

The buildings lacked ventilation and only had a single "privy" which was often "very full" and remained unemptied. As a consequence many of the residents made their own privies in the small yards fronting their houses.

By 1854 things had not improved and a public health report complained about the "filthy cesspool" in the corner of the Square from which an open drain carried the discharge off down Greyhound Lane.

Throughout the 19th century these small dwellings provided cramped, insanitary homes for itinerant labourers and factory workers and a large number of unemployed men who lodged there.

By the early 1930s Greyhound Square was widely considered to be one of the worse areas of sub-standard housing in Streatham. As a result Wandsworth Council compulsorily purchased the site, and neighbouring properties, and Sanders House was erected in their place as part of the council's slum clearance programme.

STREATHAM NOW & THEN AGAIN

PATHFIELD ROAD

Pathfield Road was laid out on an orchard which formed part of Greyhound Lane Farm. The ancient wooden farm house, with adjoining thatched barn, was situated just beyond the old co-op hall on the western junction of Pathfield Road and Greyhound Lane.

The farm was developed from the late 1700's and in 1836 was taken over by Zachariah Grout after which it became known as Grouts Farm. It eventually contained 180 acres producing vegetables, root crops and grain, which were sold in London markets.

When the harvest had been gathered in it was the custom for Mr. Grout to host a "Harvest Home" to which all the farm workers would be invited as a reward for their hard labours in bringing in the crops.

Large poles would be erected in front of the farmhouse on which would be draped tarpaulin to form a small booth. This would be decorated with large pots of flowers to provide a festive setting for the feast.

Farmer Grout would sit at the head of a long, wooden table where he would carve a massive 50lb joint of boiled beef. At the other end of the table was a 40lb rump of roast beef with huge haunches of roast and boiled mutton in the centre. The rest of the table would be covered with large bowls of vegetables and loaves of freshly baked bread.

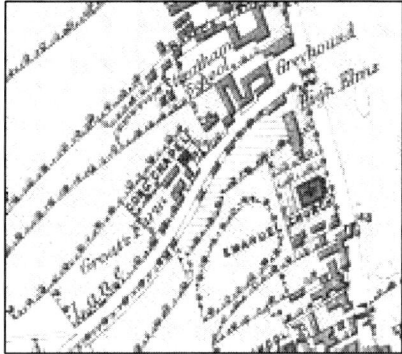

1861 Map showing Grout's Farm

After all those gathered had their fill of meat and veg they would feast on plum pudding with brandy sauce, all washed down with an abundant supply of ale.

At the conclusion of the meal tobacco and cigars would pass down the table. Farmer Grout would then take out his long Churchwarden pipe which he would suck, although it never contained any tobacco as he was a non-smoker!

Once the dishes had been cleared away preparations would be made for dancing and singing which would continue on well into the night.

When Zachariah died in 1853, his son took over the farm and on his death, in 1875, the grandson, Joseph Grout, continued farming here.

In 1888, following the death of Joseph's widow, Thomas Kidner, a local butcher, took over 43 acres of the farm on which he grazed his dairy herd. The remainder of the land became Moses Miller's market garden. By the end of the 19th century the farm had disappeared as all the surrounding fields had been covered with housing.

Left: Mr Grout and his daughter at the doorway to their old farm house, situated off Greyhound Lane, shown below.

THE RAILWAY HOTEL

The Railway Hotel stands on the eastern junction of Ellison Road and Greyhound Lane and dates from the late 1860s.

The building was started by a Mr Blenkarn who probably anticipated an influx of residents to the area following the opening of the nearby railway station in 1862. However, when this did not occur he abandoned the building and it was not finally completed until 1879 when Richard Dootson became the first publican here.

For many years the pub was known locally as Laceby's, after Coulson Laceby who took over the licence in March 1888 and was a long serving publican here. He was also Streatham's senior member on Wandsworth Council being first elected in 1901.

Coulson had very humble origins; his mother and father died when he was a boy and he then joined the railway as an engine cleaner. Through hard work he eventually became an engine driver and drove one of the first electric trains to run on London Underground.

After leaving the railway he purchased a pub in Islington which burnt down. But he recovered and rebuilt the business and at the time of his death owned pubs in Battersea, Earlsfield, and other parts of Wandsworth, as well as the Railway Hotel where he was based for many years until his son took over the license.

Coulson Laceby

Railway Hotel in the early 1900s

In late Victorian and Edwardian times the Railway Hotel had a bad reputation for the fights which occurred there. The public bar was mainly frequented by the hard-drinking working classes of the neighbourhood. Many such men were used to solving their disputes with their fists rather than the eloquence of a well-reasoned argument.

In 1905 George Newman had an argument with Herbert Cooper. Shouting out "I'm Hot Stuff" Newman then smashed a beer glass in Cooper's face. Cooper's friends moved to restrain Newman and within seconds the whole pub was the scene of one gigantic fight.

In 1915 Albert Gordon was asked to leave the pub and when he refused a fight ensued in which Coulson's son was bitten on the leg - so deep was the bite he was unable to walk. A police constable's face was so badly disfigured in ther brawl he was unable to give evidence at the following court case.

Coulson Laceby died in June 1922 aged 77 and was buried in Ely, Cambridgeshire. With his passing the locality lost a much loved publican and a highly respected councillor.

STREATHAM COMMON STATION

On 1st December 1862 the London Brighton and South Coast Railway (LB&SC) opened Greyhound Lane Station here. It was not until 1870 that the name was changed to Streatham Common Station as few travellers knew where Greyhound Lane was, so isolated a spot was it at that time.

Within 6 months of the station opening it was the scene of one of the worse train accidents in London.

On 29th May 1863, a troop train carrying a detachment of the Grenadier Guards was derailed on the curve of the track leading to the station.

As it ploughed down the embankment the carriages catapulted over the engine which exploded propelling the dome 175 yards across the fields to land in Ellison Road.

As a result of the accident 4 people were killed, including two Guardsmen, and 59 were injured of which 36 were soldiers.

This was the first rail accident involving a troop train and Queen Victoria sent a telegram of condolence to the Guards.

In February 1893 the station was to experience an accident of a different nature when a new Pullman car was on its trial run between London and Brighton. All went well until it reached Streatham Common Station when, passing under Greyhound Lane bridge, it lost 10 shining copper ventilators on the roof of the carriages due to contact with the bridge which is one of the lowest on the line.

The train continued on its journey unaware of its loss until it pulled into Brighton Station and the driver noticed the damage to his brand new train. Meanwhile the staff at Streatham Common were busy collecting the mangled ventilators from the line before other trains were due to pass under the bridge.

Streatham Common Station in the 1890s

The original station was demolished in 1902 when the much larger station we see today was erected on the site.

During the Zeppelin raid on Streatham on the night of the 23rd September 1916 a bomb fell nearby at 10-14 Estreham Road and the Station received severe blast damage to the booking office, with the shock waves of the explosion blowing off the roofs of the platforms.

A bid by the Streatham Vale Property Owners Association to get the station renamed Streatham Vale Station was rejected by the railway authorities in 1936.

However, the Association was more successful in its recent campaign for a new Eardley Road entrance to the station which was opened on Monday 14th May 2007.

Streatham Common Station c1903

Zeppelin bomb damage at Streatham Common Station in September 1916

STREATHAM NOW & THEN AGAIN

HOLY REDEEMER CHURCH

The Holy Redeemer Church stands on the southern junction of Churchmore Road and Streatham Vale and is the largest building in the neighbourhood.

The site of the church, hall and vicarage was secured by the parishioners of St. Andrew's Church, South Streatham who, together with local residents, raised around £4,500 to cover the cost of building and equipping the hall which was erected in Churchmore Road in 1927.

It was dedicated by the Bishop of Southwark in March 1928. The first vicar of the church, the Rev. C. P. Turton, conducted services in the hall until the church was built.

The parish was provisionally named St. Luke's, Streatham Vale, and incorporated the work of St. John's Church in Eardley Road and the Mission of the Good Shepherd, which operated from a large wooden hut off Marion Road.

It was not until 10th January 1931 that work began on building the church when the first turf was cut by the Rev. J. W. Crauford-Murray, the Rural Dean of Streatham.

On this occasion the dedication was changed to the Church of the Holy Redeemer as a memorial to the Clapham Sect and to reflect the evangelical nature of the new parish.

The foundations were quickly formed and on the 28th March Canon Robert Joynt, Archdeacon of Kingston, laid the foundation stone before a gathering of local dignitaries, including Sir William Lane-Mitchell, the local MP, and the Mayor of Wandsworth, Cllr. A Bellamy.

The church was designed to accommodate 500 worshippers by the architects, Martin Travers and T. F. W. Grant, who were charged with producing a building that would "possess dignity, simplicity and beauty".

The large renaissance cupola on the roof provides a landmark feature to the building and can be seen from most parts of the Vale.

The church was consecrated on 5th March 1932 at which time its cost of £12,700 had been fully raised with an excess sum of £500 being allocated towards the cost of the organ.

Interior of the church in 1932

The stone octagonal font was the gift of the children of the Sunday School who collected farthings to cover its cost. This was a remarkable feat considering that a farthing was worth one quarter of a penny. One young boy collected 1,700 farthings while another accumulated 1,500.

It was not until 1938 that building work came to a close when the Vicarage was completed.

Canon Robert Joynt blesses the foundation stone of the Holy Redeemer Church in 1931

CRUMB'S BAKERY 132 STREATHAM VALE

Crumb's Bakery, at 132 Streatham Vale, continues a tradition of selling bread and cakes at this site which dates back over 80 years.

The building was erected in the late 1920s and its first occupant was W Rock who had established his bakery business in 1905. The bake house was situated in the yard behind the property where the delivery carts in their brown livery with gold lettering were kept.

Ronnie Rock's iced cake made in 1946 on display at the London South Bank University

By 1930 a sub Post Office had been established at the back of the shop which provided a valuable service to the new residents moving into the houses erected on the Streatham Vale estate.

Rocks advertised themselves as "high-class Bakers and Pastry Cooks" with "a local reputation for high-class bread and pastries".

By 1932 the firm had opened a second outlet in Streatham at 418b Streatham High Road, opposite Streatham Common. Four years later in 1936 the firm was advertising a third shop in Manor Road, Mitcham.

W Rock had three sons, Walter, Thomas and Sidney, and a daughter, Bertha. Other members of the family worked at the bakery and Walter's son, Ronnie, became a champion cake icer and a master at producing lace net icing and piping. He taught icing at the Borough Polytechnic, now known as the London South Bank University, and a cake he iced in 1946 is on display there in the National Bakery School.

In the early 1950s Ronnie won a triple first at the British Icing Championships. One of his cakes, a magnificent creation shaped like a cathedral, was exhibted at Olympia.

During the war Ronnie's sister, Joan, was one of the famous Windmill Girls, and thier brother, Peter, worked at the theatre as an electrician. Tragically he was killed when a bomb fell on the cafe opposite the Windmill where he was having a meal at the time.

By 1953 the firm had grown to become "Wholesale and Retail Bakers and pastry cooks specialising in the manufacture of bride cakes and cakes for special occasions". It had transferred its head office from Streatham to a "Modern Hygienic Bakery" situated on the Hackbridge Mills Estate, where the firm's own brand of "sliced and wrapped bread - the 'IDEAL' loaf" was produced.

The Streatham Vale shop subsequently became Clarke's Bakers, and it was while the premises was under this ownership that it was robbed by an armed 14 year old boy in 1995.

After a period trading as KIDZ-UK, a children's clothing and gift shop, the outlet returned to its confectionary roots when it became the Crumbs Bakery in December 2008.

W. Rock & Sons

HIGH-CLASS

BAKERS AND PASTRYCOOKS

A local reputation for high-class Bread and Pastries

ROUT SEATS, TABLES & ALL TABLE REQUISITES FOR HIRE
PRICES ON APPLICATION

Head Office and Bakeries
132 Streatham Vale, S.W.16
Branches at
**HIGH ROAD, STREATHAM
MANOR ROAD, MITCHAM**

Advert for Rock's Bakery in 1936

MONTI'S DAY NURSERY

Monti's Day Nursery has occupied the hall at the bottom of Streatham Vale since 14th June 1999.

The building dates from 14th June 1930 when it was opened by Mrs George J Campbell as the Streatham Vale Citadel of the Salvation Army.

The hall measured 58ft by 36ft and could accommodate around 250 people. It incorporated an ante-room, a tea room, and a platform and cost £1,953 to build and furbish.

The venture was made possible through a grant from the Salvation Army Headquarters and a generous donation from Sir James Carmichael.

Sir James had formerly lived at Redclyffe, in Ullathorne Road, Streatham Park, and was a benefactor to a number of local charities. He was the head of a large building firm and was knighted in 1919. He died in 1934 and is commemorated by a stained glass window in Streatham United Reform Church.

The opening of the hall was a great success with the band of the Balham Corps performing throughout the day. In the first week 447 children attended meetings at the hall.

A Sale of Work was a highlight in the Army's year, which usually featured an evening music concert and special services. The event was normally opened by the wife of Streatham's MP and the

Streatham Salvation Army Band

proceeds made an important contribution towards the cost of running the citadel.

During the Second World War the Streatham Vale Corps played an important part in raising the morale of local residents and providing help and assistance to those in need.

They acquired an old builder's barrow on which they transported tea and refreshments to those sleeping in local air raid shelters. So impressed was the Mayor of Mitcham by their service he presented them with £50 from which they purchased a second hand car, together with new tea urns etc., which enabled them to extend their operations to Mitcham.

At Christmas it became the custom to provide dinners to those living by themselves. About 40 local residents would gather in the hall for a traditional Christmas dinner and an afternoon of yuletide entertainment.

In 1969 the Streatham Vale citadel was featured in a special programme about the Salvation Army made by the BBC.

Members from the Streatham Vale Corps would regularly visit local pubs to sell copies of the Salvation Army's Newspaper, *The War Cry,* and continued to minister to local residents until their citadel in Streatham Vale closed in 1993.

Mrs Campbell opens the Salvation Army Hall on 14th June 1930

STREATHAM VILLAGE

Traditionally, Streatham Village was centred around the ancient parish church of St. Leonards. As Streatham was absorbed into the sprawling southern suburbs of London the concept of "Village Streatham" disappeared.

The term came back into use in the early 1970s to describe the area encompassed by Sunnyhill and Wellfield Roads. At that time this locality was under threat of compulsory purchase by Lambeth Council which had plans to demolish most of the houses there and replace them with a massive new council estate.

Although some of the properties were run down and in need of renovation, this was in part a consequence of the uncertainty over the future of the area.

Many local residents found references by the council to their homes as "slums" offensive and were concerned at the continual decline of the area.

In 1972 they banded together to form a residents group, called the Streatham Village Community, to fight the council's plans and press for improvement rather than redevelopment. Their crusade was a success and the wholesale destruction of the neighbourhood was averted.

Lambeth Council eventually designated the area a Conservation and General Improvement Area.

Slowly over the past 40 years or so the housing stock has been renovated and repaired, rather than demolished and replaced with a massive housing estate.

Today, wandering along Sunnyhill and Wellfield Roads is indeed like strolling through a Victorian village, complete with its own cottages, shops, school, pub and chapel.

STREATHAM NOW & THEN AGAIN

48-50 SUNNYHILL ROAD

Dry Cleaning by Thomas Clark occupies this pair of Victorian semi-detached cottages. For many years this building was occupied by the Bush family whose links with Streatham date back to 1871 when John Bush came to the town where he was a coachman at Kenmure, a large house on the western side of Streatham High Road.

Kenmure was the home of Daniel Bax who made a small fortune selling Mackintosh raincoats. He was the grandfather of Clifford Bax, the playwright and Sir Arnold Bax, the composer and Master of the King's and Queen's Musicke.

In the early 1870s John left Kenmure and worked as a shop assistant. By 1877 he had established a confectionary business at 17 Leigham Lane, later to become 48 Sunnyhill Road.

When he died in 1896 his widow, Harriett, took over the business and around 1900 the family acquired the shop next door, No. 50, where they sold a range of leather goods.

After Harriet's death in 1906 her family continued the business and the leather store traded at No. 50 until 1959.

The Bush family use to take in lodgers and local legend has it that Tom Norman, the famous Victorian showman, once lodged there.

Tom is best known today for exhibiting the Elephant Man, Joseph Merrick. His character was much maligned in the film *The Elephant Man* which portrayed him as a drunken bully. This was far from the truth as Tom had signed the pledge with the Church of England Temperance Society and did not drink. Secondly he never bullied those he exhibited and Merrick himself said he was always treated with kindness by Tom.

The success of Tom's penny shows continued well into the 20th century and he was

Bush's confectionary shop in the early 1900s

still appearing at Mitcham Fair up to 1930, when in August of that year he died at Croydon Hospital.

Tom's son, George "Barnham" Norman, lived for many years at 8 Streatham Vale, where he died aged 86 in May 1993.

George's wife, Brenda, ran her antique shop beneath their home there for almost a quarter of a century from 1980 until her death in March 2004.

Known as Norman's Corner, the shop was always filled with the most unusual mixture of items.

It was not unusual to see among the delicate Edwardian bone china and Victorian furniture a large, rusty old motorbike over the seat and handlebars of which was draped an old tiger skin with the animals head resting on the headlight!

Tom King

Joseph Merrick "The Elephant Man"

J. BUSH,
WHOLESALE AND RETAIL
STATIONER & CONFECTIONER,
CIRCULATING LIBRARY,
No. 17, Leigham Lane, Streatham, S.W.
Bookseller and Binder.   Newspapers, Magazines and Periodicals regularly supplied to order.
Cheapest house in Streatham for all kinds of Fancy Goods, &c., &c.

Advert for J Bush Stationer & Confectioner March 1879

THE LEIGHAM ARMS

Locals have been supping their ale here since 1860 when George Lockwood was granted a licence to operate a beer shop on the premises.

George was born in Streatham on 13 September 1828 and his grandparents are buried in St. Leonard's churchyard together with seven of their children, all of whom died in their infancy.

Although the family were originally labourers, they prospered from their endeavours and eventually owned a number of cottages in what was then called Wells Lane. These are known today as Wellfield House, Nos. 33-39 Wellfield Road.

George split his time between pulling pints in the pub and plastering to maintain his wife, Mary, and their seven small children.

The family lived above the inn, along with four lodgers and the pub potman. At times there must have been more people living above the tavern than those sitting in the bar below drinking beer!

Another long term publican here was Walter Topliss who managed the pub for over 40 years until his death in 1927. During the term of his occupancy the pub was affectionately called "Toppo's" by local residents.

Towards the end of his life, Walter became quite infirm and the management of the house was left to his daughter. It was while she was in charge that the police raided the pub a few weeks before Christmas, 1925, and broke up a gambling syndicate which had been set up in the bar by George Roberts. Despite Walter pleading his ignorance of Robert's betting activities the magistrates still fined him £50.

A member of the family, Flight Lieutenant G. W. Topliss, was awarded the Distinguished Flying Cross (DFC) in the Second World War for his determination in successfully pressing home bombing raids against railway targets in France in December 1944.

The pub contains a fine collection of old photographs of Streatham on its walls and today drinkers can quench their thirst on a good pint whilst feasting their eyes on pictures of the area in days gone by.

On the front wall of the pub two plaques indicate the regard with which Gordon Brown and Ken Livingstone are held by the publican who has banned both politicians for life from his establishment.

Meanwhile, on the back wall of the pub, a small monkey can be seen carrying a bottle of beer up to the second floor bedroom window of the building!

The Leigham Vale Morris Men perform at the Leigham Arms on May Day 1979.

STREATHAM NOW & THEN AGAIN

BLACKWOOD MINSTRELS

The young men of the Blackwood Institute, under the leadership of Mr. H. C. Edwards, gave a minstrel entertainment at the Blackwood Hall on Thursday of last week, and the accommodation of this hall was taxed to its utmost capacity for the occasion. The interest of the audience was maintained at a high pitch from beginning to end, and the inevitable "funny" man kept them in roars of laughter. After the opening chorus, which was composed by two of the troupe (Edwards and Waterman), E. Pierce (bones) set the programme going, and was followed by A. Jordan, who put the audience on the best terms with the troupe by his antics in a song "Keep on walking." Other comic singers were H. C. Edwards (tambo) and J. Ratcliff and T. Waterman, all of whom earned hearty applause, while A. Rogers (interlocutor) sang two more "serious" songs in fine style. The choruses by the whole company (consisting of those already mentioned and F. Dibden, H. and R. Jordan, F. Stevens and J. Young) were splendidly rendered, and were repeatedly encored.

Details of a Minstrel Concert at the Blackwood Hall in 1911

WELLFIELD CENTRE

The Wellfield Centre at 16 Wellfield Road is run by the Streatham Youth and Community Trust as a community centre specialising in services for young people.

The hall was built in 1867 by Sir Arthur Blackwood as the Streatham Mission Hall.

Sir Arthur was born at Hampstead in 1832 and in 1858 he married the Dowager Duchess of Manchester. The following year they moved to Streatham where they lived at Wood Lodge, a large house on the corner of Tooting Bec Gardens and Garrads Road.

A devoutly Christian man, Arthur became increasingly concerned about the wellbeing of the poor and the large number of navvies who were constructing the railway through Streatham.

He purchased a plot of land on which he erected a tarred wooden hut which became known as the Black Chapel. Here he conducted services as well as arranging a programme of social events.

People journeyed from miles around to attend these services. In the summer months meetings would be held in the open and large crowds would gather in the street with local residents leaning out of their windows to join in the hymn singing.

The old wooden hut was replaced by the Hall we see today which Mr. Blackwood opened on 6th October 1867.

In 1874 Arthur founded the Trinity Presbyterian Church and the hall was handed over to them for use as a Mission hall.

Sir Arthur Blackwood

Sir Arthur became a Member of Parliament and had a distinguished parliamentary career. In 1880 he was appointed Postmaster General and was knighted in 1887 when he was awarded the KCB in Queen Victoria's Jubilee honours.

He remained a firm supporter of the work at the Mission Hall and on his desk in Whitehall he had a small card inscribed with the words "Pray for Streatham".

Sir Arthur died on 2nd October 1893. After his death, the Mission Hall was named the Blackwood Hall in his honour.

By the late 1970s the hall had ceased to be used and was in a semi-derelict condition. The building was subsequently converted into offices known as Grenville House.

In 1995 the Bright Sparks Theatre School, also known as Flames Academy, moved here. Several of its students appeared on the West End stage including Joe Cumi and Adam Mead, who both played the Artful Dodger in the musical *Oliver*, and Sammy Jay who played Cosette in *Les Misersbles*. Other pupils appeared in numerous TV dramas such as *The Bill*, *East Enders* and *Casualty*.

The Streatham Youth and Community Trust purchased the hall in 2003 as a base for their activities in this part of Streatham in addition to their operations at their hall in Conyers Road.

In 2011 the trust won a public vote for a £75,000 grant from Lambeth Council to repair the roof of the Centre.

THE OLD SMITHY

Standing opposite the old Blackwood Hall are Nos 17-21 Wellfield Road, formerly known as Manor Terrace. A peek down the alleyway to the side of No 17 will reveal a small, two-storey building at the end of the courtyard which was formerly an old smithy.

For many years the sound of hammer on anvil echoed out from the ancient blacksmiths shop here which was a hive of activity in the days when horse-drawn transport ruled the roads.

Around 1910 Harry Fox took over the business from James Tuck and he was still shoeing horses here at the outbreak of the Second World War.

Part of the secret of Harry's success was he embraced modern technology and equipment in the operation of his business whilst maintaining the traditional skills of his trade.

In the 1930s he was quick to use machines to cut the iron bars used in making horse shoes. This was ten times quicker than if the same act was performed by hand.

He also used modern electric drills to bore the holes in the horse shoes which they did more quickly, accurately and cleanly than could be achieved with punch and hammer.

By such innovations he was able to boost production and shoe more horses than his competitors.

Harry served his apprenticeship as a blacksmith at the old Streatham Forge that use to stand opposite

AN UP-TO-DATE FORGE
HARRY FOX
- THE FORGE -
Wellfield Road, Streatham
AND
63, Westcote Road, Streatham
HORSES CAREFULLY SHOD
On the most modern principles, by contract or otherwise

Estimates give for any kind of

JOBBING AND GENERAL SMITHS' WORK

including Springs and Tyres.

Bob Davis (third from left), a local coal man from Sunnyhill Road, having his horse shod at the old smithy before the First World War.

Sketch of the old smithy and forge in Wellfield Road

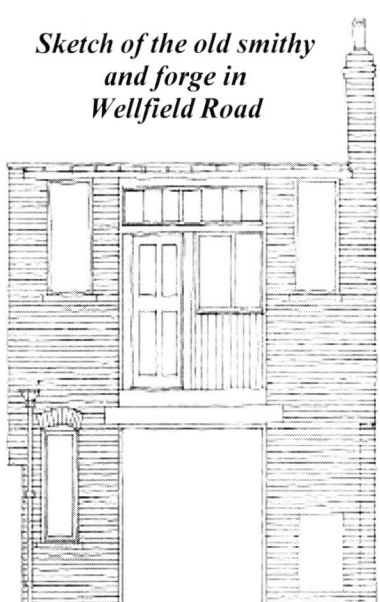

St. Leonard's Church, at the junction of the High Road and Mitcham Lane.

So successful was he at his trade that he eventually owned three other forges in Streatham, located in Estreham, Westcote and Blegborough Roads.

A great boost to his business was the lucrative contract he had to shoe all the horses used in pulling the milk carts at the United Dairies Depots in Streatham and Dulwich and as a result had around 200 horses under his care.

It was not unusual for Harry, and his three companions who worked at the Wellfield Road smithy, to put on 150 horse shoes a week.

After the Second World war the number of horses in the area dramatically declined. The United Dairies phased out their horse drawn milk carts in favour of new electric milk floats and the old forge eventually closed.

However, although the anvil lay silent the building continued in use as a warehouse and workshop until 2003 when it was converted into the self-contained house we see today.

STREATHAM NOW & THEN AGAIN

54 WELLFIELD ROAD

Number 54 Wellfield Road is the birthplace of one of Streatham's famous residents, the comedian Tommy Trinder, who was born here on 24th March 1909.

Tommy's father, Thomas Trinder, was a south London tram driver and moved to Streatham shortly before Tommy was born so as to be close to the tram depot at Streatham Hill where he worked.

The family only lived here for about a year before Mr. Trinder's work caused him to move out of the area and the family settled in Fulham.

Tommy become a keen supporter of Fulham Football Club and later became club Chairman for 21 years from 1955 to 1976. Fulham FC often formed the butt of his gags and he would tell of how as a lad he used to sneak into the club's riverside ground when the Thames was at low tide.

Tommy Trinder

Tommy made his first appearance on the stage in Collins Music Hall at the age of 12 and went on to become one of the greatest comic stars in British variety.

By the mid-1930s he was Britain's top comic and starred at the London Palladium for a record run of 2 years and 4 months.

Tommy was a particular favourite with the Queen Mother and participated in 13 Royal Variety Shows, as well as giving private shows for the Royal family at Buckingham Palace, Windsor Castle and Balmoral.

For many years he hosted the popular television programme *Sunday Night at the London Palladium* where his catch phrase of "You lucky people" was particularly apt during the "Beat the Clock" quiz that was held in the middle of the show.

Tommy used to quip he was a social climber and pretended to have been born in Sunnyhill Road, rather than Wellfield Road.

In 1961, when he was in the pantomime Jack and Jill at Streatham Hill Theatre, he visited Wellfield Road to have his photograph taken outside the house in which he was born. Unfortunately he thought this was No 57 rather than No 54 and as a result was photographed outside the wrong house!

Tommy was awarded the CBE in 1975. He died in hospital at Chertsy on July 10th 1989 aged 80.

A Streatham Society plaque to commemorate Tommy's birthplace adorns 54 Wellfield Road. This was unveiled in June 1990 by another famous local comedian, Roy Hudd. The unveiling was attended by Tommy's widow, Tonie, their daughter Jane and their granddaughter Louise.

Roy Hudd unveiling the Streatham Society plaque to Tommy Trinder in June 1990.

142-144 WELLFIELD ROAD

The former St. Paul's Mission Church & St. Peter's Church Hall stands at 142-144 Wellfield Road although it no longer serves any religious purpose having been converted into a private residence.

When St. Peters Church was formed in Leigham Court Road it originally met in a corrugated iron tabernacle located on the other side of the road to the Church building that stands today.

This temporary chapel was dedicated to St. Peter and St. Paul. When the new church was completed it was decided to call it St. Peter's and the connection with St. Paul was transferred to the church's Mission Chapel in Wellfield Road.

St. Peter's Church mainly catered for the rich and wealthy inhabitants of the large houses in Leigham Court Road, whilst the more humble St. Paul's Chapel ministered to the poor people of Wellfield and Sunnyhill Roads.

The Mission Church normally held services in the evening which were attended by around 50 people.

The building was also used as St. Peter's Church Hall for this part of the parish and accommodated various social and parochial events.

The Church Lads Brigade met here and every Sunday the boys would parade to St. Peter's Church led by their drum and bugle band.

A local Doctor's Club was also based at the hall and residents would pay their weekly subscriptions there.

In 1937 Wellfield and Sunnyhill Roads organised a street party to commemorate the Coronation of King George VI. When it started to rain the event was transferred to the hall which became the centre of the celebrations. Later that day around 180 children were packed into the Hall when the vicar of St. Peter's, the Rev Morson, and Billy Cotton, the famous band leader of "wakey, wakey" fame, visited them.

The Hall became a valuable community centre during the Second World War and a rabbit keepers club was established there to provide encouragement and advice to local residents who bred rabbits to augment their war-time rations.

Billy Cotton

The building was sold in 1945 and in 1994 it was renovated and converted into a residence with two double bedrooms, a large Shaker-style kitchen, and a 26ft x 9ft 3ins reception room. It was offered for sale in 1998 for £225,000 and again in 2002 for £395,000.

The building still retains a number of its original features, such as the ornate wrought iron cross above the side gate and the small turret on the roof complete with bell.

King George VI Coronation street party in Wellfield Road 1937

SUNNYHILL SCHOOL

Left: Sunnyhill Road School in the early 1900s.
Right: Sunnyhill Road School pupils in the late 1920s.

Sunnyhill School was built in 1900 at a cost of just over £26,000. This was the 444th school to be erected by the London School Board, whose initials LSB can be seen in various motifs on the outside of the building together with the date of the year in which it was built.

The school was opened on Thursday November 9th and was originally designed to accommodate 450 juniors and 358 infants, a total of 808 pupils.

In 1903 an Evening Institute was established here teaching book keeping, cooking and shorthand among other subjects.

In the Second World War the school was badly damaged by a V1 flying bomb which fell at the back of the building, between the gardens in Valley and Harborough Roads, at 8am on the 29th June 1944.

Fortunately the school was empty and no one was fatally injured. However, tremendous damage was done to surrounding properties and sixty families were made homeless.

In one back garden a coup containing five chickens was destroyed by the blast. One chicken was blown into the air and returned to earth "as feathers" whilst its four companions quickly recovered and sought shelter in a nearby house where they promptly laid eggs!

The school became the focus of attention on May 10th 1945 when King George IV, Queen Elizabeth, Princess Elizabeth and Princess Margaret visited the school to inspect the local ARP and Civil Defence groups.

Crowds ten deep lined the route and among the cheering throng was even a lorry full of Italian Prisoners of War whose transport had been delayed to allow the Royal car to pass by.

The King and Queen were greeted at the school gates by the Mayor of Wandsworth and the local MP, David Robinson, and were then introduced to members of the various Civil Defence organisations who had worked so hard throughout the war, particularly during the blitz and the period when some 41 flying bombs fell on Streatham.

Among the former pupils of the school is Howel Bennet, the actor, who lived nearby at 118 Valley Road. He made his first appearance on the stage as "Doc" in the school's production of *Snow White and the Seven Dwarves*.

He later gained fame starring in feature films with Hayley Mills such as *The Family Way* in 1967. He also appeared on television in such programmes as *Shelly* and *Eastenders*.

Bottom left: Visit by King George VI, Queen Elizabeth and the Princesses Elizabeth and Margaret, to Sunnyhill Road School to inspect Streatham's Civil Defence services in May 1945. Right: Howel Bennet

CENTRAL STREATHAM

Although Streatham is today a sprawling southern suburb of London, in the centre of our town can still be found the vestiges of its origins as a small Surrey village.

Our ancient parish church of St. Leonard's sits atop a hill at the heart of the local community and for centuries its spire was a well known prominent landmark which could be seen as far away as Streatham Hill until buildings obscured the view.

The Town's village Green also survives as a small public open space amid the buildings lining the northern end of Streatham High Road and Mitcham Lane. Today it provides a valuable green lung at the centre of the town and a refuge from the noise and fumes of the heavy traffic journeying along the busy A23 London to Brighton Road.

Although the 18th century Bedford House, the home of Daniel Macnamara, agent to the Duke of Bedford the Lord of the Manor of Streatham and Tooting Bec, has disappeared from the centre of the town, part of one of its wings can still be seen at the rear of the shop at 119 Streatham High Road.

While Streatham's Alms Houses, erected by the Thrale family in 1832 to provide accommodation for four poor widows or single women who had "attained an honest old age" no longer stand next to the police station, they survive having been rebuilt in a quieter, location off Polworth Road.

Some of the town's early 20th century buildings have been adapted for modern use, such as the Golden Domes Cinema and the Temperance Billiard Hall. These stand alongside modern buildings, such as Norwich House, a large office block erected in 1963, to form Streatham's present-day street-scape.

The Green in the mid 1800s

STREATHAM GREEN

Streatham Green covers an area of .47 of an acre and was originally more than twice this size having been reduced over the years by the widening of both Streatham High Road and Mitcham Lane.

This ancient open-space has occupied this site for centuries and it was here in the middle ages that villagers practised archery.

In 1792 an abandoned baby girl was found on the Green and as her parents could not be identified she was baptised Perdita Green on May 19th.

Lady Pitches caused a minor riot in 1794 when she absorbed the Green into the grounds of her house. She disliked funeral processions passing the front of her mansion as they took the short cut across the Green on their way to the graveyard.

The Lord of the Manor, the Duke of Bedford, forced her to open up the Green again, thus saving pall-bearers the hard labour of having to carry heavy coffins up the steep incline of Streatham High Road.

In the 19th century children from the parish school in Mitcham Lane used the Green as a playground. In the summer they played cricket here using one of the trees as a wicket. If the ball was hit into the road it automatically counted as a six as it would roll down the hill and take some stopping!

A row of ancient elms use to front the High Road side of the Green and when these were cut down in 1904, when the road was widened, they were found to be over 300 years old.

In 2004 the Green was landscaped at a cost of £250,000. As part of these works the last surviving public air raid shelter in Streatham was removed. This stood alongside the High Road and was filled in at the end of the Second World War and covered with earth. The Streatham Society planted daffodils on the mound which provided a magnificent display of blooms each spring.

At the centre of the Green stands the Dyce Fountain which was erected by public subscription in 1862. It was designed by William Dyce, a famous Victorian painter and a pioneer of the pre-Raphilite movement, who lived in Streatham and was Churchwarden of St. Leonard's Church from 1862-64.

The Fountain cost £250 and originally stood at the Junction of Streatham High Road and Mitcham Lane. It was moved to its present position in 1933 following the redevelopment of the site with shops and flats.

Pigeons shelter from the wind on the Dyce Fountain

Part of a wing of Bedford House at the rear of 119 Streatham High Road

BEDFORD HOUSE

Nos 119-127 Streatham High Road mark the site of one of Streatham's grandest buildings - Bedford House. This was erected in the mid-1780s by Daniel Macnamara, agent to the Duke of Bedford, the Lord of the Manor of Streatham and Tooting Bec.

This magnificent mansion was a large red-brick house of five stories the ceilings and walls of which were elaborately decorated with intricate plaster work.

One of the most impressive features of the house was an imposing double staircase which swept up from each end of the house, curving from one landing to the next.

The building had a total of 143 windows, many of which commanded breathtaking views over the surrounding countryside.

Bedford House stood in 22 acres of landscaped gardens and grounds, which incorporated a large boating lake located roughly where Ashlake Road now stands.

Daniel Macnamara persuaded the Duke of Bedford to buy the house whilst he continued to live there.

The Duke and MacNamara were friends of the Prince Regent who often visited Bedford House as their guest.

Bedford House when in use as a Furniture Depository by Pratts

In the privacy and luxury of this fine mansion the Prince enjoyed the convivial companionship of his friends, both male and female, as well as Mr. Macnamara's extensive cellar of fine French wines.

After Macnamara died in 1800, the Duke's brother, Lord William Russell, briefly lived here. The house was subsequently sold and in 1872 it was purchased by George Pratt, the owner of a large department store in Streatham.

George developed the grounds as the Bedford Park Estate and moved the entrance of the house to 9 Gleneldon Road to allow shops to be built on the High Road frontage, most of which survive today.

The house was eventually converted into a furniture depository comprising over 300,000 cubic feet of storage space.

It was sad that such a magnificent Georgian mansion should have ended its days as a warehouse, and its final demise was equally as tragic for at 1.30 am in the morning of Sunday 2nd August 1936 a fire broke out in one of the ground floor offices and rapidly spread throughout the building.

Despite the best endeavours of more than 150 fire fighters and 26 fire engines the building burnt to the ground.

The site of the house survives behind the shops as a courtyard and the end of one of the entrance wings is all that remains of this once grand building.

St. Leonard's church and Bedford House in an 1804 etching.

PRATTS & PAYNE

On the site of the Pratts & Payne public house once stood a terrace of four small almshouses erected in 1832 to provide subsidised accommodation for four poor widows or single women who had "attained an honest old age" in Streatham.

Known as the Thrale Almshouses, they were built in memory of Henry and Hester Thrale of Streatham Park by their surviving daughters, Hester Maria, Viscountess Keith; Susanna Arabella Thrale; Cecilia Margaretta Mostyn and Henry Hoare, the widower of Sophia Thrale.

In the latter half of the 18th century, Streatham Park was a fashionable centre for society and here Henry and Hester entertained the leading luminaries of the day, such as David Garrick, Edmund Burke, Oliver Goldsmith, Sir Joshua Reynolds, and Dr. Samuel Johnson.

The Thrale Almshouses in Streatham High Road in the 1930s

In 1930 the almshouse site was sold for £11,792 and the money used to build eight new almshouses in Polworth Road which were designed by Cecil M Quilter. In 1938 an additional two almshouses were errected on the site in memory of Lady Edith Robinson, the wife of the president of the Streatham Conservative Association.

The old Almshouse site was redeveloped in 1932 with a large retail unit for Montague Burton, tailors.

Burton's shop was designed by Harry Wilson of Leeds, and is typical of the company's Art Deco style. The building contains some lovely features including an elephant head frieze and other Art Deco motifs.

The shop front was dressed in black marble and two foundation stones survive laid by Stanley Howard Burton and his wife Barbara Jessie Burton.

For over half a century a dance studio has occupied the first floor of the building. In 1951 it was called the Streatham Rooms School of Ballroom Dancing.

In modern times the Starlight Dance Centre has been based here originally run by Bill and Bobby Irvine, 13 times world ballroom champions. At one time their dancing school boasted 3,000 members aged between 7 and 70 years of age. A number of champion dancers have trained here including the Japanese professional ballroom dancing champions.

Following the closure of Burtons the ground floor became the Streatham Job Centre.

In 1997 an application to open a public house on the ground floor was initially refused, but was later granted and the Furze and Firkin opened here on 12th March 1998. This closed in the autumn of 2000 and the premises were refurbished and reopened as The Goose on November 5th, 2000. This pub closed in September 2011 and after renovation was reopened as the Pratts and Payne, named after George Pratt's Streatham department store and Cynthia Payne, the town's famous "hostess".

The Thrale Almshouses in Polworth Road

St. Leonard's Terrace c1900

75-89 STREATHAM HIGH ROAD

Nos 75-89 Streatham High Road marks the site of St. Leonard's Terrace, a parade of shops with two floors of accommodation above which was erected here around 1860.

At 4 St. Leonard's Terrace, later known as 87 Streatham High Road, William Weller Nichols established a watch making business in 1860. He was the son of William Henry Nichols, the organist at St. Leonard's Church. In the 1870s the business was run by another of William's sons, Alfred Latimer Nichols, until around 1877 when William Fountain took it over.

Around 1893/4 a German watchmaker called Otto Mohr bought the business and to advertise his ownership he erected a large clock over the entrance with his name proudly displayed above it. As a consequence his shop soon became known as "The Clock House".

The clock quickly became a well-known local landmark as it was the only timepiece on public display in Streatham other than the clock on the tower of St. Leonard's church.

Otto played an active part in local affairs and in 1897 he was one of the local tradesmen who subscribed towards the cost of a special Fete for local children to celebrate Queen Victoria's Diamond Jubilee.

In 1902, he once again exhibited his patriotism to his adopted country by being a member of the Sports Committee for the Fete which formed part of the celebrations to mark the coronation of King Edward VII.

In 1909 St. Leonard's Terrace was demolished and replaced with the parade of shops we see today. Otto transferred his business to one of these new shops and continued to trade from 85 Streatham High Road.

He was still occupying these premises in 1914 at the outbreak of the First World War but his business was not to survive the conflict. Being a German by birth, albeit one who had spent many years in England, he may have experienced a downturn in trade as hostilities against Germany intensified. However, by this time he would have been in his 60s and he may have taken the opportunity to retire.

By 1917 his shop had been taken over by R Barratt and Otto is no longer listed in local directories.

Today, an example of Otto's clock making skills can still be seen in St. Peter's Church in Leigham Court Road where a time piece made by him over a hundred years ago was recently restored and still keeps perfect time.

THE CLOCK HOUSE, STREATHAM.
Established 1860

OTTO MOHR

Chronometer, Watch and Clock Maker, &c., Optician and Jeweller.
OLD GOLD AND SILVER BOUGHT AND TAKEN IN EXCHANGE.
87, HIGH ROAD (A few doors from the Tate Free Library).

Right: Otto Mohr's shop c1905 Above: 1905 Advert

CARPET RIGHT

The Carpet Right showroom at 130-132 Streatham High Road occupies the site of one of Streatham's earliest cinemas - the Streatham Picture Theatre.

Commonly known as *The Golden Domes*, after two large gold coloured domes which adorned the roof of the two wings of the building, this was the second cinema to be built in Streatham.

The building could seat 900 patrons, 200 of which were accommodated in the gallery. Seats were available at 3d, 6d and 1s, with children under the age of 10 years being admitted at half price.

Advert for Beau Geste screened at the Golden Domes in 1927

The cinema opened in September 1912 with the screening of Caesar Borgia.

To attract a good attendance for the matinee performances the cinema provided afternoon teas free of charge.

The management was at the forefront of cinema administration and pioneered Sunday opening in the area in 1913. This drew considerable criticism from the local clergy who organised a petition which attracted over 2,500 signatures and forced the directors of the cinema to abandon Sunday screenings in the light of such widespread opposition.

However, Sunday evening presentations recommenced in January 1914 with the screening of *The Vicar of Wakefield*.

In 1929 the building was extended; the balcony was rebuilt and new seating provided to accommodate an additional 400 patrons. The building was wired so that it could show the latest "talking pictures" and a stage was built in front of the screen to accommodate variety acts.

By the late 1930s the Golden Domes was unable to compete

The Streatham Picture Theatre also known as The Golden Domes

successfully with the much larger "modern" cinemas which had been opened in the area, such as the Astoria (now the Odeon) and the Gaumont at Streatham Hill, and in November 1938 it closed.

During the war the building was used by the Ministry of Food. After the war it remained empty for many years until the mid 1950s when £6,000 was spent on levelling the sloping auditorium and the building reopened in September 1955 as one of Hardy's chain of furniture shops.

This venture was short lived and in 1957 it was converted into a Thomas Wallis store.

In 1967 the building was taken over by Tesco and in April 1968 "Sir Savealot" rode his white steed through cheering crowds when Tesco opened their new 9,000 square feet store here.

The Site was subsequently occupied by Kwik Save, Land of Leather, and MFI before Carpet Right opened their shop here in 2007.

STREATHAM NOW & THEN AGAIN

Norfolk House in the 1890s

NORFOLK HOUSE ROAD

Norfolk House was one of the grand mansions of Streatham and was erected by Francis Yates around 1843.

The house had 13 bedrooms and stood in over 15 acres of landscaped grounds comprising extensive gardens and parkland laid out between the High Road and Mount Ephraim Lane.

A small lodge on the High Road guarded the entrance to the grounds and a long, sweeping carriage drive led to the front door of the property.

In the grounds were numerous greenhouses, 3 coach houses, and a stable.

At the rear of the property was a large lake with a boat house. In the centre of the lake was a small island which was a popular place for family picnics in the summer months.

In 1851 Christopher Gabriel, a wealthy timber merchant, moved into the property.

In 1833 he had married Ruth Thurston, the only daughter of John Thurston a noted Billiard table maker. John had established his business in 1799 and is considered to be the father of snooker.

His firm continues to make billiard and snooker tables today and are responsible for many improvements to the designs of the table, including the introduction of the slate bed and placing rubber within the cushions which allowed billiards and snooker to develop into the major sport it is today.

The Gabriel family were wealthy and influential timber merchants.

Christopher's grandfather, also called Christopher, had laid the foundations of his family's fortunes in 1770 when he set up in business as a plane maker.

Thomas Gabriel

Following his death in 1809 two of his sons, Thomas and Christopher, continued to run the family business and subsequently concentrated their activities on the timber importing and wholesale side of the business based at Gabriel's Wharf on the Thames.

The third generation of the family continued to run the firm, building on the success of their forefathers. In 1866 Christopher's brother, Thomas Gabriel, became Lord Mayor of London.

The family business continued under various names for the next hundred years until 1968, when as Gabriel Wade and English Ltd it was acquired by Montague L Meyer Ltd.

After Christopher Gabriel died on 2 March 1873 his widow, Ruth, continued to live on at Norfolk House until her death in 1898 after which the estate was sold for development.

The house was subsequently demolished and in 1903 Norfolk House Road was laid out through the grounds between Kingscourt and Mount Ephraim Roads.

The formal garden at the rear of Norfolk House

Belle Vue Terrace c1905

BELLE VUE TERRACE

Belle Vue Terrace, 62-70 Streatham High Road, were erected in the early 1860s as residential houses.

At 1 Belle Vue Terrace, now 66 Streatham High Road, lived Commander John Lane, a retired naval officer. He was a veteran of the Battle of Trafalgar in which as a young 15 year old midshipman on HMS Thunderer, a vessel of 74 guns, he saw action.

John was born in 1791 in Halifax, Nova Scotia and spent most of his life at sea. Following his service at Trafalgar he continued in the Navy and was promoted to Lieutenant in 1815 and finished his career as a Commander.

He was well-known for his anecdotes about his seafaring adventures and would thrill local children with his tales of the Napoleonic War and the hardships he endured during his time at sea.

A couple of doors away, at 3 Belle Vue Terrace, 70 Streatham High Road, one of the first photographic studios in Streatham was established in 1888.

Known as Belle Vue Studios, the business was run by William Henry Douglas Pym, who was born in Norwood in 1862, and had previously operated a studio at 191 Newington Butts.

William traded under the name of Douglas Pym and, after setting up in Streatham, he quickly established himself as one of the town's premier photographers. In May 1893 he won a silver medal and a diploma of honour for his portraits of children.

In 1912 Reginald Rawlings took over the business and he continued taking pictures here for a quarter of a century until 1937.

By the early 1990s the old homes of Commander Lane and Douglas Pym had been converted into rival restaurants. At 66 Streatham High Road La Pergola traded while at 70 Streatham High Road was La Caretto.

Both establishments had competed for business for 13 years when their rivalry reached new heights when they both started to feature Elvis Presley impersonators to serenade their customers.

For seven years diners were never 'lonesome at night' with a choice of Elvis performers just two doors apart. In 1997 Kim Bridges at La Pergola became the unrivalled "King" of the Streatham Presleys when the Il Carretto was sold and Lou Jordan no longer had a local venue to perform his Elvis impersonations. The Il Carretto subsequently became part of the Hogshead pub which now trades as the Five Bells.

Belle Vue Terrace c1910

TESCO EXPRESS

Streatham's Tesco Express store stands at the southern junction of Streatham High Road and Broadlands Avenue.

A large, grey brick, Victorian house called Broadlands use to occupy this site - hence the name of the adjoining road.

The house had an extensive garden and a large paddock adjoined the grounds at the rear. On the southern side of the house was a glass conservatory running the full width of the property.

John Margetson, a wealthy warehouseman, died here in 1867 leaving a bequest of £100 to his servants in appreciation for their faithful service.

The house was demolished in the 1920s and Broadlands Avenue was laid out through the grounds in 1928 with Broadland Mansions being built on the High Road frontage of the property.

On the southern junction of the road a Temperance Billiard Hall was erected in the early years of the First World War and over the following fifty years many of the great players of the day played here, including Joe Davis and his brother Fred.

By the mid-1960s players were no longer satisfied with the outdated amenities offered at the hall and the building was taken over by the Mecca organisation.

They completely refurbished the property and opened it as the Golden Q in November 1965. The hall then accommodated 17 tables, each one covered in gold baize as opposed to the traditional green cloth.

The Temperance Billiard Hall

The club also had a Black Jack table and numerous one-arm bandit machines. The success of these facilities led Mecca to convert the building into a gambling establishment, firstly known as the Craywood Club, and then the Albany. The Albany lost its license in 1970 and the club was converted into a supermarket.

In March 1998 "Page Three" model Jo Guest brought glamour back to the site when she opened the Gasoline Alley car accessory shop here. The store became a well-known local landmark as parked on its roof was a full-size replica of a red 1950s Cadillac.

The shop was dedicated to the memory of the American film star James Dean and a clock mounted on the roof rafters showed the time of his death - 5.59pm on 30th September 1955.

The 1950s theme was continued throughout the store which was decorated with period memorabilia including original petrol pumps and an old telephone of the period.

In 2007 Gasoline Alley moved accross the road to 9 Streatham High Road and the building became a Tesco Express store.

James Dean

NORWICH HOUSE

Norwich House, a seven storey office block designed by Scott Brownrigg and Turner, was built by the Norwich Union Life Insurance Company in 1963 on the site of two large, semi-detached, Victorian houses, Nos 9 and 11 Streatham High Road.

These houses were built in 1855 and were three storeys high with a basement. They stood well back from the pavement with large front gardens and circular carriageways leading to steps rising up to the entrances to the houses. Both properties had extensive rear gardens.

No. 9 was called the Chestnuts after a large horse chestnut tree that stood in the front garden. No. 11 was originally known as Sussex Villa, then Nuthurst and was renamed Auchenblae in 1919 to commemorate the Scottish birthplace of the mother of the new resident of the house, Dr. George Scott.

Both houses were identical in design although Dr. Milne, who lived at No. 9, added a rather plain two storey extension to the northern end of the Chestnuts in 1928 to provide some extra accommodation for his large family which consisted of eight children - three boys and five girls.

Dr Milne and his wife in their car

Dr. Milne's family moved to the Chestnuts during the First World War having previously lived at 129 Mitcham Lane. He was a much respected ophthalmic surgeon with consulting rooms at 59 Queen Ann Street in London. He became a director of public health and held positions with a number of London hospitals. During the First World War he served in France and Mesopotamia with the Royal Army Medical Corps.

For most of his working life Dr. George Scott was a General Practitioner in Brixton Hill and Streatham. He was also a coroner and for a quarter of century was police surgeon for the South Western division of the Metropolitan Police. During the First World War he served in the Royal Army Medical Corps in Salonika.

Dr. Scott retired to Reigate in the late 1940s and died aged 88 in 1963. Dr. Milne retired around 1950 and moved to Mount Ephraim Road where he died just before his 91st birthday in 1974.

Both houses were demolished in the early 1960s to make way for Norwich House. At the rear of the site there stands a magnificent horse chestnut tree. The conker from which it has grown having been taken from the Chestnut tree in the front garden of "The Chestnuts" by Dr. Scott's eldest daughter who planted it there almost a hundred years ago.

The Chestnuts, 9 Streatham High Road.

STREATHAM HILL

Up until the First World War, other than for the shops between Leigham Court Road and Downton Avenue erected in the mid-1890s as part of the Artisans', Labourers and General Dwelling Company's Leigham Court Estate, Streatham Hill remained chiefly a residential road with large, late Regency and Victorian houses lining the street.

The oldest of these dwellings were 20 early 19th century villas, known as the Paragon, which were erected on the western side of the road, south of the Crown and Sceptre public house. Of these only Nos 40, 42 and 44 Streatham Hill survive and are now Grade II listed buildings of historic and architectural interest. On the eastern side of the road, Nos 5, 7 and 9 Streatham Hill, survive.

Many of the houses were swept away in the inter-war years to make way for large blocks of flats with shops at ground floor level such as Telford Court, Telford Parade Mansion, Wavertree Court and Christchurch House.

Those that escaped development in the 1920s and 30s were demolished after the Second World War to be replaced by the Claremont Council Estate which spans both sides of Streatham Hill.

Opposite Streatham Hill Station, at 113-117 Streatham Hill, is a site of national retail historic significance. It was here on 26th November 1951 that Britain's first self-service supermarket was opened by Express Dairy. Later rebranded as a Premier Supermarket this store was at the forefront of retailing in the early 1950s when it generated sales ten times higher than other shops.

DAILY FRESH FOODS

Daily Fresh Foods occupies Nos 113-117 Streatham Hill. This is a site of national, historic significance as it was here the first self-service supermarket opened in Britain on 26th November 1951.

The idea was the brainchild of Patrick Galvani, an executive with Express Dairies. He had seen supermarkets in action in the USA and believed the formula would be a great success in the UK.

So it was that the Express Dairy outlet at Streatham Hill was transformed into the Express Self Service store. Gone was the traditional shop layout and in its place three long lines of shelving filled the middle of the store with more shelves along the walls. These were stocked with a variety of groceries, fruit and vegetables, fresh meat, frozen produce and even toiletries and stationery from which the housewife could make her choice.

With a floor space of over 2,500 square feet and such a wide variety of items for sale the store was indeed a "super" market compared with the other outlets on the High Street at that time.

The store was an instant success with 1,500 shoppers flocking to the premises on its first day of trading. After years of standing in queues waiting to be served the novelty of self-service was an appealing aspect of the new shopping experience.

Within a short time the level of sales achieved was phenomenal. Whereas the average grocer would be making just under £100 a week the Express Self Service outlet at Streatham was taking £1,000.

The following year the store was rebranded as a Premier Supermarket and by the mid-1960s 39 other Express Dairy outlets were trading under the same banner.

Express Dairies eventually sold off their supermarket chain to Mac Fisheries.

Three small single storey shops had originally occupied this site in the opening years of the 20th century. Larger buildings were not possible here as the railway bridge was not designed to take the weight.

Walter Gardner opened his grocery and provisions store at No 117 in 1901 and eventually expanded into the neighbouring two shops. He suffered badly with his nerves and the pressures of running his business during the First World War took their toll and he hanged himself in his home in nearby Downton Avenue in 1920.

Gardner's continued to trade until the early 1930s when P W Gray took over the business before Express Dairies acquired the premises in 1937.

113-117 Streatham Hill c1905

The staff of the Premier Supermarket in 1960

*The first self-service supermarket in Britain
The Express at Streatham Hill*

45 STREATHAM HILL

Since 6th August 2001 the Mediterranean Food Centre has been selling a variety of Greek, Turkish and Cypriot foods at 45 Streatham Hill from dawn to dusk, seven days a week. The business was established by Sabri Zamur who previously ran a Turkish restaurant in Streatham.

The building dates from November 1956 when it was opened as the head office of the South London Building Society. The four storey building cost £60,000 to erect. Below ground level is a large basement with a heavy, green, steel door through which access to the vaults was obtained. Here the Society stored some 4,500 deeds of the properties it had mortgaged.

The Society was established in 1875 as the Camberwell and South London Building Society and over the years had spread its operations south to Brighton and north to Wolverhampton and Manchester.

One of the founder directors of the Society was William Andrews, and his grandson, Herbert James Andrews, was Chairman at the time they moved their HQ to Streatham.

George Hawes Butcher's shop at 45b Streatham Hill c1920

Around 1962 the building was extended in an identical style with the Streatham Hill Post Office occupying the ground floor of the extension.

On 31st December 1966 the South London Building Society merged with the Chelsea Building Society. Although the headquarters of the combined societies were moved from Streatham to Thirlestaine Hall in Cheltenham in 1973 the Chelsea continued to operate a branch here until February 1991 when it opened a new office at 112-114 Streatham High Road.

In 1998 the property became a

R Higgs Dairy shop at 45 Streatham Hill in 1910

building society again when it was transformed into the Meridian Building Society by a film company who used it to shoot a bank robbery scene featured in the film *Greenwich Mean Time*.

The site had previously been occupied by a three storey building with shops at street level erected here sometime between 1906 and 1909.

At midnight on 24th June 1944 a V1 flying bomb fell opposite this building scoring a direct hit on Wyatt Park Mansions. The blast from the explosion blew the front and roofs off the property. An emergency first aid post was set up in Streatham Hill Theatre to treat those injured in the explosion which fortunately resulted in no fatalities.

The blast damage caused to No 45 Streatham Hill was so serious that it was not possible to repair the premises which were subsequently demolished. The plot remained a bomb site for over 10 years until the present building was erected here in 1956.

Damage to 45 Streatham Hill from a V1 flying bomb that fell close by on 24th June 1944.

TELFORD COURT

Telford Court stands between Nos 62-88 Streatham Hill and was built in 1931 to the designs of Frank Harrington.

Among the large houses demolished to make way for this development was one of Streatham's most extraordinary buildings, No. 54 Streatham Hill, which was known by a variety of different names during its lifetime.

The property was built in 1838 by John Mallcott who lived there with his family until 1850. John was a builder and erected the National Gallery which was built between 1832 and 1838 on the site of the old Royal Mews.

He was responsible for the demolition of a number of London's old landmarks and is reported to have built his Streatham home from stones reclaimed from these buildings, including masonry from the Royal Mews.

In the forecourt of his Streatham home stood some large stone balls which were identical to those featured in an old print of the Royal Mews dated 1764.

Around the top of his house was a frieze that once adorned Carlton House in Pall Mall, the former residence of the Prince Regent. Hence his Streatham house being known as Carlton Villa and later as Carlton House.

Along the front of the property ran a white stone wall made from parts of old London Bridge which was demolished when the new bridge was erected in 1823-31. This led to the house also being called the Stone House and Bridge House.

When the building was demolished in 1927 much of the old stonework ended up as hardcore and was used in the laying down of Streatham Hill roadway and the foundations for Telford Court.

However, some small relics of Mallcott's house survived. A part of the London Bridge wall was used by Edward Wates to make a garden seat for his home at No 4 West Drive, Streatham Park, which is now Yew Tree Lodge.

Another stone from the house was taken to America where it was incorporated into Rotary Lodge constructed by the Rotary Club of Greenville, South Carolina, from rocks donated by Rotary Clubs from all over the world.

When the house was being demolished in 1927 a workman believed the building was at least 400 years old and said the stone work "would last for ages and was equal to many of the important public buildings in London and elsewhere". A shrewd judgement by a labourer who at the time was unaware of the origins of the building!

The stone balls on the Royal Mews which stood in the garden of No. 54 Streatham Hill.

Carlton House the residence of the Prince of Wales

Royal Mews

TELFORD PARADE MANSIONS

Telford Parade Mansions stands at the southern junction of Streatham Hill and Telford Avenue and were built in 1934/5. They were designed by the architects Frank Verity, Beverley, Horner, and built by Barnard & Childs for the Telford Property Trust Ltd.

It occupies the site of a large and impressive Georgian house called Telford Lodge which was erected here in 1833. The house was one of the grandest fronting Streatham Hill and stood amid large gardens.

In December 1869, William Henry Ryder, a rich New Bond Street jeweller, moved into the property and it was he that christened the house Telford Lodge. He died in 1882 bequeathing much of his £128,000 fortune to various charities.

George W Ryder, his nephew, occupied the house until around 1899 after which the property appears to have been empty, or let on short leases.

In the autumn of 1904 Thomas Whitford, the headmaster of Montrose College, transferred his school here.

The College had been founded in 1884 when it probably was originally known as the South London High School. It provided education to boys seeking careers in commerce and the professions.

Montrose College

As an indication of the college's success, in 1894 a special Masonic Lodge was formed by its old boys which survives today as Old Boys Lodge No. 2500. In 1898 it had 35 members rising to 55 in 1903.

Among the founding members was Thomas Whitford, the Senior Warden of the Lodge, who was described then as a Schoolmaster of Montrose College, Brixton Hill.

In 1909 a thief, Wallace Harry Wagner from Germany, was jailed for 9 months for breaking into Whitford's home at 15 Wavertree Road, Streatham Hill, and stealing his masonic apron, sash and ceremonial "jewels" among other items.

As well as being head of the College, Whitford played an active part in local politics and was a member of the Streatham Conservative Club.

By 1912 Charles Hinton had taken over the headship of the College but it would appear that at this time the college was experiencing difficulties and it subsequently closed.

In 1913, in response to requests from parents of many of the boys attending the College, the principal master, John Henry Bray, established a new school nearby at 225 Brixton Hill called Belmont College.

Telford Lodge was then converted into a hotel and subsequently became the Telford Court Residential Hotel until 1934 when it was demolished to make way for the building of Telford Parade Mansions.

Junction of Streatham Hill and Telford Avenue c1912

STREATHAM NOW & THEN AGAIN

WAVERTREE COURT

Wavertree Court stands at the southern junction of Wavertree Road and Streatham Hill and comprises three, four-story high blocks of flats, containing a total of 83 apartments.

Between the blocks are two secluded gardens, incorporating water fountains, trees, shrubs and flower beds.

The development dates from 1933/4 and was designed by Frank Harrington.

One of Wavertree Court's former residents is the actor Roger Moore. He started his career as an extra on the film *Caesar and Cleopatra* after which he attended RADA, where he met Lucy Woodard, the 16 year old daughter of a Streatham taxi driver. They married in 1946 and lived with Lucy's parents in Buckleigh Road before moving to Wavertree Court.

Lucy, a professional skater and actress, changed her name to Doorn Van Steyn, but her marriage was short lived and ended in 1952 after Roger's affair with the singer Dorothy Squires, whom he married in 1953.

Roger's big break came with his television role as Ivanhoe in 1957 which ran for 39 episodes. This led to him playing The Saint for seven years from 1962-9, with international stardom following after he took over the role of James Bond.

Wavertree Court in 1934

Wavertree Court stands on the site of two large Georgian houses, Nos. 37 and 39 Streatham Hill which were built c1823. Dennis Wheatley, the famous novelist, spent his early boyhood at No. 37, then known as Wooton Lodge.

Dennis was born on January 8 1897 and moved to Wooton Lodge in 1904, where he stayed until 1910 when his family moved to *Friars Croft*, No 1 Becmead Avenue. It is interesting to note that today this former home of Britain's most famous occult author is now the residence of the Rector of Streatham!

In 1914 the family moved to their third and final residence in Streatham, at Clinton House, No 1 Palace Road.

In 1927 Dennis's father died and in 1930 his mother became Lady Newton when she married Colonel Sir Louis Newton, a former Lord Mayor of London in 1923, and the family's association with Streatham came to an end.

Dennis sold the family-owned wine business in 1930 to concentrate on his writing, producing a series of popular novels, a number of which were filmed, such as *The Devil Rides Out* and *To the Devil a Daughter*.

At the height of his popularity his books were translated into 31 languages. He died in November 1977 and is buried in West Norwood Cemetery.

Roger Moore

Dennis Wheatley

Wooton Lodge

BRIXTON HILL BUS DEPOT

The old London County Council (LCC) Tram Depot, erected in 1923, is now used to provide extra parking facilities for buses operating out of the neighbouring Brixton Hill Bus Garage.

The building was originally designed as a tram trailer shed but by the time it was completed the decision had been made to no longer use trailers and it therefore provided extra accommodation for 31 tram cars operating out of the Telford Avenue Tram Depot.

After the last tram trundled through Streatham on 7th April 1951 the Telford Avenue Depot was demolished and the Bus Garage was erected on the site.

As there was no need for the tram shed it was sold for commercial use and was, for many years, Stratstone Garage which operated a Daimler and Jaguar dealership from the building.

Stratstones closed in the late 1990s and in 2003 the building was acquired by Arriva to supplement facilities at their adjacent bus garage.

In September 2010 a serious fire occurred here which took 50 firemen with 10 appliances over three hours to bring under control.

The fire is believed to have been caused by an overheated engine in a bus parked here. The roof of the building was badly damaged and two buses were destroyed in the blaze.

Trams in the Brixton Hill Tram Depot in the 1930s

Aspen House

The building stands on the site of Aspen House which was built in 1839 as the family home of the Roupells. John, and his son Richard Palmer Roupell, were north Lambeth lead merchants. They developed the Roupell Park Estate between 1840-60 which includes Christchurch and Palace Roads.

William Roupell, one of Richard's four illegitimate children, became MP for Lambeth in 1857 at the age of twenty seven. His election campaign was described as one of the most corrupt in London's history.

By 1862 he was on the verge of bankruptcy having squandered a vast sum of money and losing most of the land it had taken his father and grandfather fifty years to acquire.

At trials in 1862 and 1863 he confessed to forging his father's will and was sentenced to 14 years imprisonment.

Following Mrs. Roupell's death in 1878, Aspen House was purchased by William Yeats Baker, the maternal grandfather of Dennis Wheatley, the famous Streatham author.

William was an avid collector of objects d'art, and after his death on 11th August 1916 it took Christies 6 days to auction off the contents of his house.

STREATHAM HILL RESERVOIR

Standing behind Pullman Court and Brixton Hill Bus Garage is a large area of grassed over land which backs onto Daysbrook Road. However, this is not a public park or open space but the turfed roof to Streatham's underground water reservoir.

This dates back to 1832 when a reservoir covering an area of over 3 acres was built here by the Lambeth Water Works Company. Originally the reservoir was open to the elements providing a large expanse of water, like a huge lake, amid the fields which then covered this part of Streatham Hill.

After the construction of the reservoir its owners, listed in the parish rate books as the Directors of the South Lambeth Water Works, had the honour of owning the largest single rated property in the parish valued at £200.

Originally water was pumped to the reservoir directly from the Thames where the company drew their supplies from their works under Hungerford Bridge.

By the early 1850s the Lambeth company had moved their extraction plant to Seething Wells near Kingston.

This meant that although the water still came from the Thames it was abstracted 15 miles upstream from Hungerford Bridge where the water was much cleaner than that to be found in the river in central London.

As demand for fresh water increased the reservoir at Streatham Hill was doubled in size and then roofed over to protect the drinking water stored there from becoming contaminated by the notoriously polluted London air.

Streatham MP, Keith Hill (right), inspects the underground reservoir at Streatham Hill in May 2003.

Leading from the Reservoir, and running under Brixton Bus Garage to Streatham Hill, is a specially constructed tunnel which carries two 18 inch pipes which connect the reservoir to the water mains.

Streatham Hill Reservoir in 1897

The pipes were originally buried in the ground with no special protection.

However, when the Bus Garage was built in 1951-3 their position posed serious problems to London Transport as they lay where the foundations of the garage were to be constructed.

The only solution was to build a special tunnel around the pipes to protect them from the weight of the building above.

By the start of the 21st century the reservoir was no longer in use. However, in order to boost London's water supplies during the summer months the reservoir was brought back into use by Thames Water in 2003.

A special membrane was installed to provide added protection to the 4.3 million gallons of water now stored there.

Aerial view of Streatham Hill Reservoir between Pullman Court and Daysbrook Road

CHRISTCHURCH HOUSE

Christchurch House stands on the site of Boylands Oak, a large detached Victorian mansion which was the home of the famous music hall star Kate Carney (1869-1950).

Kate was born in Southwark in 1869. Her father had been on the stage as an old time comedian who worked with a partner as "The Brothers Raynard".

She spent a lifetime in the music halls, appearing under the names of Kate Raynard, Kate Paterson and "The Cockney Sweetheart".

She started her career at the age of 10, and although she later gained fame for her cockney character songs, she made her first stage appearance in 1890 as a singer of Irish ballads. Her first hit song was *Here's My Love to Old Ireland*, but it was *A Donkey Cart Built for Two*, *Sarah* and *Three Pots a Shilling*, which established her at the top of the bill, and earned for her a reputation as the "Cockney Queen".

In 1885, at the age of 15, Kate married 17 year old George Barclay (1867-1944), a comedian then working in the double act of "Barclay and Perkins - the Brewers of Mirth".

By 1905, Kate and her husband were both well known and wealthy performers. It was in this year that they moved to Boylands Oak, at 241 Brixton Hill on the northern junction with Christchurch Road. From these premises, her husband ran a very successful variety agency where he booked many of the well known stars of the day.

In 1912 Kate attended her first Royal Command Performance. She made a second appearance in 1935, the year of her golden wedding, when she appeared in the finale "Cavalcade of Variety".

In 1937, Kate and George moved to Brockenhurst, 50 Aldrington Road. Following George's death in 1944, Kate moved to Tring, later returning to London to reside in various hotels.

In 1948 she went to live with her son, George Barclay, at his flat in Christchurch Road. She died at the Whitlands Nursing Home at 38 Palace Road on New Years Day 1950, aged 80.

After Kate vacated Boylands Oak it was demolished, along with adjacent houses, to make way for Christchurch House which was built on the corner of Brixton Hill and Christchurch Road in 1938.

This 6 storey building containing 110 flats was designed by Couch and Copeland, and is considered to be a fine example of pre-war contemporary architecture.

Boylands Oak, Brixton Hill.

BRIXTON HILL UNITED REFORM CHURCH

Brixton Hill United Reform Church was built in 1993. A place of worship has occupied this site since 7th October 1829 when a Union Chapel was opened here for use by Anglicans and Non-conformists living in this part of Streatham parish.

It was described as a "handsome detached building" and cost £2,387. The church could accommodate 530 worshippers and had a small vestry which opened into the chapel.

The Anglicans left to form Christ Church in 1837 and the Baptists departed in 1842 to form their own congregation in New Park Road.

The chapel then became a Congregational Church which was rebuilt in 1871 at a cost of £7,836.

The new church stood on the summit of Brixton Hill some 300ft above the Thames placing it level with the foot of the golden cross on the dome of St. Paul's Cathedral.

The church was built of Kentish ragstone with Bath stone dressings and had a magnificent organ, the pipes of which formed a major focal point behind the central pulpit. The building opened for worship on 12th December 1871 and could accommodate 1,050 people.

In 1982 the church was declared unsafe and as the congregation could not afford the £40,000 required for repairs it was demolished. Church meetings were then held in a small prefabricated hut at the rear of the site until a new church was built. This opened for services on Valentine's Day 1993.

Behind the church stands a mid-Victorian Sunday School, a Grade II listed building. It was designed by Rowland Plumbe and the foundation stone was laid by Mr. James Spicer JP in December 1878. The building cost £3,707 and could accommodate 400 children. After standing derelict for a number of years the school was restored in 2001-2 and converted to provide residential units and now forms part of Rush Common Mews.

During the Second World War the Sunday School was used as a First Aid Post. In July 1941 Lady Louis Mountbatten visited the Post and met the members of the British Red Cross and the St. John Ambulance Brigade based there. During the Blitz this post, together with that at Streatham Baths, treated over 1,000 casualties as well as 760 victims of street accidents.

Also at the rear of the church was a small burial ground in which the first minister of the Union Chapel, the Rev. John Hunt, was interred in 1856.

Streatham Hill Congregational Church c1910

Interior of Streatham Hill Congregational Church in 1932

Henri Marie
avec la participation de
Georges Bernage et Wolfgang Schneider

Villers-Bocage

dédié aux victimes civiles

HEIMDAL

– Ouvrage conçu par Georges Bernage.

– Ecrit par Henri Marie, décédé le 14 septembre 2007.

– Traduction : John Lee.

– Cartes : Bernard Paich.

– Maquette : Erik Groult qui a réalisé les reportages sur le terrain.

– Composition et mise en pages : Annick Le Coquet, Christel Lebret.

– Photogravure : Christian Caïra, Philippe Gazagne.

– Infographie : Philippe Gazagne.

– Iconographie : IWM, BA.

Editions Heimdal
Château de Damigny - BP 61350 - 14406 BAYEUX Cedex
Tél. : 02.31.51.68.68 - Fax : 02.31.51.68.60 - E-mail : Editions.Heimdal@wanadoo.fr

Copyright Heimdal 2003. Seconde édition 2010. La loi du 11 mars 1957 n'autorisant, aux termes des alinéas 2 et 3 de l'article 4, d'une part, que les « copies ou reproductions strictement réservées à l'usage privé du copiste et non destinées à une utilisation collective » et, d'autre part, que les analyses et les courtes citations dans un but d'exemple et d'illustration, « toute reproduction ou représentation intégrale, ou partielle, faite sans le consentement de l'auteur ou de ses ayants droit ou ayants cause, est illicite. Cette représentation, par quelque procédé que ce soit, constituerait donc une contrefaçon sanctionnée par les articles 425 et suivants du code pénal.

ISBN 978-2-84048-292-5

Avant-Propos
Foreword

La **bataille** de **Villers-Bocage** est une bataille ignorée de beaucoup, une bataille oubliée. Pourtant, c'est une bataille qui aurait pu précipiter la défaite de l'armée allemande en Normandie. Le voyageur qui traverse la Normandie ne s'étonne pas de rencontrer, sur sa route, des villages, des bourgs, de petites villes aux clochers modernes, aux maisons bien alignées, peu dissemblables dans leur architecture d'époque. Il ne doit pas ignorer qu'en 1944, de nombreuses agglomérations prises dans la tourmente de la guerre ont été détruites et que, la paix revenue, elles ont été reconstruites. Lorsqu'il traverse Villers-Bocage, il comprend rapidement que la cité fait partie de ces villes mais, à moins qu'il ne soit particulièrement averti, il ignore sans doute à quelle occasion Villers a été détruite.

A-t-il entendu parler de la bataille de Villers-Bocage ? Connaît-il la part importante qu'a eu cet affrontement dans la guerre après le débarquement ? Sait-il que le résultat, s'il avait été autre, aurait pu changer le cours de la bataille de Normandie et en précipiter le déroulement ? Mis à part les anciens qui ont vécu cette période et quelques plus jeunes intéressés par l'histoire locale et le débarquement, les Villérois ignorent pour la plupart ce qu'a été cette partie de l'histoire de leur commune.

Dans de nombreux ouvrages écrits sur la guerre, on peut trouver quelques pages, quelques lignes sur la bataille de Villers, souvent des ouvrages relatent une action précise, la plupart ne s'attardent que sur quelques épisodes, souvent les mêmes, et passent les autres sous silence. Parfois ces épisodes sont décrits d'une façon plus ou moins romancée sans souci de vérité historique et parfois dans un but de propagande. C'est après avoir parcouru ces livres et y avoir relevé des contradictions que m'est venue l'idée de rassembler écrits et photos puis de recueillir des témoignages de civils qui ont connu cette époque, des témoignages de soldats anglais qui ont combattu dans nos murs, de fouiller les archives municipales et d'essayer d'en faire une synthèse.

J'avais écrit en 1993 une histoire de la bataille de Villers dans un livre paru chez le même éditeur sous le titre « Tigres au Combat » mais, depuis cette date, à la suite des manifestations qui se sont déroulées à l'occasion du cinquantième anniversaire du débarquement, j'ai pu recueillir de nombreux témoignages d'anciens combattants britanniques et de civils. J'ai pu ainsi, tout en conservant le récit initial, l'étoffer, en préciser certains points, le modifier même, rapporter quelques controverses, le compléter et ajouter un rappel des combats qui se sont déroulés dans le Pré-Bocage après le 13 juin jusqu'au début du mois d'août, c'est-à-dire au moment de la libération de notre région. J'espère ainsi m'être rapproché le plus possible de la vérité historique.

Pour beaucoup de villes, le jour de la libération marque une date importante qui est célébrée partout avec respect. Nous verrons que, pour l'histoire de Villers-Bocage, une autre date est importante, le 13 juin 1944. Ce jour-là, les soldats alliés entraient dans la ville pour livrer une bataille indécise, la bataille de Villers-Bocage, alors que le 4 août, ils libéraient une ville détruite, abandonnée de ses habitants et des troupes allemandes.

Selon de nombreux commentateurs, la bataille de Villers-Bocage a été un « pitoyable épisode de la guerre » et, si l'attaque hardie du général Montgomery avait réussi, le résultat aurait pu être « fantastiquement différent ». Nous verrons que la « bataille perdue » a eu des prolongements positifs et qu'elle a apporté de nombreux enseignements aux Britanniques. L'armée allemande a vu une de ses contre-attaques stoppée net. Elle a subi dans la bataille du 13 juin et du 14 juin dans la région de Villers des pertes importantes, surtout en blindés, qui l'ont affaiblie dans cette partie du front, ce qui permet de penser que les sacrifices consentis par les soldats britanniques n'ont pas été vains.

Henri Marie

The Battle of Villers-Bocage is a battle unknown to many, a forgotten battle. And yet it is a battle which might have brought about the early defeat of the German Army in Normandy. The traveller crossing through Normandy is not surprised to see, along the way, villages, boroughs, small towns with modern church steeples, neat rows of houses, all very much alike in their period architecture. He ought to know that in 1944, many of these towns and villages were caught up in the tumult of war and destroyed, and how they were rebuilt when peace returned. Passing through Villers-Bocage, he soon understands that this was one such town but, unless he is particularly well-informed, no doubt he does not know in what circumstances Villers was destroyed.

Has he ever even heard of the Battle of Villers-Bocage? Is he aware that this was a major battle in the campaign after D-Day? Does he know that had the result been different, it might have changed the course of the Battle of Normandy and bring it to an early close? Apart from the older folk who lived through this period and a few younger ones who are interested in local history and the D-Day landings, the people of Villers are mostly ignorant of this chapter in the history of their town.

In many books written about the war, one finds a few pages, a few lines on the Battle of Villers, such books often give an account of some specific action, most only have room for one or two episodes, often the same ones, and have nothing to say about the rest. Sometimes these episodes are described in a more or less fictionalized way with no care for historical truth and occasionally for propaganda purposes. After perusing

these books and noted their contradictions, it occured to me to collate all these writings and photographs and then to collect eye-witness accounts from civilians familiar with the time, accounts of British soldiers who fought inside the town, sift through the municipal archives and to try to bring everything together.

In 1993, I wrote a history of the Battle of Villers in a book with the same publisher entitled "Tigres au Combat" ("Tigers in Battle") but, since that time, following the ceremonies to commemorate the fiftieth anniversary of D-Day, I was able to collect a good many eye-witness accounts from British ex-servicemen and civilians. In this way, working from my initial account, I have contrived to expand on it, clarify certain points, even modify it, tackle some controversial issues, supplement it and add an account of the battle which was fought in the Pré-Bocage after June 13th and until the beginning of August, i.e. up to the moment when the area was finally liberated. I hope this has enabled me to get as close as possible the historical truth.

For many towns, the day of liberation is a big day which is celebrated everywhere with great respect. We will see that there is another important date in the history of Villers-Bocage: June 13th, 1944. On that day, Allied soldiers entered the town to fight a finely balanced battle, the Battle of Villers-Bocage, while the town they liberated on August 4th was in ruins, abandoned by both its inhabitants and the German troops.

According to many commentators, the Battle of Villers-Bocage was a "pitiful episode in the war" and, had General Montgomery's bold attack succeeded, the outcome would have been "fantastically different". However, we will see that the "lost battle" had positive repercussions, as it taught the British many a lesson. The German Army saw one of its counter-attacks stopped in its tracks. During the battle of June 13th and 14th it sustained heavy losses in the Villers, especially in tanks, which weakened it along this section of the front, which leads one to think that maybe the sacrifices made by the British soldiers were not in vain.

Henri Marie

CETTE PREMIERE PIERRE DE LA RECONSTRVCTION DE VILLERS-BOCAGE,
DETRVIT AV COVRS DES COMBATS DE CHARS ET DES BOMBARDEMENTS
DV 13 JVIN AV 5 AOVT 1944, A ETE POSEE LE DIMANCHE 7 MARS 1948 PAR
MONSIEVR ALEXANDRE STIRN PREFET DV CALVADOS
MAIRE: JEAN LEVEQUE.
PRESIDENT DE L'ASSOCIATION DE RECONSTRVCTION
MAURICE LEPIETRE.
PRESIDENT DE L'ASSOCIATION DE REMEMBREMENT:
ANDRE DOUBLET.
ARCHITECTE EN CHEF: PIERRE DUREUIL
ARCHITECTE EN CHEF ADJOINT: LEON REME

Sommaire/*Summary*

Avant-propos/*Foreword* 3

1. Villers-Bocage en juin 1944/*Villers-Bocage in June 1944* 6

2. *SS-Panzer-Abteilung 101* 14

3. La bataille de Villers-Bocage/*The Battle of Villers-Bocage* 36

4. Tragique dénouement/*Tragic ending* 118

Annexe : Evalution de l'engagement de la *SS-Panzer-Abteilung 101*
d'un point tactique/*Assessment of the commitment of SS-Panzerabteilung 101
in Normandy from the tactical viewpoint* 154

Bibliographie 160

Le texte est de Henri Marie, l'iconographie est présentée et légendée par Georges Bernage, l'annexe est de Wolfgang Schneider.

Ch. MILNER
Major
A Coy IRB

D. GORDON
Maj. Gen.
Lt. Col. 7Q

T. Pearson
General
2RB

Joe LEVER
RTR

Comte ONSLOW
fils de Lord Col.
CRANLEY 4CLY

FORESTER
GQ's

Pat DYAS
QG
4CLY

A. BURN
5 RHA

Tim LANYON
Adjutant
Rugard

1   Villers-Bocage en juin 1944
Villers-Bocage in June 1944

C'est une commune rurale de 1 200 habitants, chef-lieu de canton situé au sud-ouest de Caen. La commune s'étire sur le flanc ouest d'une colline entre le sommet, la cote 217, le haut de la côte des Landes de Montbroc, la « côte des Landes », et une vallée où coule une petite rivière paisible, la Seuline, affluent de la Seulles. La ville est entourée de vallons alignés en amphithéâtre, au sud, le Mont-Pinçon, point culminant de la Normandie, et les bois d'Aunay, à l'ouest, les bois de Jurques, au nord-ouest, Caumont-l'Eventé.

Villers est un important nœud de communications où convergent les routes de Caen, de Tilly sur Seulles et Bayeux, de Caumont et Saint-Lô, d'Evrecy, de Vire, d'Aunay sur Odon, Thury-Harcourt et Condé sur Noireau. C'est la porte du Bocage et du Bessin. Finis les grands espaces éternellement monotones de cultures que le voyageur qui vient de Caen a pu voir dans la plaine. Ils laissent la place, au fur et à mesure qu'il approche, à des zones d'herbages clairs ou plantés de pommiers et de petits chemins creux et ombragés. Dans le Bocage, cultures et zones d'élevage alternent selon la nature du sol et la disposition du terrain.

Villers, c'est essentiellement une grande rue et deux rues longues et étroites parallèles à cet axe, la rue Saint Germain, au sud mène de l'église à la place du marché, la rue Saint Martin au nord, relie la place Jeanne d'Arc à la place Richard-Lenoir. Ces trois rues sont entrecoupées de rues transversales, rue Montebello, boulevard Joffre, rue Jeanne Bacon, rue du pied fourchu, rue Curie, rue aux Bouchers et de ruelles étroites. La rue principale, la RN 175, est constituée de deux rues, la rue Pasteur, le « bas du bourg », la rue Georges Clemenceau, le « haut du bourg ». Elles longent des places, la place Jeanne d'Arc à l'ouest, puis en montant le bourg, la place de la Liberté, la place Richard-Lenoir. Là, se trouvent plus de quatre-vingts commerçants, artisans, membres de professions libérales, cafés, restaurants, hôtels, qui constituent la vitrine de la commune. Tout voyageur qui traverse la ville doit s'étonner qu'elle ne soit pas plus importante. Grâce à sa position géographique, la ville accueille chaque mercredi un important marché. En plus du marché forain se tient aussi un marché aux bestiaux ; la plus grande partie des animaux transite par la gare.

La voie ferrée à voie unique a été construite autour de 1885. Elle relie Caen à Vire puis Saint-Lô, elle passe par le bourg de Villers, puis entre Aunay et Saint-Georges d'Aunay, dessert ensuite Jurques, Saint-Martin des Besaces et Bény-Bocage avant d'atteindre Vire. Depuis la guerre, un service voyageurs interrompu en 1938 a été rétabli. Acheteurs, vendeurs, courtiers sont souvent dans l'obligation de séjourner dans les hôtels et « cafés-restaurants » de la commune où la plupart des affaires sont réglées autour d'un verre ou d'une tasse, ce qui explique le nombre anormal de débits de boisson pour un bourg de 1 200 habitants.

Depuis le début du XXᵉ siècle, un nouveau quartier s'est créé surtout après la Première Guerre mondia-

It was a country town of 1,200 inhabitants, the main town in this area south-west of Caen. The town extended along the western flank of a hill between its summit (Point 217), the top of the Montbroc moors (the "Côte des Landes"), and a valley through which gently flowed a small tributary of the Seulles, the River Seuline. The town was surrounded by small valleys, like an amphitheatre; to the south, Mont Pinçon, the highest point in Normandy, and the woods of Aunay; to the west, the woods of Jurques, to the north-west, Caumont-l'Eventé.

Villers is a major crossroads where the roads to Caen, Tilly-sur-Seulles and Bayeux, Caumont and Saint-Lô, Evrecy, Vire, Aunay-sur-Odon, Thury-Harcourt and Condé-sur-Noireau all meet. It is the gateway to the "Bocage" (hedgerow country) and the Bessin area around Bayeux. This marks the end of the endless expanses of cropland which the visitor to Caen will have seen in the plain, as they make way for pastureland maybe planted with apple-trees, and small shadowy sunken lanes. In the Bocage, crops alternate with stock farms according to the nature and lie of the land.

Villers was primarily a main street and two long, narrow streets running parallel to it; the Rue Saint Germain to the south ran from the church to the marketplace; the Rue Saint Martin to the north linked the Place Jeanne d'Arc and the Place Richard-Lenoir. Intersecting these three streets were the Rue Montebello, Boulevard Joffre, Rue Jeanne Bacon, Rue du Pied Fourchu, Rue Curie, Rue aux Bouchers, and some narrow back lanes. The main road, the N175 highway, comprised two streets: the Rue Pasteur, "down the hill", and the Rue Georges Clemenceau, "up the hill". They ran alongside some squares, the Place Jeanne d'Arc to the west, then further up, the Place de la Liberté and the Place Richard-Lenoir. Here were over eighty storekeepers, craftsmen, professional people, cafes, restaurants and hotels, making up the town's showcase. Any traveller passing through would have been surprised that it was no bigger. With its geographical position, the town hosted a big market every Wednesday. In addition to the open market there was a cattle market; most of the stock was brought in via the station.

The single track railroad was built in c. 1885. It went from Caen to Vire then Saint-Lô, passing through Villers, then between Aunay and Saint-Georges d'Aunay, with stops at Jurques, Saint Martin-des-Besaces and Bény-Bocage before coming to Vire. Since the war, a passenger service that was discontinued in 1938 has been restored. Buyers, sellers and brokers are often obliged to stay in the town's hotels and "bar-restaurants" where a great deal of business is settled over a glass or cup of something, hence the unusual number of bars for a population of just 1,200.

At the turn of the 20th century and particularly after World War I, a new district was created. The Town Hall was built in 1896, and the station was opened in 1886.

A boulevard was constructed, the Boulevard Joffre south of the main street, round a new district in which private housing was built, the first low-rent apartment

Villers-Bocage au début du XXe siècle :

1. Villers, c'est essentiellement une grande rue. Nous voyons ici un aspect du haut du bourg (rue G. Clemenceau), à l'est, vers Caen.

2. Grâce à sa position géographique, la petite ville accueille chaque mercredi un important marché. Il s'y tient aussi un marché aux bestiaux dont nous voyons ici un aspect.

3. Le bas du bourg, à l'ouest, au niveau de la place Jeanne d'Arc. L'église se trouve à proximité.

(Collection Henri Marie.)

Villers-Bocage in the early 20th century:

1. *Villers was mostly just the high street. Here we see a view of the top of the town (rue G. Clemenceau), on the east side towards Caen.*

2. *Thanks to its geographical position, it had a big market each Wednesday for such a small town. There was also a cattle market, of which we see a view here.*

3. *The bottom end of town, to the west, on a level with the Place Jeanne d' Arc.*
The church is not far away.

1. Les commerces se succèdent dans la grande rue où passe toute la circulation de la RN 175.

1. Shops are strung along the main street where all the N 175 highway traffic passes.

le. L'Hôtel de Ville a été construit en 1896, la gare a été inaugurée en 1886.

Un boulevard a été tracé, le boulevard Joffre qui délimite au sud de la rue principale un nouveau quartier où ont été édifiées des maisons particulières, les premiers H.L.M (loi Loucheur de 1928), de petites industries, bonneterie Philips, la « Tricoterie » et, près de la gare, scierie Rivière qui fabrique pour l'instant du combustible pour gazogène.

Villers est une commune rurale de petite superficie. Les exploitations agricoles y sont peu nombreuses et se trouvent à la périphérie du bourg mais surtout aux alentours, aux Hauts-Vents, à Montbrocq, à la queue de Renard, au moulin, au château. Dans les grands herbages qui entourent la ville, paissent des bovins qui appartiennent pour la plupart à des négociants en bestiaux, les herbagers. Les animaux y demeurent plus ou moins longtemps, le temps d'être mis en état avant d'être menés au marché. Une plaine céréalière s'étend dans la partie sud de la commune, elle est exploitée par les fermes du château.

Le **château** appartient depuis 1929 à Monsieur Amédée de Clermont-Tonnerre. Il a été construit au XVIIe siècle sous le règne de Louis XIII et n'était lors de sa construction qu'un relais de chasse. Il a été agrandi d'une aile droite vers 1770 et depuis lors s'est constitué autour un domaine avec un grand parc, des bois, des étangs, des pièces d'eau. Depuis la guerre, les Allemands se sont installés dans des baraquements sous les grands arbres et un corps de garde en contrôle les allées et venues depuis l'entrée..

---

blocks (the Loucheur Act of 1928), along with small factories, the Philips hosiery, the "Tricoterie" (knitwear) and, near the station, the Rivière sawmill which currently manufactures fuel for gas generators.

Villers is a small country town. There were very few farms, and they were on the outskirts of the town, or more usually in the neighboring countryside, the Hauts-Vents, Montbrocq, the Queue de Renard, the mill, the château. In the large meadows around the town were grazing cattle belonging mostly to stock-traders and graziers, where they stayed for however long it took to fatten them up for market. A plain of cereal crops stretching out to the south of the town was operated by the castle farms.

The château had been the property of Monsieur Amédée de Clermont-Tonnerre since 1929. When it was first built during the reign of Louis XIII in the 17th century, it was no more than a hunting lodge. A right wing was added onto it in c. 1770 and since then a domain has grown up around it, with a large park, woods, ponds and lakes. Since the outbreak of war, the Germans had settled in quarters under the big trees, and there was a guard at the entrance to monitor comings and goings.

2. La rue du pied fourchu. Le pied fourchu est un pied constitué de deux sabots ongulés comme celui des bovins qui empruntaient cette rue lorsqu'ils étaient amenés au marché.

2. « Rue du pied fourchu ».

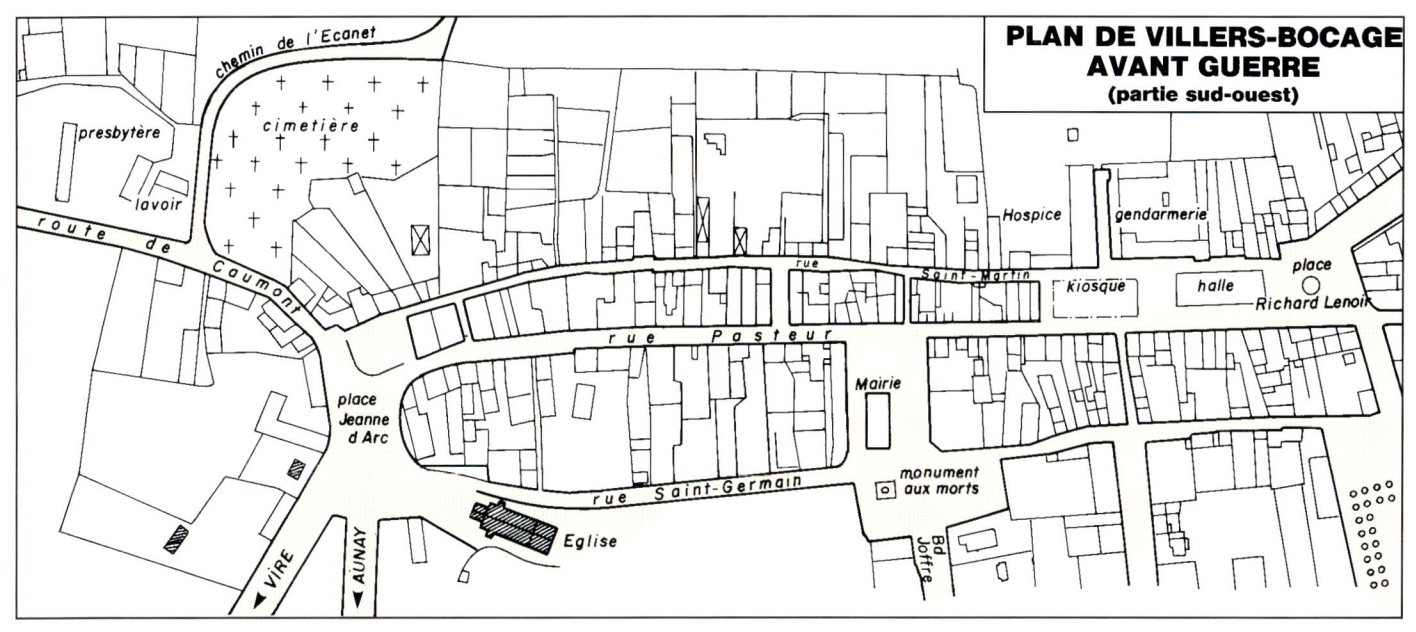

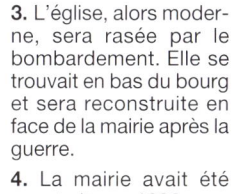

PLAN DE VILLERS-BOCAGE AVANT GUERRE
(partie sud-ouest)

3. L'église, alors moderne, sera rasée par le bombardement. Elle se trouvait en bas du bourg et sera reconstruite en face de la mairie après la guerre.

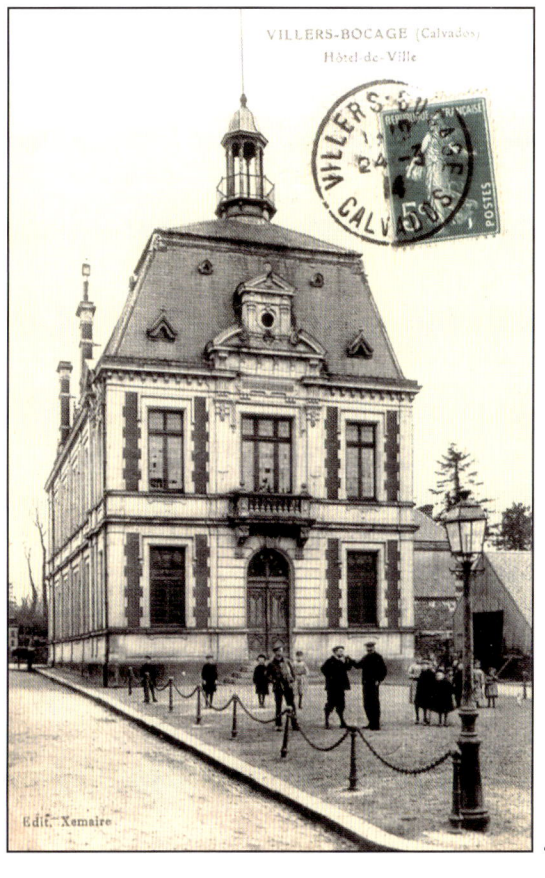

4. La mairie avait été construite en 1896.

5. A proximité se trouvait la poste. Tous les édifices présentés dans ces deux pages ont été détruits en 1944.

6. Le monument aux morts se trouvait entre la mairie et la poste.

(Documents coll. H. Marie.)

3. The church, modern at the time, was razed in the bombardment. It was at the bottom end of town and was rebuilt opposite the Town Hall after the war.

4. The Town Hall was built in 1896.

5. The Post Office was close by. All the buildings presented in these two pages were destroyed in 1944.

6. The war memorial was between the Town Hall and the Post Office.

(Documents coll. H. Marie.)

1. Marché, place de la liberté devant le kiosque et l'hospice.

2. La gendarmerie située face à cette place et face aux halles.

3. Maquette du centre ville avant guerre situant les divers édifices.

4. Les halles.

5. La place Richard-Lenoir, à l'est des halles.

6. La grande rue et les halles.

7. En continuant vers l'est, après les halles, la grande rue et la route de Caen à l'angle de la place Richard-Lenoir. Cette vue est prise dans le prolongement de celle de la photo « 5 ». La pharmacie est aussi visible sur les photos « 4 » et « 6 ».

(Coll. H. Marie sauf **3** : photo Vandevorde.)

1. The market, the Place de la Liberté in front of the bandstand and the hospice.

2. Gendarmerie located opposite this square and opposite the market.

3. Model of the town center prewar locating the various buildings.

4. The market.

5. The Place Richard-Lenoir, east of the market.

6. The main street and market.

7. Still further east, after the market, the high street and the Caen road on the corner of the Place Richard-Lenoir.

5

6 Villers-Bocage — Grande Rue et les Halles

Hôtel de ville

VILLERS-BOCAGE - La Route de Caen

▲ Rue Pasteur

◀ Rue Georges Clemenceau

7

11

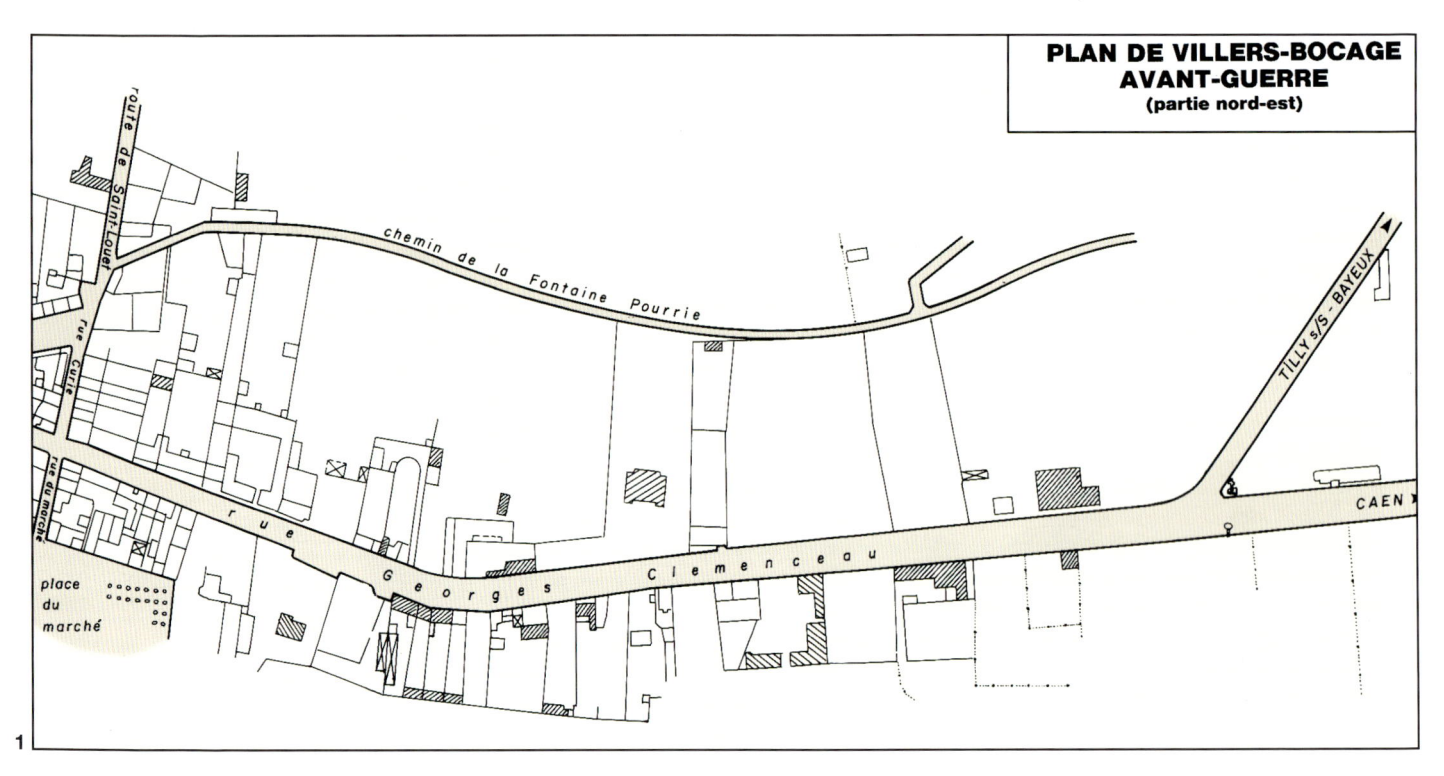

PLAN DE VILLERS-BOCAGE AVANT-GUERRE
(partie nord-est)

1. Partie nord-est du bourg, en direction de Caen. (Plan Heimdal.)

2. La gare au début du XXe siècle.

3. L'avenue de la gare.

4. Le quartier de la gare, d'après la maquette.

5. Le quai pour l'embarquement des bestiaux.

6. Le quartier de la gare créé surtout après la Première Guerre mondiale, rue Foch.

7. Le marché aux bestiaux est situé entre le quartier de la gare et la grande rue.

8. Le château au sud-ouest du bourg.

9. Le parc du château.

(Coll. H. Marie, sauf **4** et **7**, photo Vandevorde.)

1. North-eastern section of the town, out towards Caen. (Heimdal map.)

2. The station in the early 20th century.

3. The avenue to the station.

4. The station quarter, after the model.

5. The quay for loading cattle.

6. The station quarter created mostly after World War I, rue Foch.

7. The cattle market located between the station quarter and the high street.

8. The castle in the south-west of town.

9. The château grounds.

(coll. H. Marie, except 4 and 7, photo Vandevorde.)

2   *SS-Panzer-Abteilung 101*

Le chef de la Ire section, de la 2e compagnie (la 2/101 commandée par Wittmann) est assis en manteau sur la tourelle du Tiger « 232 ». Cette photo a été prise alors que ce Tiger remorque le « 231 » (voir page 156) le long de la RN 175 à l'est de Villers-Bocage. (BA 738/207/18a.)

1. *Wearing an overcoat and seated on the right on the turret of the Tiger "232", the commander of the 1st Platoon 2nd Company (2./101) commanded by Wittmann. This picture was taken while the tank was towing away Tiger "231" east of Villers-Bocage along Highway 175 (see page 156).*

Pourvu de puissants chars Tiger I, la SS-Panzer-Abteilung 101, arrivée sur le front de Normandie le 13 juin 1944, va jouer un rôle important en bloquant une offensive britannique. Ses puissants chars Tiger joueront ensuite un rôle défensif dans le secteur de Caen. Nous vous la présentons ici à l'entraînement dans le secteur de Beauvais un mois avant le débarquement puis lors de sa montée en ligne. Elle a alors bénéficié d'une importante couverture photographique, à des fins de propagande, ainsi qu'après la bataille de Villers-Bocage. Ces documents sont une importante contribution à l'histoire d'une des pages de la Bataille de Normandie.

With its powerful Tiger I tanks, the SS-Panzer-Abteilung 101, which arrived at the front on June 13th 1944, was to play a key role in blocking a British offensive. Its powerful Tigers went on to play a defensive role in the Caen sector. Here we see it in training a month before D-Day in the Beauvais sector and again when it moved up into the front line. It then came in for substantial photographic coverage, for propaganda purposes, and again after the battle for Villers-Bocage. These documents are a major contribution to one page of the Battle of Normandy.

2. Cette tenue est la plus courante de celles portées par les tankistes de la SS-Panzer-Abteilung 101. Nous voyons ici le treillis de protection deux pièces taillé dans de la toile camouflée *(Erbsenmuster)* mais de la même coupe que la tenue noire des tankistes de la Waffen-SS. Il a été adopté en janvier 1944 et est très courant lors de la bataille de Normandie. Les passants noirs de pattes d'épaule portent les lettres brodées « LAH » identifiant un tankiste du *SS-Pz-Rgt. 1*, ou un ancien de ce régiment (nombreux dans la *SS-Panzer-Abteilung 101*). Le calot noir du modèle 1940 est celui porté par la troupe. (Documents Heimdal.)

2. *The two-piece protective outfit is of the same cut as the black version of the tank uniform of the Waffen-SS, but made of camouflage cloth (Erbsenmuster). It was issued from January 1944 and was commonly worn in Normandy. Thee black loops on the shoulder straps have the letters "LAH" embroidered on them showing that the wearer was a tank crew member of the 1st SS-Panzer-Regiment or a veteran of that unit (as were many in the SS-Panzer-Regiment-Abteilung 101). The 1940 pattern black cap is that worn by the troopers. (Heimdal.)*

3. Marquages de Tiger I de la *SS-Panzer-Abteilung 101*. Chaque compagnie avait une manière bien particulière de marquer les signes tactiques à l'avant et à l'arrière de ses Tiger. **1.** 1re compagnie, à gauche un losange avec un S pour *schwere* (lourde) suivi d'un « 1 » et à droite l'écusson du corps blindé légèrement pointu. **2.** 2e compagnie, l'écusson du corps est à gauche, arrondi dans le bas. **3.** 3e com-

14

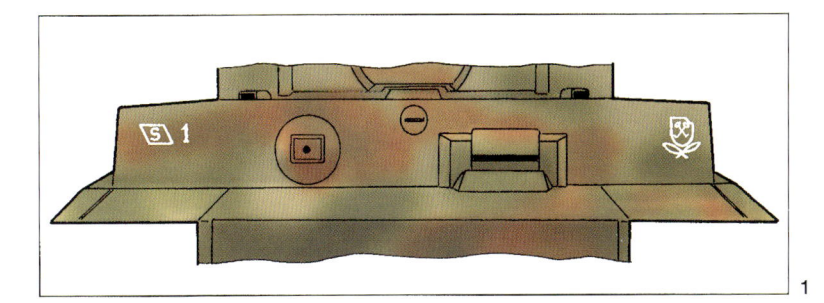

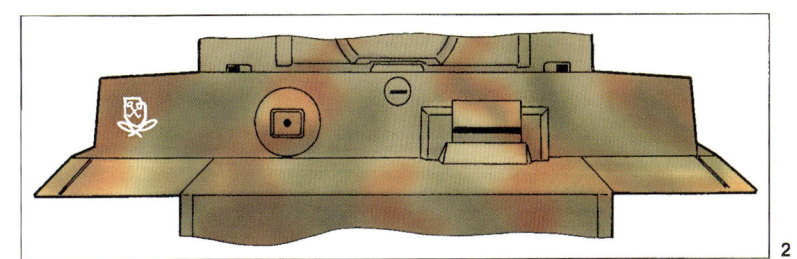

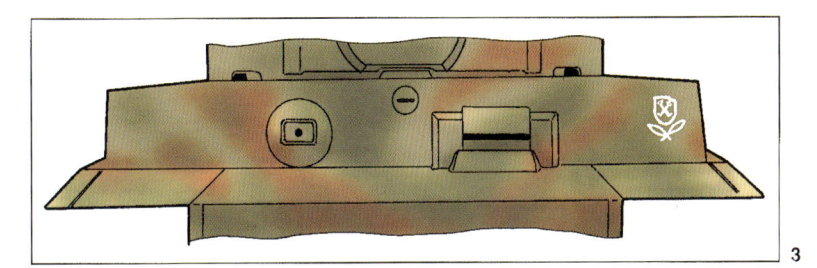

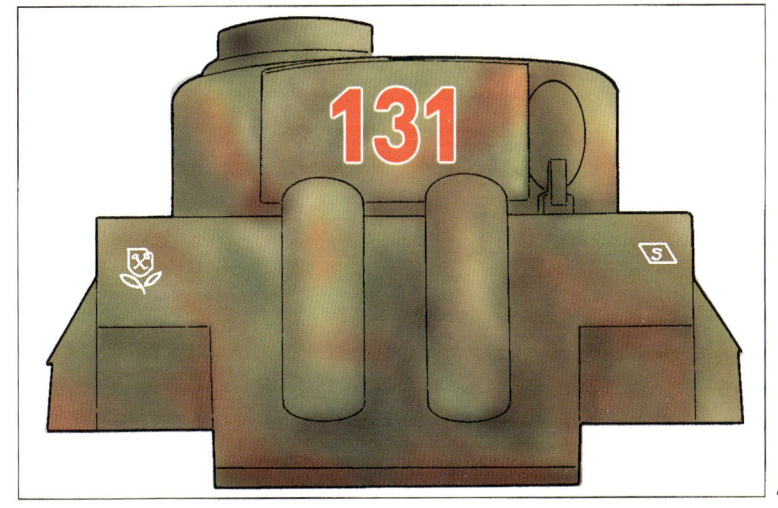

pagnie, même écusson mais placé à droite. **4.** 1ʳᵉ compagnie, avec l'écusson pointu à gauche et le losange avec le « S » à droite. **5.** 2ᵉ et 3ᵉ compagnies, avec un écusson arrondi à gauche. Les chiffres sont peints en rouge avec un liseré blanc. Le Tiger « 122 » était le char de l'Uscha. Arno Salamon, détruit à Villers-Bocage. Le « 213 » était celui de l'Oscha. Jürgen Brandt qui a détruit trois chars à Villers-Bocage le 13 juin. (Heimdal.)

3. *Specific markings on the Tiger Is of the heavy SS-Panzer-Abteilung 101, of which each company had its own way of marking the tactical identifications on the front and rear of its Tigers. The 1st Company **(1)** had on the left a rhombus shape with an "S" for schwere (heavy). On the right was the insignia of the corps slightly pointed at the base. For the 2nd Company **(2)** the corps insignia was on the left, rounded at the base. The 3rd Company **(3)** had the same insignia, but placed on the right. Rear identification markings of the 1st Company **(4)** and of the 2nd and 3rd Companies **(5)**. The numbers were painted in red outlined in white. Tiger "122" was Uscha. Arno Salomon's tank (destroyed in Villers-Bocage) and "223" was commanded by Oscha. Jürgen Brandt, who destroyed three tanks at Villers-Bocage on June 13th. (Heimdal.)*

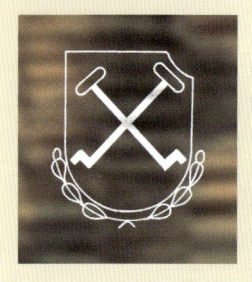

Mai 1944, manœuvres de la *SS-Panzer-Abteilung 101* dans le secteur de Beauvais.

Alors que les Alliés préparent le débarquement, un bataillon de chars lourds Tiger, la *SS-Panzer-Abteilung 101* s'entraîne dans le secteur de Beauvais, au nord-est de la Normandie. Quelques semaines plus tard, il rejoindra le front de Normandie. Disposant, lors de la montée en ligne, de 51 chars Tiger, seules ses 1re et 2e compagnies auront rejoint le secteur de Villers-Bocage le 13 juin.

1. Ici, les principaux officiers du bataillon discutent du déroulement de la manœuvre devant le Tiger « 311 ». De gauche à droite : le *SS-Untersturmführer* Willi Iriohn, en veston de cuir hérité de la Kriegsmarine via la marine italienne, en pantalon de treillis et avec la *Schirmmütze* (casquette plate) de la tenue de service ; le *SS-Sturmbannführer* Hein von Westernhagen, le chef du bataillon, il porte la combinaison camouflée modèle 1943 et le calot noir d'officier modèle 1940 à passepoils de fil d'aluminium ; et le *SS-Obersturmführer* Hanno Raasch, qui porte l'imperméable de motocycliste gris-vert et le même style de calot. L'officier de droite n'est pas identifié.

2. Autre photo du reportage où l'on retrouve les officiers Hein von Westernhagen et Hanno Raasch. Le troisième à partir de la droite, regardant l'objectif, est le *SS-Unterscharführer* Otto Blase, chef de char de la 3e compagnie.

3. Cette dernière photo nous montre plus nettement l'équipage du « 311 », celui de Hanno Raasch. Appuyé au canon, son conducteur, le *SS-Unterscharführer* Hofmann.

(Photos Bundesarchiv.)

May 1944, SS-Panzer-Abteilung 101 exercising in the Beauvais sector.

While the Allies prepared for the landing, a heavy Tiger tank battalion, SS-Panzer-Abteilung 101 was training in the Beauvais sector, north-east of Normandy. A few weeks later, it joined the front in Normandy. At the time of moving up to the front line, it had 51 Tiger tanks, but only the 1st and 2nd Companies had reached the Villers-Bocage sector by June 13th.

1. *Here the battalion's senior officers discuss how the exercise went in front of Tiger "311". Left to right: SS-Untersturmführer Willi Iriohn, wearing a leather jacket handed down from the Kriegsmarine via the Italian navy, drill trousers and the service uniform Schirmmütze (flat cap); SS-Sturmbannführer Hein von Westernhagen, the battalion commander, wearing the 1943 model camouflage tunic and the 1940 model officer's black cap with aluminum wire piping; and SS-Obersturmführer Hanno Raasch, who is wearing a gray-green motorcyclist's raincoat and the same type of cap. The officer on the right is unidentified.*

2. *Another photograph from the reportage, again with the officers Hein von Westernhagen and Hanno Raasch. Third from the right looking towards the camera is SS-Unterscharführer Otto Blase, a tank commander with the 3rd Company.*

3. *This last photograph gives a better picture of the crew of Hanno Raasch's tank "311". Leaning on the gun is the driver, SS-Unterscharführer Hofmann. (Bundesarchiv photographs.)*

2

3

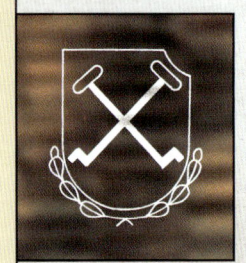

Mai 1944, secteur de Beauvais, manœuvres de la *SS-Panzer-Abteilung 101*.

1. Nous voyons ici une compagnie formée en bataille, sur large front. Une compagnie de Tiger est constituée de quatorze engins, soit trois sections de quatre Tiger chacune et deux Tiger pour la section de commandement. Les tankistes du bataillon n'auront jamais l'occasion d'utiliser cette formation lors de la Bataille de Normandie.

2. Le Tiger « 321 » de la 3ᵉ compagnie, il s'agit d'un type E.

3. Le « 304 » est un char d'état-major, celui du *Kompanietruppführer*. Le 6 juin, il sera commandé par le *SS-Unterscharführer* Heinrich Ritter. Le jour où cette photo est prise, il est sous le commandement du *SS-Unterstumführer* Amselgruber.

4. Au premier plan, le Tiger « 321 ».

(Photos Bundesarchiv.)

May 1944, Beauvais sector, SS-Panzer-Abteilung 101 exercising.

1. *Here we see a company in battle formation, along a broad front. A Tiger company had fourteen tanks, i.e. three troops of four Tigers each and two Tigers for the command troop. During the Battle of Normandy the battalion's tank crews never got a chance to use this formation.*

2. *The 3rd Company's Tiger "321", an E type.*

3. *The "304" is an HQ tank, that of the Kompanietruppführer. On D-Day, its commander was SS-Unterscharführer Heinrich Ritter. The day this photograph was taken, its commander was SS-Untersturmführer Amselgruber.*

4. In *the foreground, Tiger "321".*

(Bundesarchiv photographs.)

19

1

Mai 1944, secteur de Beauvais.

1. Le Tiger « 232 » de la 2ᵉ Compagnie (la *2./SS-Pz.-Abt. 101*). Le 6 juin 1944, il sera sous le commandement du *SS-Unterscharführer* Kleber.

2. Un équipage de la 1ʳᵉ compagnie vient d'adopter une corneille comme mascotte. Nous voyons ici l'un des membres de cet équipage, le conducteur, et la corneille.

3. Trois autres membres de l'équipage avec la corneille. Ces tankistes portent le treillis camouflé modèle 1944, vêtement de protection sensiblement le plus répandu dans les régiments de chars de la *Waffen-SS* durant l'été de 1944. Le deuxième tankiste à partir de la gauche est le chef de char, le *SS-Junker* Erwin Asbach. Le 6 juin, il commandera le Tiger « 124 ».

(Photos Bundesarchiv.)

May 1944, Beauvais sector.

1. Tiger "232" of the 2nd Company (2./SS-Pz.-Abt. 101) on June 6th, 1944, its commander was SS-Unterscharführer Kleber.

2. A crew of the 1st Company has just adopted a crow as its mascot. Here we see one of the crew, the driver, and the crow.

3. Three other crew members with the crow. These men are wearing 1944 model camouflage drill, definitely the most widespread protective clothing among the Waffen-SS tank regiments during the summer of 1944. The second man from the left is the tank commander, SS-Junker Erwin Asbach. On D-Day, he commanded Tiger "124".

(Bundesarchiv photographs.)

Mai 1944, Michael Wittmann et cinq commandants de Tiger.

Le 2 juin 1944, le journal « Wacht am Kanal », destiné aux soldats du front de l'Ouest, publie un reportage du correspondant de guerre Scheck pris au mois de mai, dans le secteur de Beauvais. Le titre de l'article était « Six commandants de Tiger ont détruit près de trois brigades de chars ».

1. Voici le plus prestigieux de ces six commandants, le SS-Obersturmführer Michael Wittmann qui est déjà une « vedette » pour son score obtenu sur le front de l'Est. La Croix de chevalier avec feuilles de chêne lui a été remise par Hitler le 30 janvier 1944 pour avoir détruit 117 chars soviétiques en l'espace de sept mois.

2. Sur cette photo, nous retrouvons les six commandants de Tiger. Tout d'abord, en tenue noire, l'*Ostuf.* Michael Wittmann (Tiger « 205 », il commande la 2ᵉ compagnie du bataillon), puis cinq chefs de char de sa compagnie : l'*Uscha* Karl-Heinz Warmbrunn (à demi caché derrière Wittmann, Tiger « 214 ») et, de gauche à droite, le *Hscha* Hans Höflinger (de face, Tiger « 213 »), l'*Oscha* Georg Lötzsch (à moitié caché, Tiger « 223 ») et l'*Uscha* Kleber (Tiger « 212 »).

3. De gauche à droite, Kleber, Wittmann et Woll (Tiger « 212 »).

4 à 7. Le correspondant de guerre réalise plusieurs portraits de Wittmann, pour la propagande…

(Photos Bundesarchiv.)

May 1944, Michael Wittmann and five Tiger commanders.

June 2, 1944, the newspaper "Wacht am Kanal", intended for the soldiers on the front in the West, published a reportage by war correspondent Scheck taken in the Beauvais sector in May. The title of the article was. "Six Tiger commanders destroy nearly three tank brigades".

1. Here is the most prestigious of the six commanders, SS-Obersturmführer Michael Wittmann, who was already an "ace" for the number of his kills on the Eastern front. The Knight's Cross with Oak Leaves was presented to him by Hitler on January 30th, 1944 for destroying 117 Soviet tanks in the space of seven months.

2. In this photograph, we again see the six Tiger commanders. First, in a black uniform, Ostuf. Michael Wittmann (Tiger "205", he was commander of the battalion's 2nd Company), then five of his company tank commanders: Uscha Karl-Heinz Warmbrunn (half hidden behind Wittmann, Tiger "214") and, from left to right, Hscha Hans Höflinger (facing the camera, Tiger "213"), Oscha Georg Lötzsch (half hidden, Tiger "223") and Uscha Kleber (Tiger "212").

3. Left to right, Kleber, Wittmann and Woll (Tiger "212").

4 to 7. The war correspondent took several portraits of Wittmann, for propaganda purposes.

(Bundesarchiv photographs.)

23

Voici la suite du reportage consacré aux six commandants de chars Tiger titulaires de très nombreuses victoires sur le front de l'Est.

1. L'*Uscha* Balthasar Woll, l'ancien pointeur de Michael Wittmann, Sarrois âgé de 22 ans, titulaire de la Croix de Chevalier pour avoir détruit 81 blindés, 107 canons antichars, deux batteries d'artillerie, un blindé de reconnaissance, deux tracteurs et une position de mortiers, commande maintenant un Tiger, le « 212 ». On voit derrière lui le conducteur de son char, l'*Uscha* Jupp Sälzer.

2. Karl-Heinz Warmbrunn, 19 ans, 1,87 mètre, originaire de Nuremberg, le plus jeune chef de char de la compagnie. Il a détruit 51 chars et 68 pièces antichars.

3. Autre photo de Woll, image de propagande. Celui-ci ne participera pas à la Bataille de Villers-Bocage.

4. Hans Höfflinger, 26 ans, Bavarois, a rejoint l'arme blindée en 1942.

5. Balthasar Woll et Sälzer.

6. Georg Lötzsch, surnommé « le général des panzers » par ses camarades. 30 ans, originaire de Dresde, il a rejoint la *Waffen-SS* en 1933.

7. Autre image de propagande de l'*Uscha* Woll.

(Photos Bundesarchiv.)

Here the next part of the reportage on the six Tiger tank commanders who had notched up an impressive number of kills on the Eastern front.

1. Uscha Balthasar Woll, Michael Wittmann's old gun layer. A 22 year-old from the Saar, awarded the Knight's Cross for destroying 81 tanks, 107 anti-tank guns, two artillery batteries, a reconnaissance tank, two prime movers and a mortar position, now commanded a Tiger, no. "212". Behind him is his tank driver, Uscha Jupp Sälzer.

2. Karl-Heinz Warmbrunn, aged 19, 1.87 meters tall, from Nuremberg, the company's youngest tank commander. He destroyed 51 tanks and 68 anti-tank guns.

3. Another photograph of Woll, a propaganda picture. He did not take part in the Battle of Villers-Bocage.

4. Hans Höfflinger, aged 26, a Bavarian, joined the tank arm in 1942.

5. Balthasar Woll and Sälzer.

6. Georg Lötzsch, called "the panzer general" by his comrades. Aged 30, from Dresden, he joined the Waffen-SS in 1933.

7. Another propaganda picture of Uscha Woll.

(Bundesarchiv photographs.)

25

SS-Panzerabteilung 101

007
Kommandeur
Stubaf. Heinz von Westernhagen

008
Adjutant
Ustuf. Eduard Kalinowsky

009
Nachrichtenoffizier
Ustuf. Helmut Dollinger

Stabs-und Versorgungskompanie/SS-Panzerabteilung 101
(companie de commandement et de ravitaillement)

Etat : 6 juin 1944

Kompaniechef
Ostuf. Paul Vogt

Betriebsstoffkolonne

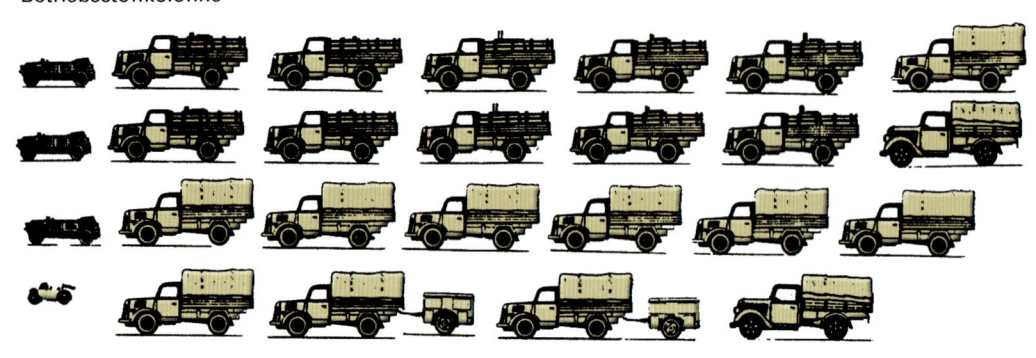

Munitionskolonne

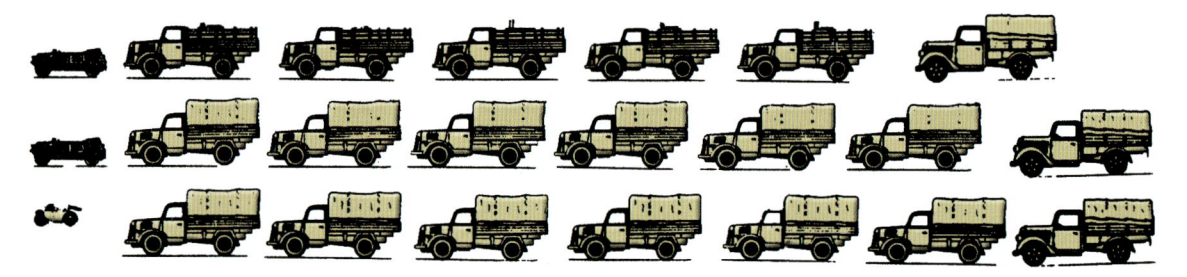

Nachrichtenzug
Ustuf. Helmut Dollinger

Sanitättsstaffel
Hstuf. Dr. Wolfgang Rabe

Strm. Artur Bergmann
Strm. Jochen Borchert
Rttf. Herbert Debusmann
Strm. Alfred Bahlo.
Oscha. Robert Bardo
Ustuf. Peter Harsche
Pz.Schtz. H. Rudolf Schneider
Pz.Schtz. Albert Habenicht
Strm. Karl. Heinz Heim
Pz.Schtz Franz Krippel

Strm. Kurt Krötzsch
Uscha. Lothar Kühn
Rttf. Edmund Laule
Pz. Schtz. Johann Müller
Pz. Schtz. Lothar Krüschel
Uscha. Hartwig
Rttf. Otto Hahn
Strm. Künast
Strm. Bruno Grabowsky
Strm. Günther Kruschwitz

Strm. Adolf Leuschner
Hstuf. Dr. Erich Hausamen
Uscha. Otto Piehler
Uscha. Otto König
Strm. Gerhard Kaschlan
Rttf. Horst Uhlig
Oscha. Erich Schmitz
Schtz. Carlo Pangallo
Schtz. Dino Bandini
Rttf. Giovanni Sala

1./SS-Panzerabteilung 101
Etat : 6 juin 1944

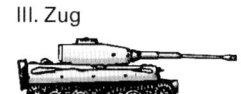

105
Kompaniechef
Hstuf. Rolf Möbius

104
Kompanietruppführer
Uscha. Sepp Franzl

I. Zug

111
Ostuf. Hannes Philipsen

112
Uscha. Cap

113
Oscha. Heinrich

114
Uscha. Willi Otterbein

II. Zug

121
Ustuf. Fritz Stamm

122
Uscha. Arno Salamon

123
St. Jk. Franz Staudegger

124
Jk. Erwin Asbach

III. Zug

131
Ustuf. Walter Hahn

132
Uscha. Werner Wendt

133
Oscha. Fritz Zahner

134
Uscha. Helmut Dannleitner

Pointeurs/Gunners
Uscha. Georg Przibylla
Rttf. Helmut Steinmetz
Strm. Friedel Fischer
Rttf. Alfons Ahrens

Pourvoyeurs/Loaders
Pz.O.Schtz. Alfred Weyel

Chefs de char/Tank commanders
Oscha. Hans Swoboda
Oscha. Hein Bode
Oscha. Karl Müller
Uscha. Günther Kunze
Uscha. Josef Franzl

Radios/Radio operators
Strm. Helmut Schrader
Uscha. Fritz Belbe
Strm. Lorenz Mähner

Conducteurs/Drivers
Strm. Theo Janekzek
Rttf. Walter Bingert
Uscha. Gerd Beutel
Oscha. Walter Sturhahn
Uscha. Paul Berendt
Rttf. Kurt Koch
Rttf. Lemaire
Rttf. Anesi
Oscha. Helmut Fritzsche
Strm. Hans Stecher

Fonctions précises/Combat functions unknown
Hscha. Kurt Michaelis
Hscha. Fritz Hibbeler
Uscha. Gerhard Langer
Uscha. Heinz Stoye
Uscha. Robert Zellmer
Rttf. Horst Luley
Rttf. Walter Kahl
Rttf. Rudolf Foege
Rttf. Horst Daniel
Rttf. Gerhard Koelbl
Rttf. Hans Hermani
Strm. Siegfried Hruschka
Strm. Walter Bausch
Strm. Ernst Wedehin
Pz.Schtz. Arno Laupsin

Adjudant de compagnie/CSM : Hscha. Günter Lueth

Spécialiste radio/Radio Tec. : Oscha. Quenzer

Secrétaire/Secretary : Uscha. Karl Mollenhauer

Trésorier/Treasurer (Pay Corps) : Uscha. Peter Schnitzler

Chef mécanicien/Chief engineer : Oscha. Seifert

Equipe de dépannage/Breakdown gang : Uscha. Heinrich Wölfel

Armurier/Armorer : Uscha. Bernhard Bauer

Cuisine/Kitchen : ?

Agent de liaison/Liaison agent : Strm.

2./SS-Panzerabteilung 101

Etat : 6 juin 1944

205
Kompaniechef
Ostuf. Michael Wittmann

204
Kompanietruppführer
Uscha. Seifert

I. Zug

211
Ostuf. Jürgen Wessel

212
Uscha. Balthasar Woll

213
Hscha. Hans Höflinger

214
Uscha. Karl-Heinz Warmbrunn

II. Zug

221
Ustuf. Georg Hantusch

222
Uscha. Kurt Sowa

223
Oscha. Jürgen Brandt

224
Uscha. Ewald Mölly

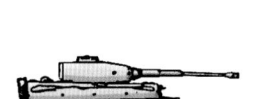

III. Zug

231
St. O. Jk. Heinz Belbe

232
Uscha. Kurt Kleber

233
Oscha. Georg Lötzsch

234
Uscha. Herbert Stief

Pointeurs/Gunners
Uscha. Karl Wagner
Rttf. Johann Kern (211)
Rttf. Rudi Lechner
Strm. Walter Lau (234)
Rttf. Friedrich Faltlhauser
Strm. Erich Tille
Strm. Werner Knocke
Strm. Max Gaube
Strm. Deutschwitz

Pourvoyeurs/Loaders
Rttf. Gustav Grüner
Rttf. Gerhard Bückner
Strm. Werner Irrgang
Strm. Willibald Schenk (205)
Strm. Günther Weber
Strm. Günter Boldt (222)
Strm. Paul Sümnich
Strm. Peter Mayer
Strm. Julius Luscher
Strm. Alfred Bernhard
Strm. Günther Braubach
Strm. Helmut Hauck
Strm. Harald Henn
Strm. Erich Ditl

Radios/Radios
Uscha. Ernst Wohlleben
Rttf. Heinz Stuß
Rttf. Paul Bender
Rttf. Gerhard Waltersdorf
Strm. Rudolf Hirschel
Strm. Franz Rausch
Strm. Josef Rößner (205)
Strm. Günter Jonas (222)
Strm. Herbert Werner
Strm. Aribert Wideburg (214)
Strm. Hubert Heil (213)
Strm. Friedhelm Zimmermann

Conducteurs/Drivers
Rttf. Eugen Schmidt
Rttf. Franz Elmer
Rttf. Kurt Kämmer
Rttf. Erlander (221)
Rttf. Fritz Jäger
Uscha. Arthur Sommer
Uscha. Jupp Sälzer (211)
Uscha. Heinrich Reimers
Uscha. Hans Focke
Uscha. Kurt Hühnerbei
Uscha. Herbert Stellmacher
Uscha. Walter Müller (222)
Uscha. Ludwig Eser

Uscha. Bernhard Ahlte
Strm. Karl Piper
Strm. Werner Hepe
Strm. Augst

Chefs de char/Tank commanders
Oscha. Otto Augst
Uscha. Joachim Willhelmi

Fonctions précises inconnues/Combat functions unknown
Oscha. Lange
Uscha. Alfred Stubenrauch
Uscha. Werner Licht
Uscha. Herbert Boden
Uscha. Willy Wils
Uscha. Günther Jacob
Rttf. Rolf Brandenburg
Rttf. Johann Schöppler
Rttf. Bruno Ryll
Rttf. Willhelm Brock
Rttf. Kurt Lange
Rttf. Hermann Schmitz
Strm. Siegfried Kriebisch
Pz.Schtz. Wilhelm Dahlmann
Strm. Johann Schmidtbauer

Strm. Max Wosnitza
Strm. Heinrich Wagenfeld
Strm. Willy Martschausky
Strm. Wilhelm Longkamp
Strm. Otto Gollan
Strm. Ottmar Günther
Strm. Otto Koch
Strm. Christel Diemens
Strm. Goerg Eckly
Strm. Harry Teubner
Strm. Max Schulze
Strm. Heinz Borck

Adjudant de compagnie/CSM : Hscha. Georg Konradt
Infirmiers/Medics : Uscha. Adolf Schmidt
Equipe de dépannage/Breakdown gang : Uscha. Adolf Frank
Agents de liaison/Liaison agent :
Rttf. Adolf Becker
Strm. Hannes Winkelmann
Rttf. Hans Schmitt
Armurier/Armorer : Uscha. Richard Heib
Cuisine/Kitchen : Uscha. Günter

3./SS-Panzerabteilung 101
Etat : 6 juin 1944

305
Kompaniechef
Ostuf. Hanno Raasch

304
Kompanietruppführer
Uscha. Heinrich Ritter

I. Zug

311
Ustuf. Alfred Günther

312
Oscha. Peter Kisters

313
Uscha. Schöppner

314
Uscha. Otto Blase

II. Zug

321
Hscha. Max Görgens

322
Uscha. Heimo Traue

323
Hscha. Hermann Barkhausen

324
Uscha. Jürgen Merker

III. Zug

331
Ustuf. Thomas Amselgruber

132
Uscha. Albert Leinecke

333
Uscha. Waldemar Warnecke

334
Oscha. Rolf von Westernhagen

Pointeurs/Gunners :
Strm. Alfred Lünser (314)
Strm. Siegfried Ewald (321)
Uscha. Kurt Diefenbach (305)
Pz.Schtz. Günter Wagner (333)
Rttf. Otto Garreis
Strm. Heinz Bannert
Strm. Leopold Aumüller (331)

Pourvoyeurs/Loaders :
Strm. Ewald Graf (331)
Pz.Schtz. August-Wilhelm Belbe (333)
Pz.Schtz. Lund
Pz.Schtz. Krupp (321)

Chefs de char/Tank commanders :
Uscha. Richard Müller
Uscha. Herbst

Radios/Radio :
Strm. Gerhard Jäsche (313)
Strm. Richard Garber (333)
Strm. Werner Dörr
Strm. Duwecke
Strm. Ernst Kufner (305)
Strm. Mitscherlich
Pz.Schtz. Jonny Heuser (314)
Pz.Schtz. Frahm (321)
Strm. Willi Hagen

Conducteurs/Drivers :
Uscha. Ludwig Hofmann
Uscha. Gerhard Noll (333)
Rttf. Paul Rohweder
Strm. Sippel
Uscha. Bernhard Ahlte
Strm. Joseph Heim (311)

Rttf. Herbert Bölkow
Rttf. Konrad Peuckert (321)
Strm. Ulrich Kreis
Rttf. Müntrat (314)

Fonctions précises inconnues/ Combat functions unknown :
Uscha. Werner Albers
Uscha. Sepp Engshuber
Uscha. Toni Zietz
Strm. Paul Pilz
Strm. Georg Christian
Strm. Walter Kolanus
Pz.Schtz. Heinz Becker
Pz.Schtz. Reinhold Bytzeck
Pz.Schtz. Rudolf Stallmann
Pz.Schtz. Heinrich Eiselt

Adjudant de compagnie/CSM :
Hscha. Wilhelm Hack

Spécialiste radio/Radio Tec. :
Uscha. Maier

Secrétaire/Secretary :
Uscha. Robert Bofinger

SDG (infirmier)/Medic :
Rttf. Gerhard Scherbarth

Mécanicien/Engineer :
Oscha. Herbert Tramm

Trésorier/Treasurer : Uscha. Martens

Cuisine/Kitchen :
Uscha. Hüsken,
Rttf. Käse

Equipe de dépannage/Breakdown gang :
Uscha. Foko Ihnen
Uscha. Georg Sittek

4.(leichte)/SS-Panzerabteilung 101
Etat : 6 juin 1944

Kompaniechef : Ostuf. Wilhelm Spitz
Kompanietruppführer : Oscha. Gerhardt Klett

Pionierzug : St. O. Jk. Walter Brauer

Uscha. Richard Ackermann   Uscha. Heinz Fiedler   Uscha. Siegfried Thomsen

Gepanzerter Aufklärungszug : Hscha. Benno Poetschlak

Erkundungszug : Ustuf. Rolf Henniges

Uscha. Konrad Mankiewitz   Uscha. Manfred Krebs   Uscha. Richard Heidemann

Flakzug : Uscha. Kurt Fickert

Uscha. Gottlob Braun   Uscha. Heinrich Hölscher   Uscha. Werner Müller

Schtz. Willi Gerstner
Schtz. Anton Hriberscheg
Schtz. Horst Kahlfeld
Section de Flak/AA Platoon :
Strm. Dörr
Rttf. Valentin Roth
Rttf. Hans Gaiser
Kan. Rolf Bergmann
Rttf. Hans-Adalbert Gürke
Rtff. Eduard Hofbauer
Rttf. Ewald Mletzko
Uscha. Walter Frisch
Rtff. Viktor Bolduan
Strm. Willy Jagschas
Strm. Gustav Look
Strm. Herbert Turck
Strm. Karl Schwab
Strm. Erich Will
Strm. Manfred Blumberg
Uscha. Franz Tilly

Section du Génie/Pioneer Platoon :
Schtz. Stefan Hartmann
Schtz. Harmann
Schtz. Janni Corrado

Adjudant de compagnie/CSM : Hscha. Fritz Müller
Chef mécanicien/Engineer : Oscha. Heinz Pfeil

Pionierzug : section du Génie/Pioneer Platoon
Gepanzerter Aufklärungszug : section de reconnaissance blindée/Reconnaissance Platoon
Erkundungszug : section d'éclairage
Flakzug : section de Flak/AA Platoon

Werkstattkompanie/SS-Panzerabteilung 101

Etat : 6 juin 1944

Kompaniechef : Ostuf. Gottfried Klein

Werkstattzug : Ustuf. Oskar Glaser

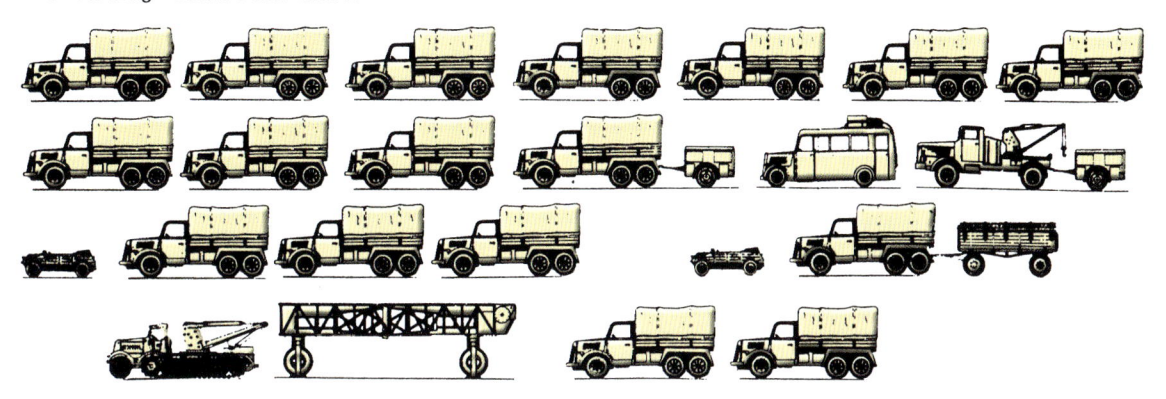

Bergezug : Ustuf. Reinhold Wichert

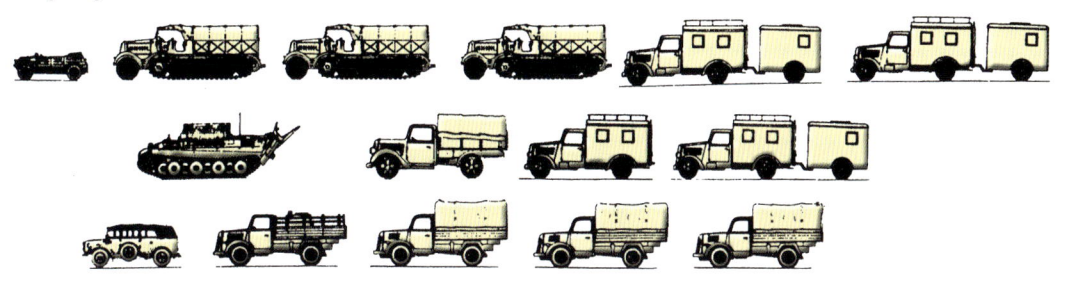

Waffenmeisterzug : Oscha. Reichert

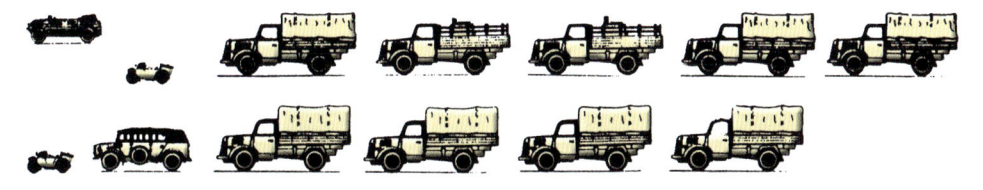

Rttf. Franz Gilly	*Strm.* Paul Müller	**Adjudant de compagnie/*CSM* :** *Hscha.* Seidel
Strm. Oskar Ganz	*Strm.* Eduard Kastl	**Trésorier/*Treasurer (Pay Corps)* :** *Oscha.* Walter Havemann
Uscha. Willi Seibert	*Schtz.* Robert Oswald	**Chef mécanicien/*Chief engineer* :** *Oscha.* Michael Heimes
Oscha. Benno Bartel	*Rttf.* Erwin Reisch	
Uscha. Werner Freytag	*Oscha.* Lehmann	
Uscha. Heinrich Roth	*Strm.* Fehrmann	
Strm. Lothar Bartholomeus	*Uscha.* Heinz Flebig	
Strm. Otto Büchner	*Rttf.* Pitt Roland	
Strm. Heinz Feldstedt	*Rttf.* Langholz	
Uscha. Max Freisenegger	*Osch.* Sepp Hafner	
Schtz. Franz Janski	*Rttf.* Ludwig Schulz	*Werkstattzug* : section atelier/*Workshop*
Schtz. Willi Kalender	*Rttf.* Karl-Heinz Fetz	*Bergezug* : section de dépannage/*Repairs*
Strm. Erich Kleinschmidt	*Rttf.* Walter Rudolf	*Waffenmeisterzug* : section d'armurerie/*Weapons*

31

Morgny, 7 juin 1944.

1. L'alerte a été donnée et la *SS-Panzer-Abteilung 101* marche vers le front de Normandie. Cette photo a été prise dans la matinée du 7 juin 1944 sur la RN316 entre Bézu-la-Forêt et Morgny, dans le nord-est de la Normandie. Nous voyons sur cette photo Wittmann à la tourelle de son Tiger « 205 ». (BA.)

2. On aperçoit ici, montant les lacets de la route nationale 316, les Tiger de la I⁰ section qui suivent le « 205 ». En tête se trouve le Tiger « 211 » commandé par le chef de section, le *SS-Obersturmführer* Jürgen Wessel. Il est suivi par le « 212 » du *SS-Unterscharführer* Balthasar Woll et par le Tiger « 213 » du *SS-Hauptscharführer* Hans Höflinger. (BA.)

3. Les Tiger continuent de gravir la pente. On aperçoit ici le *SS-Oberscharführer* Jürgen Brandt à la tourelle de son Tiger « 223 ». Il n'y a pas encore d'avions dans le ciel mais les mitrailleuses de caisse ont été démontées et placées sur les tourelleaux des chefs de char en protection antiaérienne. Les housses de protection sont placées sur les freins de bouche des canons pour éviter que la poussière n'y entre. (BA.)

Morgny, June 7th, 1944.

1. The alert was given and SS-Panzer-Abteilung 101 marched towards the front in Normandy. This photo was taken on the morning of June 7th, 1944 on the N316 highway between Bézu-la-Forêt and Morgny, north-east of Normandy. On this photograph we see Wittmann at the turret of his Tiger "205". (BA.)

2. Here we see, climbing up the hairpin bends of Highway 316, the Tigers of no. 1 Troop following "205". Leading the way is the troop commander, SS-Obersturmführer Jürgen Wessel's Tiger "211". It is followed by SS-Unterscharführer Balthasar Woll in "212" and SS-Hauptscharführer Hans Höflinger's Tiger "213". (BA.)

3. The Tigers continue their way up the hill. Here we see SS-Oberscharführer Jürgen Brandt at the turret of his Tiger "223". There are as yet no planes in the sky but the hull machine-guns have been dismounted and placed on the anti-aircraft protection tank commanders' cupolas. The protective covers are placed over the guns' muzzle brakes to prevent dust getting in. (BA.)

4. The 57 ton Tigers climb round the bends.

5. The rear of tank "221" belonging to Ustuf Georg Hantusch, commander of no. 2 Troop, 2nd Company. Notice this company's special markings.

(Bundesarchiv photographs.)

4. Les Tiger de 57 tonnes grimpent les lacets.
5. L'arrière du « 221 » de l'*Ustuf* Georg Hantusch, chef de la IIe section de la 2e compagnie. On remarquera le marquage propre à cette compagnie.
(Photos Bundesarchiv.)

33

Morgny, 7 juin 1944.

1. Le 7 juin 1944, la *schwere SS-Panzer-Abteilung 101* rejoint le front de Normandie et traverse Morgny, un village situé à l'est de Rouen. Nous voyons ici un Tiger de la 1re compagnie. A droite, dans le Schwimmwagen, véhicule amphibie, on aperçoit le *SS-Unterscharführer* Willi Röpstorff. (BA.)

2. Le Tiger « 131 », commandé par le *SS-Untersturmführer* Walter Hahn, qu'on aperçoit assis au sommet de la tourelle du char, passe devant l'église de Morgny. (BA.)

3. Il est suivi par le Tiger « 132 » de l'*Unterscharführer* Werner Wendt avec, à l'avant-gauche, le *SS-Sturmmann* Schader (radio) et, à l'avant droit, le *SS-Sturmmann* Fischer (pointeur). (BA.)

4. Arrive ensuite le Tiger « 133 » commandé par l'*Unterscharführer* Fritz Zahner, en haut de la tourelle, avec sa casquette. (BA.)

Morgny, June 7th, 1944.

1. On June 7th, 1944, schwere SS-Panzer-Abteilung 101 came up to the front in Normandy, passing through Morgny, a village located east of Rouen. Here we see a Tiger of the 1st Company. On the right, SS-Unterscharführer Willi Röpstorff can be seen in the Schwimmwagen, an amphibious vehicle.

2. Tiger "131", commanded by SS-Untersturmführer Walter Hahn, whom we see sitting on the top of the tank turret, passes by Morgny church.

3. It is followed by Unterscharführer Werner Wendt's Tiger "132" with, in the foreground on the left, SS-Sturmmann Schader (radio-operator) and SS-Sturmmann Fischer (gun layer) in the foreground on the right.

4. Then comes Tiger "133" commanded by Unterscharführer Fritz Zahner, in the top of the turret, wearing the cap.

35

3 La Bataille de Villers-Bocage
The Battle of Villers-Bocage

La résistance pendant l'occupation

Depuis 1940, comme dans toutes les communes de France occupée, les Français souffrent moralement et physiquement, beaucoup ne peuvent supporter cette présence étrangère et tous subissent les restrictions imposées par l'occupant. Dans toute cette partie de la France, une résistance s'organise, une résistance passive ou une résistance active. Une résistance civile et passive pour aider tous ceux qui sont recherchés par la police allemande ou française, pour camoufler les prisonniers de guerre évadés d'Allemagne, les aviateurs alliés victimes de la défense aérienne, les civils qui n'obéissent pas aux réquisitions, les jeunes qui refusent d'aller travailler en Allemagne dans le Service du Travail Obligatoire, le S.T.O., les familles de confession juive. Cette résistance passive aide aussi la résistance active, celle du renseignement et de l'action « terroriste ». Il fallait pour tous ces gens, des cartes d'identité, des cartes de ravitaillement que les mairies pouvaient délivrer irrégulièrement. Trois Villérois travaillaient dans ce sens, Madame Marguerite, secrétaire de mairie, et Monsieur Auguste Briard qui servait de boîte à lettres. Quant à Monsieur Ernest Huet, représentant, il était chargé de se procurer des timbres fiscaux lors de ses déplacements. Après un vol simulé à la mairie, Mme Marguerite et M. Briard étaient arrêtés le 8 décembre 1943.

Dès juin 1940, un réseau s'était constitué, le réseau Hector, dont le rôle était de renseigner les alliés sur la présence des unités allemandes, des champs d'aviation, vrais ou faux, et de transmettre les renseignements. Dans la région, son chef était Monsieur Falcoz-Vigne, d'Aunay. Il contacta M. André Doublet, minotier, fils du maire de Villers, M. Marcel Chiron, quincaillier, M. Jean Lebaron, agent d'assurances. Mais le réseau inexpérimenté fut découvert en 1942 et démantelé. Seul M. Falcoz-Vigne fut arrêté le 28 avril 1942 et faute de preuves formelles condamné à 10 ans de forteresse. Il sera libéré le 8 août 1945.

Il s'était créé dans toute la France des réseaux importants, l'un d'eux était le réseau « Alliance ». Le secteur normand avait pour rôle de surveiller la côte, de Ouistreham à Isigny-sur-mer. Depuis mai 1942, l'agent radio est Jean Caby, électricien à Villers. Il recrute Julien Thorel, Marcel Chiron, Jean Lebaron, organise dans la région un véritable réseau de renseignement et s'assure plusieurs points de transmission autour de Villers, à Villy chez Joseph Langeard, à Amayé chez M. de Saint-Pol, à Epinay chez Marcel Marié, à Longvillers chez André Robert, son poste émetteur est chez Marcel Chiron.

Tous les renseignements recueillis ne pouvaient pas être transmis par radio. Croquis et plans ne pouvaient être acheminés que par coursiers vers Paris. C'est lors d'un de ces voyages que, le 12 mars 1944, l'agent de liaison « Tadorne » est surpris par la Gestapo. Torturé, il aurait donné quelques noms. Cinq jours plus tard, le 17, une rafle s'opère entre Paris et Cherbourg. Sont arrêtés entre autres, M. de Saint-Pol, Jean Caby et son épouse Marcelle. Le 4 mai,

The Resistance during the occupation

Since 1940, as in all the towns and villages of occupied France, the French were suffering both morally and physically; many could not bear this foreign presence, and everyone was subjected to restrictions imposed by the occupying Germans. Throughout this part of France, there was organized, both passive and active resistance. A passive civilian resistance to help all those wanted by the German or French police, to conceal prisoners-of-war who had escaped from Germany, Allied pilots shot down by flak, civilians refusing to comply when requisitioned, young people refusing to go and work in Germany on Compulsory Labor Service, and also families of the Jewish faith. This passive resistance also gave help to active resistance involving intelligence and "terrorist" activities. All these people needed identity cards, ration cards issued illegally by the town halls. Three people of Villers contributed to this effort: Madame Marguerite, a secretary at the Town Hall, and Monsieur Auguste Briard, who served as a mailbox. As for Monsieur Ernest Huet, a salesman, his task was to procure revenue stamps as he moved from place to place. Madame Marguerite and Monsieur Briard were arrested on December 8th 1943, following a simulated theft at the Town Hall.

As of June 1940, a network had been set in place, the Hector network, whose role was pass on intelligence to the Allies relating to the presence of German units and of airfields, whether real or fake, and to supply information. In the area, it was headed by Monsieur Falcoz-Vigne of Aunay. He contacted Monsieur André Doublet, a miller and son of the mayor of Villers, hardware merchant Monsieur Marcel Chiron, and insurance agent Monsieur Jean Lebaron. But the inexperienced network was uncovered and dismantled in 1942. Only Monsieur Falcoz-Vigne was arrested, on April 28th 1942, and in the absence of any formal proof sentenced to 10 years imprisonment in a fortress. He was released on August 8th 1945.

Major underground networks were set up throughout France; one of these was the "Alliance" network. Its Normandy sector's task was to monitor the coast, from Ouistreham to Isigny-sur-Mer. Its radio operator since May 1942 had been Jean Caby, an electrician at Villers. He took on Julien Thorel, Marcel Chiron, Jean Lebaron, organized a proper intelligence network in the area and made sure of several transmission points around Villers, in Villy at Joseph Langeard's house, Amayé at Monsieur de Saint-Pol's, Epinay at Marcel Marié's, Longvillers at André Robert's, and his wireless set was at Marcel Chiron's.

All the information collected could not be transmitted by radio. Sketches and plans had to be taken by couriers to Paris. It was during one of these trips, on March 12th 1944, that liaison officer "Tadorne" was captured by the Gestapo. He is thought to have given a few names under torture. Five days later, on the 17th, a raid took place between Paris and Cherbourg. Among others, Monsieur de Saint-Pol, Jean Caby and his wife Marcelle were arrested. On May 4th, acting on a tip-off by a double agent who had infil-

suite à la dénonciation d'un agent double infiltré dans le réseau, la Gestapo lance un grand coup de filet autour de Villers. Sont raflés, Chiron, Thorel, Lebaron, Marié, Langeard, Robert et même André Aubin bien qu'il soit concerné par une affaire différente. Sont arrêtés avec eux, René Loslier de Jurques, Ernest Margerie d'Anctoville et Octave Loisel qui, complètement étranger à ces événements, est arrêté à la place de son fils André.

Ce dernier était employé chez Jean Caby et son patron lui avait demandé au début de son engagement dans le réseau s'il ne pouvait pas lui rendre quelques petits services. Ces services, c'était tout simplement d'aller porter à Bayeux de petits bouts de papier qu'il dissimulait dans les tubes de sa bicyclette. Un jour, il porta même à Bayeux, un poste émetteur sur son porte-bagages. Le 4 mai, chez son père, prévenu par une voisine de l'arrivée d'une voiture suspecte, il a le temps de s'échapper. Vu son âge, il aurait dû être requis en Allemagne au STO mais Jean Caby avait réussi à le maintenir dans la région et l'employait tout en le camouflant.

Le **lundi 5 juin**, Villers est calme. Pour l'instant les Villérois s'interrogent sur le sort de tous ceux qui ont été arrêtés et n'ont pas été relâchés par la Gestapo après leurs interrogatoires. Le Maire, M. Albert Doublet recommande à chacun de mesurer ses paroles et de ne pas trop manifester sa réprobation. Ils savent que le 11 mai, Mme Caby et M. Chiron, le 13 mai M. Thorel et Loisel sont rentrés chez eux très amaigris. Les Allemands ont écouté Jean Lebaron qui a réussi à les persuader qu'ils ne participaient pas à ses activités de renseignement.

Par contre, on est sans nouvelles de Mme Marguerite et d'Auguste Briard. Ils sont en Allemagne depuis janvier. Les autres doivent être à Caen. Il faudra attendre la libération pour savoir ce qu'ils sont devenus. Caby, Lebaron, de Saint-Pol, Langeard, Marié, Robert, Loslier et Margerie sont en effet à la prison de Caen où ils seront fusillés le 6 juin. M. Briard et Mme Marguerite sont en Allemagne, déportés à Mauthausen. M. Briard y mourra à son arrivée en mars, Mme Marguerite, libérée le 2 mai 1945 reprendra son poste à la mairie de Villers mais, très marquée par sa déportation, elle décédera le 28 avril 1955. André Aubin, agent de l'OCM, Organisation Civile et Militaire eut plus de chance. D'abord condamné aux travaux forcés à l'île d'Aurigny, il s'évadera le 4 août à Paris après avoir passé 28 jours dans un wagon à bestiaux d'un train appelé « le train de la puanteur ».

On ne sait pas exactement ce qu'il se passe sur les lieux des combats, on ne peut que se contenter de rumeurs. Les articles des journaux, l'*Ouest-Eclair, la Presse Caennaise, le Bonhomme Normand, le Petit Parisien, le Soir*, sont contrôlés et censurés. Les postes de T.S.F. ont été raflés voici quelques mois mais de rares auditeurs qui en possédaient plusieurs en ont conservé au moins un et courent le risque d'être arrêtés. Avec un poste à galène, des bricoleurs réussissent à capter les informations de Radio Londres sur les grandes ondes mais pour entendre quelque chose, il faut pouvoir disposer d'une antenne importante. Les avions alliés lancent des tracts, tel le « courrier de l'air » de la R.A.F. mais il est prudent de ne pas être surpris à en prendre connaissance. Les troupes d'occupation n'ont pas trop le moral. Elles ne connaissent pas exactement la position des leurs, sachant que les Allemands reculent. Chacun s'attend à un débarquement massif sur les côtes françaises. Les Allemands n'y croient pas trop ou feignent d'y croire et se contentent, sur la côte, de scruter la mer, à l'abri dans les bunkers de l'Organisation Todt et derrière les défenses côtières. Les troupes de Normandie ne sont pas les meilleures,

trated the network, the Gestapo launched a major roundup in the Villers area. Chiron, Thorel, Lebaron, Marié, Langeard and Robert were all taken in, even André Aubin, although he was actually working on something quite different. Arrested along with them were Rene Loslier from Jurques, Ernest Margerie from Anctoville, and Octave Loisel who had had nothing to do with these events and was mistaken for his son André.

This André Loisel was employed at Jean Caby's and on first taking him on in the network, his boss asked him if he would go on a few small errands for him. These errands just involved taking some tiny small pieces of paper to Bayeux concealed in the tubes of his bicycle. One day, he even took a wireless set to Bayeux on his luggage rack. On May 4th, at his father's, he was warned by the woman next door that a suspicious-looking car had just arrived, and he slipped away in the nick of time. At his age, he should have been doing compulsory labor in Germany, but Jean Caby had contrived to keep him in hiding and employ him locally as well.

On Monday June 5th, Villers was quiet. For the time being, the townsfolk were wondering what had become of all those who had been arrested and not been released after being interrogated by the Gestapo. The Mayor, Monsieur Albert Doublet, recommended that everyone should hold their tongues and avoid voicing their disapproval too loudly. They did know that on May 11th, Madame Caby and Monsieur Chiron, and Messrs Thorel and Loisel on May 13th, had returned home very much thinner. The Germans believed Jean Lebaron, who managed to persuade them that they had had no part in his intelligence activities.

On the other hand, there was no news of Madame Marguerite or Auguste Briard. They had been in Germany since January. The others were presumably in Caen. It was not until after the liberation that it emerged what had become of them. Caby, Lebaron, de Saint-Pol, Langeard, Marié, Robert, Loslier and Margerie had indeed been detained at Caen prison, where they were shot on June 6th. Monsieur Briard and Madame Marguerite were in Germany, having been deported to Mauthausen. Monsieur Briard died there on arrival in March; released on May 2nd, 1945, Madame Marguerite went back to her job at Villers Town Hall, but the trauma of her deportation never left her, and she died on April 28th 1955. Andre Aubin, an agent of the Civil and Military Organization (OCM), was luckier. Initially condemned to forced labor on the island of Aurigny, he contrived to escape in Paris on August 4th after spending 28 days in a boxcar on what was known as "the foul-smelling train".

Nobody was quite sure exactly what was going on on the battlefield, and everyone had to make do with rumors. The newspapers Ouest-Eclair, la Presse Caennaise, le Bonhomme Normand, le Petit Parisien and le Soir all had their articles vetted and censored. All wireless sets had been confiscated only a few months earlier, but the rare listeners who had several sets still had at least one and ran the risk of being arrested. Using crystal sets, handymen managed to collect information from London over long-wave radio, but you could not hear much without a large aerial. Allied planes dropped leaflets, like the R.A.F.'s "air mail", but it was not a good idea to be caught reading them. Among the occupying troops, morale was not too high. Given that the Germans were on the retreat, they never knew quite where their own people were. Everyone was expecting a massive landing along the French coast. The Germans only half-believed it or pretended to believe, and those on the coast were content to scan the sea from the shelter of the bunkers of the Todt Organization behind the coastal defenses. These were by no means crack troops in

elles sont sous équipées et dépourvues de blindés. Le Maréchal Rommel qui les a inspectées au mois de mai s'en est aperçu. Si l'état-major allemand ne croit pas au débarquement, la troupe, elle, y croit.

6 juin 1944, la bataille de Normandie commence

Le **2 juin**, des fusées lancées par des avions alliés ont éclairé la plage de Sainte-Marie-du-Mont, sur la côte est du Cotentin et le **4**, un soldat qui en revenait déclarait à des civils « *j'ai laissé mon capitaine, il est à la bonne place* ». Le **5**, vers 16 heures, un autre soldat dit à la fermière qui lui servait son lait quotidien, « *vos amis Anglais débarquent cette nuit* ». Mais les civils sont quand même sceptiques.

Durant la **nuit du 5 au 6**, le sommeil des habitants de Villers est troublé par un grondement sourd et continu qui vient de la mer. « Des obus de marine » pensent les anciens qui ont fait leur service militaire dans l'artillerie.

Le jour venu, des avions sillonnent le ciel en tous sens, prêts à lâcher leurs bombes sur un quelconque objectif. Deux avions anglais font crépiter leurs mitrailleuses route de Vire et route d'Aunay. Vers 16 heures, des bombes tombent près de la gare des marchandises, route de Canchères mais ne touchent pas la voie.

Aucun doute, c'est le débarquement en Normandie ! On l'espérait depuis quatre ans mais... en d'autres lieux.

Chacun commence à s'inquiéter tout en se réjouissant, souhaitant qu'il ne s'agît pas d'une simple tentative et que l'opération réussisse.

On pense à des parents, à des amis qui habitent sur le bord de la côte, à des enfants partis assister le dimanche précédent à une communion à Courseulles qui ont prolongé leur séjour et ne sont pas encore rentrés. « *Ils vont être en première ligne et seraient mieux chez nous, loin de la bataille, on ne risque rien* » pense-t-on.

Et pourtant, les parents, les amis seront libérés dans les jours qui suivront le débarquement alors que les habitants de Villers et de la région subiront des bombardements, des mitraillages, des tirs d'artillerie, seront atteints dans leurs biens et quelquefois dans leur chair, devront tout quitter, partir sur les routes dans des conditions pénibles et dangereuses pour fuir les combats et revenir chez eux pour déblayer leurs ruines et reconstruire leur ville.

Dans la **journée**, la nouvelle se confirme, les alliés ont réussi à prendre pied. L'heure de l'espoir est arrivée. Beaucoup ont du mal à dissimuler leur joie. Quelques-uns s'extériorisent d'une façon trop visible : aux fenêtres de l'Hôtel « le Vieux puits », rue Georges Clemenceau, chez Madame Guillois, apparaissent des drapeaux britanniques. Ils resteront en place un petit moment puis disparaîtront... par prudence. L'après-midi, quelques enfants, parmi lesquels Pierre Quéruel, partent à pied en direction de Bayeux, au-devant des Anglais, croyant qu'un avion britannique a atterri ! Ils iront ainsi jusqu'à Fains et, à leur retour, ils se verront sermonnés énergiquement par leurs parents naturellement inquiets.

Dans la **soirée**, les Allemands qui occupaient le parc du château entre autres, se replient. La *Felgendarmerie*, la police militaire allemande, abandonne l'immeuble de l'ancienne poste qu'elle occupe rue Pasteur, ferme les volets et quitte Villers. Pour beaucoup, Villers risque de ne pas devenir sûr. Nombreux sont ceux qui préfèrent abandonner leur maison pour se réfugier dans les villages voisins moins exposés, croit-on, aux attaques aériennes,

Normandy, they were under-equipped and had no tanks. Field-Marshal Rommel was aware of this for having inspected them in May. The German staff may not have believed the landing was coming, but the men certainly did.

June 6th 1944, the Battle of Normandy begins

On **June 2nd**, the beach at Sainte-Marie-du-Mont, on the east coast of the Cotentin peninsula, was lit up by flares dropped from Allied planes, and on the **4th**, a soldier on his way back declared to some civilians, "I've left my captain behind, he's better off where he is". At around 4 p.m. on the **5th**, another soldier told the farmer's wife as she served him his daily milk, "Your English friends are going to invade tonight". However, the civilians were not so sure.

During the night of **June 5-6th**, the sleep of the people of Villers was disturbed by a deafening noise coming continuously from the sea: "naval shells", surmised the older ones who had done their military service in the artillery.

When day broke, there were aircraft criss-crossing the sky in every direction, ready to drop their bombs on some target or other. There were bursts of machine-gun fire from two British planes on the Vire road and the road to Aunay. At around 4 p.m., some bombs fell close to the goods station, on the Canchères road, missing the railway lines.

No mistake, this was the D-Day landing! This was the moment everyone had been waiting for, for four years, only... not in my backyard in Normandy!

It was met with mixed feelings of joy and concern, wishing that this was the real thing and that the operation would succeed.

People's thoughts went out to family and friends living near the coast, for their children who the previous Sunday had gone to a first communion ceremony at Courseulles, had stayed on and not gotten home yet. "They are going to be in the front line, and they would be better off here, out of the battle, where there is no danger", or so they thought.

And yet these family and friends were to be liberated in the days following the landing, whereas the people in and around Villers would come under bombardment, machine-gun and artillery fire, suffering damage to their property and sometimes to their person, and have to leave everything behind them, take to the roads under difficult, dangerous conditions, to flee the battle, later to return home to clear the ruins and rebuild their town.

During the **day**, the news was confirmed that the Allies had succeeded in gaining a foothold. The time for hope had come. Many people could not conceal their joy. Some expressed it rather too obviously: at Madame Guillois' "le Vieux Puits" Hotel in the Rue Georges Clemenceau, British flags appeared in the windows. They remained in place for a little while but were soon removed, just in case. That afternoon, Pierre Quéruel and some other children set off on foot for Bayeux to meet the English, thinking that a British plane had landed! They got as far as Fains, but once back home they were given a good dressing-down by their naturally anxious parents.

That **evening**, the Germans occupying the castle grounds and elsewhere withdrew. The German military police, the Felgendarmerie, abandoned the old post office building which it had been occupying in the Rue Pasteur, closing the shutters and leaving Villers altogether. Villers looked to many people like becoming rather unsafe. Many preferred to leave their homes and take shelter with family, friends or

chez des parents, des amis, des connaissances, à Amayé, Saint-Louet, Tracy, Villy, Maisoncelles, Epinay, à Canchères, aux Hauts-Vents... Ils reviendront travailler demain et le soir regagneront leur refuge pour passer la nuit.

Les Villérois ne se doutent pas qu'ils sont passés auprès d'une tragédie. **Le 6 juin**, l'important nœud de communication que constitue Villers devait faire partie des villes bombardées dès les premières heures de la matinée ce jour-là. 528 bombardiers lourds américains, forteresses volantes B 17 et Libérators B 24, sont partis d'Angleterre pour lâcher leurs bombes sur Villers, Thury-Harcourt, Saint-Lô, Coutances, Falaise et Caen, mais le plafond étant trop bas pour agir efficacement, ils ont fait demi-tour et sont rentrés à leurs bases avec leur chargement. Dans la journée, le temps s'est éclairci et les quadrimoteurs ont pu appliquer leur programme pour ralentir le plus possible l'arrivée des renforts allemands. Les cibles étaient Valognes, Carentan, Vire, Condé-sur-Noireau, Flers, Saint-Lô de nouveau et plus loin Pont-l'Evêque et Lisieux, contribuant à faire les 6 et 7 juin **autant de morts dans la population civile que dans les rangs des combattants alliés** sur les plages du débarquement. Que serait-il advenu à Villers si le temps avait été plus clément ? Ce mardi, les Villérois étaient chez eux, dans les rues, commentant les événements, s'interrogeant sur la nature des explosions qu'ils entendaient au loin. Beaucoup auraient sûrement été victimes de bombes. Pas d'abris, peu de caves, peu de tranchées où se réfugier. Les voies de communication et les carrefours auraient peut-être été touchés, mais combien d'immeubles se seraient-ils écroulés sur leurs occupants ? Combien de maisons auraient-elles brûlé ?

Sur la côte, si le débarquement a réussi, la tête de pont n'a pas l'importance escomptée. Elle aurait dû s'étendre de la Vire à la Dives dans le département. L'objectif primordial à partir des plages de **Gold, Juno, Sword** était de conquérir Caen et sa région afin de pouvoir lancer depuis le sud-est de la ville des attaques de chars vers le centre de la France.

A l'ouest, les Américains, depuis *Omaha Beach* auraient dû atteindre la Vire dans la baie des Veys et, vers le sud, dépasser la Route nationale 13 et la ville de Bayeux.

En fait, **la zone libérée à l'aube du 7 juin** comprend dans le département une zone autour de Ouistreham, de Luc-sur-mer à Ranville à l'est de l'Orne et une zone plus étendue qui va de Saint-Aubin à Arromanches et atteint, au sud, les portes de Bayeux. Les Américains, à l'ouest, ont réussi à prendre pied au prix de pertes importantes à Vierville Saint-Laurent et à la pointe du Hoc.

Dans le département de la Manche, la tête de pont de *Utah Beach* est elle aussi moins importante que prévu mais des unités dispersées couvrent quand même une partie des objectifs du premier jour. Si le débarquement des troupes américaines sur les plages du Calvados avait échoué, c'est cette zone qui aurait servi d'axe d'attaque. D'ailleurs devant les difficultés rencontrées à Vierville, le général Bradley qui commande la première armée U.S. a même envisagé le 6 juin vers 10 heures d'abandonner *Omaha Beach* pour se reporter sur *Utah Beach*.

Les raisons de ce retard dans le bon déroulement des opérations sont multiples :

- le mauvais temps et la tempête qui ont empêché à **Gold Beach** d'utiliser les chars amphibies D.D. (Duplex Drive) débarqués trop loin de la côte,

acquaintances in the nearby villages - Amayé, Saint-Louet, Tracy, Villy, Maisoncelles, Epinay, Canchères, les Hauts Vents - thought to be less exposed to air raids. They went back to work the next day, returning in the evening to their place of shelter to spend the night.

Little did the local people know just how close they had come to tragedy. **On June 6th**, the major crossroads of Villers was one of the towns marked out to be bombed in the early hours of D-Day morning. 528 American heavy bombers, B17 Flying Fortresses and B24 Liberators, left England to drop their bombs on Villers, Thury-Harcourt, Saint-Lô, Coutances, Falaise and Caen, but as the cloud ceiling was too low to take effective action, they turned back and returned to their bases with their load of bombs. During the day, the weather cleared up and the four-engined planes were able to implement their plan to slow down the arrival of German reinforcements as much as possible. The targets were Valognes, Carentan, Vire, Condé-sur-Noireau, Flers, Saint-Lô again, and further away, Pont-l'Evêque and Lisieux, **causing as many deaths among the civilian population on June 6th and 7th as among the Allied troops** fighting on the landing beaches. What would have happened to Villers had the weather been fine? That Tuesday, the local people were in their homes, on the streets, commenting on the events, wondering what the explosions they had heard in the distance meant. Many would surely have been casualties of the bombing. There were no shelters, very few cellars or trenches in which to hide. Maybe communications would have been cut off and crossroads hit, but how many buildings would have collapsed on top of their occupants? How many houses would have been burned down?

On the coast, while the landing had succeeded, the beachhead was not as broad and deep as planned. In the Calvados, it ought to have extended from the River Vire to the River Dives. The vital objective on breaking out from Gold, Juno and Sword beaches was to take Caen and the vicinity so as to be able to launch tank attacks into central France from the southeast of the city.

To the west, the Americans landing at Omaha Beach were supposed to have reached the Vire River in the Baie des Veys and moved on southwards across Highway 13 to the town of Bayeux.

In actual fact, the area of the department liberated by dawn on June 7th comprised a zone around Ouistreham, from Luc-sur-Mer to Ranville east of the Orne River, and a broader sector stretching from Saint-Aubin to Arromanches as far south as the outskirts of Bayeux. The Americans in the west had gained a foothold at the price of significant losses at Vierville Saint-Laurent and at Pointe du Hoc.

In the department of La Manche, the beachhead at Utah Beach was again smaller than planned, however there were units scattered over part of the D-Day objectives. If the landing of American troops on the beaches of the Calvados had failed, the line of attack would have been in this sector. At around 10.00 hours on June 6th, faced with the difficulties encountered at Vierville, U.S. First Army commander General Bradley had actually even contemplated transferring the Omaha Beach landings to Utah Beach.

There were several reasons for the delay in pressing ahead with the operation:

- the bad weather and the storm which prevented the amphibious D.D. (Duplex Drive) tanks from being used on **Gold Beach**, as they were launched too far out from the coast,

39

- l'action de l'artillerie allemande qui détruit au fur et à mesure de leur débarquement, matériel et ravitaillement,
- les obstacles souvent minés érigés sur la plage,
- la résistance des troupes allemandes.

Heureusement pour les alliés, l'aviation allemande dirigée quelques jours plus tôt vers le nord de la France est pratiquement inexistante, et se contente d'intervenir la nuit d'une façon sporadique.

Le commandement allemand n'a pas cru à l'invasion alliée

Il est persuadé qu'il ne s'agit que d'une opération de diversion et que le véritable débarquement va avoir lieu sur les plages du Pas de Calais. Lors de la préparation de la défense côtière, deux stratégies s'étaient affrontées. Le maréchal Rommel voulait placer des forces blindées le plus près possible de la côte pour combattre les chars alliés dès leur apparition et les empêcher de débarquer. Par contre, von Schweppenburg, confiant dans la supériorité de ses panzers, était partisan de laisser débarquer les blindés alliés et de provoquer des combats de chars dans les plaines de l'intérieur.

Ce n'est que dans la journée du 6 juin que les Allemands se décident à ramener en renfort une vingtaine de divisions vers la Normandie, surtout des divisions blindées. Ces troupes sont retardées par les attaques de l'aviation et certaines manquent même de carburant. Le résultat est qu'elles ne rejoindront le front que les 7 et 8 juin. Pendant ce temps les alliés continuent de débarquer sur les plages les réserves qui ont suivi la première vague d'assaut. C'est ainsi que la *12. SS-Panzer-Division Hitlerjugend*, division puissante et bien entraînée, arrive à Caen à la fin de la nuit du 6 au 7 et ses éléments avancés sont stoppés les 8 et 9 juin dans les environs de Bretteville-l'Orgueilleuse, par la 3ᵉ Division canadienne qui essayait d'atteindre Carpiquet et son aérodrome.

Le 6, une seule division blindée allemande est sur le front, à l'est de Caen, la *21. Panzer-Division* que son commandant le général Feuchtinger, lance dans la bataille. Dans l'après-midi, les hommes d'un de ses régiments de grenadiers, le 192ᵉ, réussissent à atteindre la côte entre les plages de **Sword** et de **Juno**.

Le 7, la *50th Infantry Division* anglaise a libéré Bayeux et le général Montgomery, qui installera son quartier général le lendemain à Creully, envisage de pousser vers Tilly-sur-Seulles, Villers-Bocage, Evrecy et Falaise.

Pendant ce temps, les Allemands commencent à évacuer leurs réserves vers l'arrière. C'est ainsi que, le 8, un convoi d'une vingtaine de chariots à quatre roues traînés par des chevaux, traverse Villers. Ils viennent du « chemin de la fontaine pourrie » où ils s'étaient abrités sous les arbres, franchissent la place Richard-Lenoir et descendent le bourg par la rue Pasteur. C'est un convoi de farine qu'ils évacuent depuis le château de Villy-Bocage. Les soldats sont âgés, peut-être des « territoriaux » comme on appelle en France les soldats de la deuxième réserve. Ils paraissent joyeux, ils pensent que, pour eux, la guerre est finie et qu'ils entament le chemin qui va les ramener chez eux en emportant vers l'arrière un convoi de ravitaillement. La guerre va se terminer pour eux mais pas de la façon qu'ils l'imaginent. Sur la route d'Aunay-sur-Odon, après le passage à niveau, sur le plateau de la « campagne » de Maisoncelles-Pelvey, ils vont être repérés par des avions anglais et

- the action of the German artillery, which destroyed equipment and supplies as they were brought ashore,
- the beach obstacles, often loaded with booby-traps,
- the resistance of the German troops.

Fortunately for the Allies, the German air force had been sent up to northern France a few days earlier and was practically non-existent, making just a few sporadic sorties at night.

The German command did not believe this was the Allied invasion

It was convinced that it was no more than a diversionary operation and that the real landing was to take place on the beaches of the Pas-de-Calais. The preparation of the coastal defenses had seen a clash between two strategies. Field-Marshal Rommel wanted to bring his armored forces as close as possible up to the coast to fight the Allied tanks as soon as they arrived and prevent them from getting a foothold. Von Schweppenburg, on the other hand, trusting in the superiority of his Panzers, was in favor of letting the Allied tanks come ashore and setting up a tank battle on the plain further inland.

It was only during the day on June 6th that the Germans decided to recall some twenty divisions, mostly armored, to Normandy in reinforcement. These troops were held up by air attacks and some even were short of fuel. As a result they did not reach the front until June 7th and 8th. Meanwhile, the Allies continued to bring their reserves onto the beaches behind the first assault wave. Thus 12. SS-Panzer-Division Hitlerjugend, a powerful, well-trained division, arrived in Caen late in the night of June 6-7th, and its leading elements stopped the Canadians and were stopped in turn on June 8th and 9th in the Bretteville-l'Orgueilleuse sector by the Canadian 3rd Division, which was on its way to Carpiquet and the airfield there.

On the 6th, only one German armored division reached the front, east of Caen; this was 21. Panzer-Division, launched into the battle by its commander, General Feuchtinger. That afternoon, the men of one of its grenadier regiments, the 192nd, managed to reach the coast between Sword and Juno beaches.

The British 7th Infantry Division liberated Bayeux, with General Montgomery, who set up his headquarters the next day at Creully, planning to press on towards Tilly-sur-Seulles, Villers-Bocage, Evrecy and Falaise.

Meanwhile, the Germans began to pull out their reserves. Thus, on the 8th, a convoy of some twenty horse-drawn four-wheel carts passed through Villers. They came from the "chemin de la fontaine pourrie" (rotten fountain lane) where they had been sheltering under the trees, crossing the Place Richard-Lenoir and passing down through the town along the Rue Pasteur. It was a flour convoy being evacuated from the château at Villy-Bocage. The soldiers were old, possibly "territorials" as the soldiers of the second line of reserves were known in France. They seemed cheerful, considering that for them the war was over and that they were on their way home as they took a supply convoy to the rear. The war was indeed over for them, but not in the way they figured. On the Aunay-sur-Odon road, after the level crossing, in the flat country at Maisoncelles-Pelvey, they were spotted and massacred by British planes. Some of the panic-stricken horses made their way back to Villers trailing their wrecked vehicles behind them, and galloped through the town before heading off towards Villy.

massacrés. Quelques chevaux affolés, traînant derrière eux des débris de véhicules vont revenir vers Villers, traverser le bourg au galop et prendre la direction de Villy.

La bataille approche de Villers

A partir du **8**, la résistance allemande s'organise autour de Tilly et le **soir du 9** arrive la célèbre *Panzer-Lehr-Division*, division école commandée par le général Bayerlein qui a reçu l'ordre de Rommel de se battre sur la défensive à l'ouest de la *12. SS Panzer-Division*.

La *Panzer-Lehr-Division* est une formation blindée. Elle est dotée de 94 *Panzer IV*, 89 *Panther*. Elle a été rappelée de Chartres et les attaques des chasseurs-bombardiers qu'elle a subies lui ont fait perdre en cours de route, 90 camions, 5 chars, 84 véhicules semi-chenillés et canons portés, de nombreux véhicules blindés, 40 camions citernes remplis de carburant.

Sur le front, les trois divisions blindées allemandes, la *Panzer-Lehr-Division* à l'ouest, la *12. SS Panzer-Division* au centre et la *21. Panzer-Division* forment avec le reste des divisions qui ont combattu depuis le 6 juin, un véritable bouclier qu'il est difficile d'enfoncer en l'attaquant de front. Maintenant, l'armée allemande ne lance pas d'attaques sérieuses, elle se contente de se tenir sur la défensive, d'installer ses unités blindées, de réunir toutes les troupes disponibles et de rassembler les unités qui, retardées, arrivent enfin une par une.

Côté allié, il se forme ainsi :

- au nord et au nord-est de Caen une tête de pont un peu plus étendue que celle qui s'était constituée le soir du 6 juin. Le seul progrès réalisé est la suppression de la poche de Douvres par suite de la jonction des forces débarquées à **Sword Beach** et à **Juno Beach**.

- au centre, au nord de la route Caen-Caumont, le front forme une ligne Bretteville-l'Orgueilleuse, Brouay, Cristot, Bucéels, le château de Verrières, le château de Bernières, la Belle Epine, Torteval.

- Plus à l'ouest, dans le Bessin, les Américains qui ont éprouvé les pires difficultés devant Vierville, Saint-Laurent-sur-Mer et Colleville-sur-mer le 6 juin, ont progressé rapidement vers le sud, les renforts allemands n'étant pas encore arrivés. Le Ve corps a franchi l'Aure le 8 juin et dépassé la RN 13, la *29th I.D.* a traversé Isigny-sur-mer et, le 9 juin, atteint la Vire au pont du Vey et l'Elle à Neuilly-la-Forêt. Le 10, la *2nd I.D.* et la *1st I.D.* foncent vers le sud, la gare de Lison, Balleroy, Caumont.

Le **10**, une dizaine de bombes tombent sur Villers et deux sur la ferme du château. Il n'y a pas de victime mais des dégâts matériels et, un peu partout dans le pays, des vitres n'ont pas résisté aux déflagrations. Le **11**, les Allemands reviennent au château. Deux officiers se présentent. Ils veulent y établir un hôpital où pourront être opérés les premiers blessés. Leur première action sera de faire installer une grande Croix Rouge sur le toit.

Sur la ligne de front, la situation évolue lentement. Ne pouvant enfoncer les lignes allemandes, Montgomery projette d'utiliser ses deux anciennes divisions de la 8e Armée en Afrique du Nord, la *51st Highland Division* et la *7th Armoured Division*, **les « rats du désert »**, afin d'opérer autour de Caen des attaques en « pince ». La *51st Highland Division* passerait à travers la tête de pont tenue par la *6th Airborne Division* à l'est de l'Orne et la *7th A.D.* ferait un crochet vers le sud-ouest.

The battle approaches Villers

From the **8th**, the German resistance was organized around Tilly and the **evening of the 9th** saw the arrival of the famous Panzer-Lehr-Division, a demonstration division under General Bayerlein, who was ordered by Rommel to go onto the defensive to the west of 12. SS Panzer-Division.

The Panzer-Lehr-Division was an armored formation. It had 94 Panzer IVs and 89 Panthers. It was recalled from Chartres and on the way, coming under attack by fighter-bombers, it lost 90 trucks, 5 tanks, 84 half-tracks and mounted guns, numerous armored vehicles, and 40 tankers filled with fuel.

In the front line, the three German armored divisions, the Panzer-Lehr-Division to the west, 12. SS-Panzer-Division in the center and 21. Panzer-Division, formed with the other divisions that had fought since June 6th a veritable shield that was difficult to penetrate by frontal attack. By now, the German army was making no serious attacking movements, being content to be pinned down on the defensive, to set up its armored units, join together all available troops and gather the units that had been held up and were finally arriving one by one.

On the Allied side, the lineup was as follows:

- to the north and north-east of Caen the beachhead had been slightly widened since D-Day evening. The only progress made had been in removing the pocket at Douvres as the forces landed at **Sword Beach** and at **Juno Beach** linked up.

- in the center, north of the Caen-Caumont road, the front formed a line through Bretteville-l'Orgueilleuse, Brouay, Cristot, Bucéels, château de Verrières, château de Bernières, La Belle Epine and Torteval.

- further west, in the Bessin area, the Americans, who had faced the most difficult situation at Vierville, Saint-Laurent-sur-Mer and Colleville-sur-Mer on June 6th, made rapid progress south before the German reinforcements arrived. V Corps crossed the Aure on June 8th and passed the N13 highway; 29th I.D. passed through Isigny-sur-Mer and on June 9th reached the Vire River at the Vey bridge and the Elle River at Neuilly-la-Forêt. On the 10th, 2nd I.D. and 1st I.D. raced south to Lison station, Balleroy and Caumont.

On the 10th, a dozen bombs fell on Villers and two on the château farm. There were no casualties, but damage to property and throughout the area many window panes were blown out.

On the 11th, the Germans returned to the château. Two officers appeared, wishing to set up a hospital there where the first of the wounded could be operated on. They began by placing a large Red Cross on the roof.

In the front line, the situation developed slowly. Being unable to penetrate the German lines, Montgomery planned to use his two old Eighth Army divisions from North Africa, the 51st Highland Division and the 7th Armoured Division, the "Desert Rats", to attack in a pincer movement around Caen. The 51st Highlanders were to cross the beachhead held by the 6th Airborne Division on the eastern bank of the Orne while 7th A.D. swung round to the south-west.

The operation failed on the eastern flank. The German forces, the 346th and 711th divisions flanked by elements of 21. Panzer-Division, inflicted heavy losses on the 6th Airborne Division although it had massive support from the artillery, and on June 11th crushed the 51st Highland Division's attack in under an hour. Against the German forces in the middle of the front, west of Caen, the Allies set the Canadian 3rd Infantry Division and the British 50th I.D. The latter formation came up against German defenses, and

A l'est du front, la manœuvre échoue. Les forces allemandes, les 346ᵉ et 711ᵉ divisions flanquées d'éléments de la 21. *Panzer-Division* infligent des pertes importantes à la *6th Airborne Division* pourtant appuyée massivement par l'artillerie et écrasent le 11 juin en moins d'une heure l'attaque de la *51st Highland Division*. Au centre du front, à l'ouest de Caen, les Alliés opposent aux forces allemandes, la *3rd Canadian Infantry Division* et la *50th I.D.* anglaise. Cette dernière se heurte à la défense allemande et si elle parvient à s'infiltrer dans Tilly, elle ne peut tenir le bourg faute d'appui blindé. Reste alors la tentative d'encerclement des forces allemandes par l'ouest avec, pour premier objectif, Villers-Bocage.

La 1ʳᵉ Division d'Infanterie Américaine, débarquée à *Omaha Beach*, qui avait fait sa jonction avec la *50th Infantry Division* le 9 au sud de Port-en-Bessin, n'a plus rencontré que de petites unités de soldats allemands peu armés. Fonçant à travers la forêt de Cerisy, la division commandée par le général Huebner est ainsi parvenue à Balleroy. Le **11**, elle est aux portes de Caumont.

Mis au courant de cette offensive « éclair », le général Dempsey, commandant de la 2ᵉ armée anglaise et le général Bucknall, commandant du XXXᵉ Corps concluent qu'il doit exister une « certaine faiblesse » dans les lignes allemandes et songent à déplacer la *7th Armoured Division* du général Erskine vers l'ouest puis à lui faire décrire un large arc de cercle qui l'amènera à Villers-Bocage puis au sud de Caen derrière les lignes allemandes.

Le projet conçu par le commandement allié est le suivant :

Le 12 juin, partir de Bucéels et du Douet de Chouain, puis par Juaye-Mondaye, Saint-Paul-du-Vernay et Cahagnolles, derrière les lignes tenues par la 50ᵉ Division, atteindre le secteur américain, se diriger vers le sud en suivant une route parallèle à celle de la 1ᵉ Division U.S. jusqu'à Livry et enfin prendre la direction de Villers-Bocage vers l'est par la route de Caumont-Villers. Le général Montgomery met ainsi toute sa confiance en une division qu'il connaît bien, le 7ᵉ Division Blindée, pour réussir une attaque surprise derrière les lignes allemandes. **Le 12 juin, la « bataille pour la conquête de Villers-Bocage est lancée. »**

Ce **12 juin**, **Aunay-sur-Odon**, subit son premier terrible bombardement. La veille déjà les routes de Thury-Harcourt et de Condé-sur-Noireau ont été coupées. Le **12**, dès 7 heures du matin, trois vagues de 12 avions vont écraser la ville pour arrêter l'arrivée des renforts allemands vers Villers. Les bombes tombent en plein cœur du bourg, quartier de l'église, quartier de la poste, place de la Liberté. Les avions continuent leur vol jusqu'à Roucamps, reviennent et touchent les immeubles des rues d'Harcourt et du Docteur Tillais. Le résultat est terrifiant, certes les immeubles abattus obstruent les routes mais **120 morts** seront retrouvés sous les décombres. Villers a été épargné, demain, ce sera la bataille de Villers.

La 7th Armoured Division, les « rats du désert »

Sa composition

La 7ᵉ Division Blindée britannique, ayant à sa tête le *Major General* G.W.E.J. Erskine fait partie du XXXᵉ Corps d'Armée britannique du *Lieutenant General* Bucknall inclus dans la 2ᵉ Armée anglaise commandée par le *Lieutenant General* Sir Miles Dempsey.

Elle est composée de :

I - **La 22nd Armoured Brigade** (brigade blindée) forte de 200 chars environ sous les ordres du *Brigadier*

Le *Major General* G.W.E.J. Erskine. (Tank Museum.)

although it managed to infiltrate at Tilly, its lack of armored support prevented it from holding the town. This left the alternative of attempting to encircle the German forces from the west, with Villers-Bocage as the prime objective.

The U.S. 1st Infantry Division, which had landed at Omaha Beach and linked up on the 9th with the 50th Infantry Division south of Port-en-Bessin, now encountered only small units of Germans with little weaponry. Racing through Cerisy Forest, the division, commanded by General Huebner, reached Balleroy. By the **11th**, it had reached the outskirts of Caumont.

When informed of this "lightning" offensive, British Second Army commander General Dempsey and XXX Corps commander General Bucknall came to the conclusion that there must be a "certain weakness" in the German lines, and considered moving General Erskine's 7th Armoured Division west and then have it make a broad sweeping movement to bring it round to Villers-Bocage and then south of Caen behind the German lines.

The scheme worked out by the Allied command was as follows:

On June 12th, to start from Bucéels and Douet de Chouain, then via Juaye-Mondaye, Saint-Paul-du-Vernay and Cahagnolles, behind the lines held by the 50th Division, to reach the American sector, head off southwards along a road parallel to that taken by the U.S. 1st Division as far as Livry, and finally to veer east towards Villers-Bocage along the Caumont-Villers road. This was General Montgomery placing all his confidence in a division he knew well, 7th Armoured, to pull off a surprise attack behind the German lines. **On June 12th, the battle for Villers-Bocage began.**

On this **June 12th**, Aunay-sur-Odon came in for its first tremendous bombardment. Already the day before, the Thury-Harcourt and Condé-sur-Noireau roads had been cut off. On the **12th**, starting at 7 in the morning, three waves of 12 aircraft pounded the town to prevent German reinforcements getting through to Villers. The bombs fell right in the town center, where the church and the post office stood, and the Place de la Liberté. The planes flew on to Roucamps, turned back and hit buildings in the Rue d'Harcourt and the Rue du Docteur Tillais. The result was horrifying, for of course the bombed buildings blocked the roads but **120 dead** were found under the rubble. Villers was spared, but the next day was to be the Battle of Villers.

The 7th Armoured Division, "the Desert Rats"

Composition

The British 7th Armoured Division, under Major-General G.W.E.J. Erskine, belonged to British XXX Corps, commander Lieutenant General Bucknall, which was a part of the British Second Army commanded by Lieutenant-General Sir Miles Dempsey.

It was made up of:

I – The **22nd Armoured Brigade**, with a strength of around 200 tanks, under its commander Brigadier Robert Hinde. A British brigade was equivalent in size to a German or American regiment, while the regiment corresponded to a German battalion. (Its special tactical symbol was the number 50).

The brigade comprised:

1. In command, a Brigade H.Q.Sqn armed with 7 Cromwell tanks, 8 artillery observation Shermans, 2 Crusader AA (anti-aircraft) tanks.

42

Robert Hinde (général de brigade). Une brigade britannique est l'équivalent en effectifs à un régiment allemand ou américain, le régiment correspond à un bataillon allemand. (Son signe tactique spécifique est le numéro 50).

La brigade est composée de :

1. A sa tête, un quartier général (Brigade H.Q.Sqn) armé de 7 chars Cromwell, 8 Sherman d'observation d'artillerie, 2 Crusader AA (anti-aircraft ou DCA).

2. 3 régiments de chars, 4th C.L.Y., County of London Yeomanry, les « Sharpshooters » du lieutenant-colonel Lord Cranley, 1st R.T.R., Royal Tank Regiment du lieutenant-colonel Mike Carver, 5th R.T.R. commandé par le lieutenant-colonel Joe Lever.

Chaque régiment de chars a la composition théorique suivante :

a - A sa tête, l'état-major du régiment (RHQ) qui dispose de 4 chars Cromwell entre autres véhicules,

b - 1 escadron d'état-major (HQSqn) avec 17 chars, 6 Cromwell avec canon de (DCA) (Anti-Aircraft troop), 11 chars de reconnaissance Honey et 7 scoutcars, 31 véhicules d'administration et de liaison,

c - 3 escadrons blindés A, B, C de 19 chars chacun ainsi répartis, état-major de l'escadron (HQSqn), 3 chars, 4 sections de 4 chars, 3 Cromwell et 1 Firefly, 1 section d'administration sur divers véhicules.

3. 1 bataillon d'infanterie motorisée. Le 1ᵉ Bataillon (the Motor Battalion) détaché de la Rifle Brigade (RB) en réalité, la Rifle Brigade n'est pas une brigade mais un régiment appelé Rifle Brigade.

Le bataillon est composé : d'un état-major de bataillon (RHQ) avec 12 véhicules, d'une compagnie d'état-major (HQ/Coy), 20 véhicules pour le renseignement et l'administration, 3 compagnies motorisées, une par régiment blindé, A, C, I. La compagnie A est rattachée au 4 C.L.Y., elle est composée d'un état-major de compagnie (CoyHQ), de 3 sections de 4 half-tracks, d'une section de reconnaissance (absente à Villers), d'une compagnie de support composée :

- d'un état-major de compagnie,
- de 3 sections antichars armées de 2 canons de 57 (6 pounds) chacune, avec 4 Loydcarriers à raison d'une section détachée auprès de chaque compagnie motorisée,
- de 2 sections de mitrailleuses lourdes sur 8 carriers.

Outre les mitrailleuses lourdes et les canons de 57, les fantassins sont armés de P.I.A.T. (Projectile Infantry Anti-Tank), de fusils, de mortiers, de fusils-mitrailleurs (brenguns), de pistolets-mitrailleurs (stenguns).

II - **La 131ᵉ brigade d'infanterie portée** (Brigadier Ekins), sur camions et brencarriers, est constituée de 3 bataillons du Queen's Royal Regiment, les « Queen's. »

- le 5ᵉ Bataillon, le Surrey,
- le 6ᵉ Bataillon, le Bermondsey, lieutenant-colonel Michael Forrester,
- le 7ᵉ Bataillon, le Southwark, lieutenant-colonel Desmond Gordon. (3)

Forrester et Gordon termineront leur carrière militaire, Major General, équivalent de général de division. Seul le 7ᵉ a combattu à Villers le 13 juin 1944.

Chaque bataillon est formé : - d'un état-major et d'une compagnie d'état-major, - de 3 compagnies, A, B, C, de 5 officiers et 122 hommes divisées en 3 sections de 3 groupes et armées d'un P.I.A.T. et d'un mortier de 60 mm (2,3 inches) par section, d'un fusil-mitrailleur par groupe et des armes individuelles,

2. 3 tank regiments: 4th C.L.Y., the County of London Yeomanry, Lieutenant-Colonel Lord Cranley's "Sharpshooters"; 1st R.T.R., the Royal Tank Regiment, commander Lieutenant-Colonel Mike Carver; 5th R.T.R., commander Lieutenant-Colonel Joe Lever.

Each tank regiment theoretically comprised the following:

a - In command, the regiment HQ (RHQ), with, among other vehicles, 4 Cromwell tanks,

b - 1 H.Q.Sqn with 17 tanks, 6 Cromwells with Anti-Aircraft troop guns, 11 Honey reconnaissance tanks and 7 scoutcars, 31 command and liaison vehicles,

c - 3 armored squadrons, "A", "B" and "C", each with 19 tanks, as follows: H.Q.Sqn 3 tanks, 4 troops with 4 tanks, 3 Cromwells and 1 Firefly, 1 command troop with various vehicles.

3. 1 motorized infantry battalion. The 1st Battalion (the Motor Battalion) detached from the Rifle Brigade (RB), which was actually not a brigade at all, but a regiment.

The battalion comprised: an HQ battalion (RHQ) with 12 vehicles, an HQ company (HQ/Coy), 20 intelligence and command vehicles, 3 motorized companies, one per armored regiment, "A", "C", "I". "A" Company was attached to the 4th C.L.Y., it comprised an HQ company (CoyHQ), 3 platoons with 4 half-tracks, a reconnaissance platoon (absent from Villers), and a reserve company comprising:

- an HQ company,
- 3 anti-tank platoons each armed with 2 6-pdr guns and 4 Loydcarriers with one platoon assigned to each motorized company,
- 2 heavy machine gun platoons with 8 carriers.

In addition to the heavy machine guns and the 57 mm guns, the infantry were armed with P.I.A.T.s (Projectile Infantry Anti-Tank), rifles, mortars, bren guns and sten guns.

II - The **131st Infantry Brigade** (Brigadier Ekins), carried on trucks and bren carriers, comprised 3 battalions of the Queen's Royal Regiment, the "Queen's".

- 5th Battalion, the Surrey,
- 6th Battalion, the Bermondsey, Lieutenant-Colonel Michael Forrester,
- 7th Battalion, the Southwark, Lieutenant-Colonel Desmond Gordon. (3)

Forrester and Gordon ended their military careers with the rank of major-general. Only the 7th Battalion fought in Villers on June 13th 1944.

Each battalion comprised: - an HQ and an HQ company, - 3 companies, "A", "B", "C", with 5 officers and 122 men divided into 3 platoons of 3 squads and armed with one P.I.A.T. and one 60 mm mortar per platoon, one light machine-gun per squad and individual weapons, rifles and sten guns - one reserve company, "D", consisting of a mortar platoon armed with six 60 mm mortars (2 per company), a 57 mm gun troop (2 per company), an engineers platoon, and a reserve platoon with 13 stretcher-bearers.

Artillery

- 2 regiments each equipped with 24 25-pdr (87.4 mm) guns, the 1st R.H.A. (Royal Horse Artillery) supporting the 131st Infantry Brigade were armed with towed howitzers, and the 5th R.H.A. with the 22nd Armoured Brigade had Sexton self-propelled howitzers on a Sherman chassis.

- 1 anti-aircraft regiment, the 15th AAReg, armed with 54 40 mm Bofors self-propelled guns.

- 1 anti-tank regiment, the 65th ATReg., the Norfolk Yeomanry, armed with twelve 17-pdr (76.2 mm) guns on a turretless tank chassis.

Lieutenant-colonel Lord Cranley.

Lieutenant-colonel Gordon. (IWM.)

(3) Le lieutenant-colonel Desmond Gordon a terminé sa carrière militaire Major-General (général de Division). Il présidait la délégation anglaise venue à Villers le 13 juin 1994.

(3) Lieutenant-Colonel Desmond Gordon ended his military career as a major-general. He led the British delegation which came to Villers on June 13th 1994.

7th Armoured Division

40
- **QG divisionnaire** (Major General Erskine)
7 Cromwell
8 Sherman (observation)
2 Crusader AA

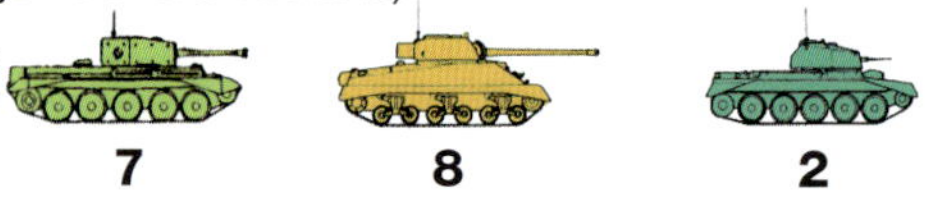

7   8   2

Régiment de Reconnaissance Blindé

44
Régiment de reconnaissance blindé
- **11th Hussars**
40 Daimler
13 Scout cars

40   13

45
Régiment de reconnaissance blindé
- **8th King's Royal Irish Hussars**
40 Cromwell
30 Honey/Stuart
6 Firefly
5 Crusader Anti-Aircraft
8 Scout cars

40   30

6   5   8

Insigne du *8th King's Royal Irish Hussars*, régiment de reconnaissance blindé de la *7th Armoured Division*

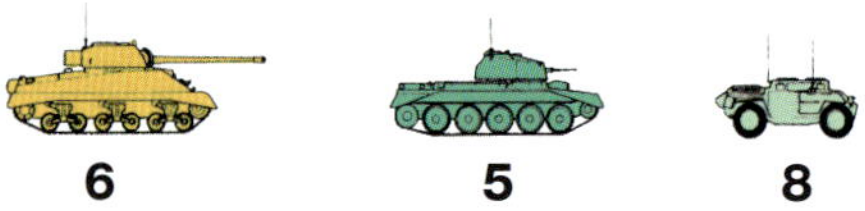

Royal Artillery

76
- **3rd RHA**
24 obusiers de 25 livres

- **5th RHA**
24 obusiers automoteurs Sexton

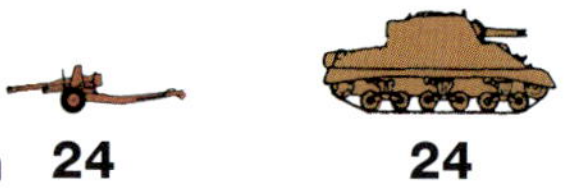

24   24

77
- **65th Anti-Tank Rgt**
18 canons de 17 livres

- **15th Light AA Rgt**
54 canons Bofors de 40 mm motorisés

18   54

Ci-contre : insigne métallique et dessous du *Norfolk Yeomanry* unité de tradition constituant le *65th Anti-Tank Regiment*.

Independant machine Gun Company (R. Northumberland Fusiliers)

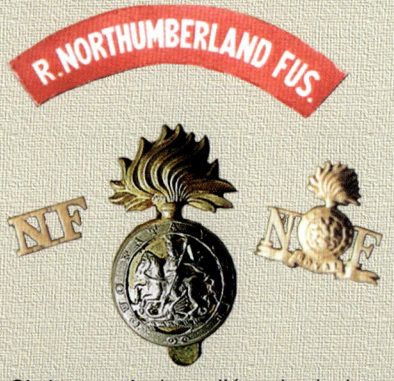

64
- **Compagnie indépendante de mitrailleuses**
12 MG Vickers (Bren Carriers)
6 mortiers/mortars 4,2 inch. (Loyd Carriers)

12   6

Ci-dessus : insigne d'épaule ; insigne de béret entouré par des insignes de collet ; insigne de manche de tradition.

et services :
Royal Engineers
Army Service Corps
Signal Corps
RAOC
REME.

131st Infantry Brigade (Brigadier Ekins)

61
- *1/5th Queen's Royal Regiment*
6 canons de six livres
6 mortiers de 3 pouces
  6   6 

62
- *1/6th Queen's Royal Regiment*
6 canons de six livres
6 mortiers de 3 pouces
30 Bren Carriers 
  6   6   30

63
- *1/7th Queen's Royal Regiment*
6 canons de six livres
6 mortiers de 3 pouces
30 Bren Carriers 
  6   6   30

Ci-contre, en haut et de haut en bas : insigne d'épaule ; insigne de béret troupe (à gauche) et officier (à droite) ; passant d'épaule 1938 toujours recherché par la troupe en 1944. (Tous objets : collection Musée Mémorial de Bayeux.)
Ci-contre : calot de tradition du *Queen's Royal Regiment*.

22nd Armoured Brigade

Ci-dessus : béret du Brigadier Hinde.
Ci-contre : insigne de la 22e brigade blindée *(collection Musée Mémorial de Bayeux).*

50
- *22e Brigade Blindée* (Brigadier Hinde)
7 Cromwell
8 Sherman d'observation d'artillerie
2 Crusader AA 
  7   8   2

Ci-contre : bande d'épaulette avec couleurs régimentaires ; badge d'épaule ; insigne de béret ; insigne de patte d'épaule pour tenue N° 1. (Musée Mémorial de Bayeux.)

51
- *4th County of London Yeomanry (Sharpshooters)* (lieutenant-colonel Lord Cranley)
55 Cromwell
6 Sherman
11 Honey
6 Crusader Anti-Aircraft
8 Scout cars Humber 
  55   6   11   6   8

52
- *1st Royal Tank Regiment* (lieutenant-colonel Mike Carver)
42 Cromwell
16 Sherman
11 Honey
6 Crusader Anti-Aircraft
8 Scout cars Humber 
  55   16   11   6   8

53
- *5 th Royal Tank Regiment* (lieutenant-colonel Joe Lever)
42 Cromwell
16 Sherman
11 Honey
6 Crusader Anti-Aircraft
8 Scout cars Humber
  55   16   11   6   8 

First Rifle Brigade

54
- *1st Rifle Brigade Regiment*
68 Half-tracks M3
20 Carrier Loyd/Bren 
  68   20

Badge d'épaule et insigne de béret. Grade de fusilier.

fusils et pistolets-mitrailleurs, - d'une compagnie de soutien D, constituée d'une section de mortiers armée de 6 mortiers de 60 mm (2 par compagnie), d'une section de canons de 57 à raison de 2 par compagnie, d'une section de génie, d'une section d'appui avec 13 brancardiers.

Artillerie

- 2 régiments équipés de 24 canons de 25 livres chacun (87,4 mm), le *1st R.H.A. (Royal Horse Artillery)* qui appuie la 131ᵉ brigade d'Infanterie est armé d'obusiers tractés et le *5th R.H.A.* avec la 22ᵉ brigade blindée possède des obusiers Sexton automoteurs à châssis *Sherman*.
- 1 régiment de Défense Contre Avion, le *15th AAReg. (Anti-Aircraft Regiment)* armé de 54 canons Bofors de 40 mm motorisés.
- 1 régiment *Anti-Tank*, le *65th ATReg.*, le *Norfolk Yeomanry* qui est armé de 12 canons de 17 livres (76,2 mm) sur châssis de char sans tourelle.
- 1 compagnie indépendante de mitrailleuses, *Royal Northumberland Fusilier*, armée de 12 MG Vickers sur *brencarriers* et de 6 mortiers de 4,2 pouces (100 mm) sur *Loydcarriers*.

Régiments de reconnaissance

- Le *8th Hussars*, the *King's Royal Irish Hussars* du lieutenant-colonel Goulburn, fort de 40 *Cromwell*, 30 *Honey*, 6 *Fireflies*, 5 *Centaur AA*, 8 *Scoutcars*.
- Le *11th Hussars*, sur 40 voitures blindées *Daimler (armed cars)* armées de canons de 75, avec un blindage de 30 mm, et d'une vitesse de 50 km heure. Il est constitué du *HQ squadron*, composé d'une section de DCA de 5 canons de 20 mm, d'une section de liaison de 13 *scoutcars* et d'une section d'administration, de 4 escadrons composés chacun d'un état-major, de 5 sections de 2 *Daimlercars* et 2 *scoutcars*, d'une section lourde de 2 AEC, d'une section d'infanterie portée sur *half-track*. Ce régiment a été détaché du XXXᵉ corps pour l'opération Perch (1) mais il faisait partie de la 7ᵉ Division Blindée lors des campagnes d'Afrique et d'Italie. Créé en 1939, il avait la particularité de faire porter à ses hommes un uniforme de cérémonie à pantalon couleur cerise d'où le surnom de « Cherry Pickers » qui lui était donné. (cueilleurs de cerises)

La *7th Armoured Division* était principalement armée de chars *Cromwell IV* et de chars *Firefly*.

Le **Cromwell IV** était un char moyen mis en service pour le débarquement. Ses principaux avantages étaient d'être maniable et rapide. Son équipage était de 5 hommes, son poids de 28 tonnes, sa vitesse de 30 à 64 km/h selon la nature du terrain et son autonomie de 270 km. Son blindage était de 76 mm d'épaisseur mais de 8 seulement aux endroits les plus vulnérables et il était armé d'un canon de 75 mm et de 2 mitrailleuses lourdes.

Le **Cromwell VI** était armé d'un obusier de 95 mm à canon court.

Le **Firefly**, était un char *Sherman* équipé d'un canon long de 76,2 mm de diamètre qui expédiait des obus de 17 pounds (soit 7,7 kilos) capables de percer des blindages de 19 cm à une distance de 1 000 mètres.

Son histoire

La 7ᵉ Division Blindée britannique, la division des « rats du désert » qui s'était illustrée en Afrique du Nord et en Italie avait été rapatriée en Angleterre pour participer au débarquement en Normandie.

Partie d'Angleterre en décembre 1940 pour l'Egypte, elle avait connu, face aux Italiens d'abord et aux Allemands ensuite en 1941 et 1942, des succès et

- 1 independent machine-gun company, the Royal Northumberland Fusiliers, armed with 12 Vickers MGs on bren carriers and six 4.2 inch (100mm) mortars on Loydcarriers.

Reconnaissance regiments

- 8th Hussars, Lieutenant-Colonel Goulburn's King's Royal Irish Hussars, with 40 Cromwells, 30 Honeys, 6 Fireflies, 5 Centaur AAs and 8 Scoutcars.
- 11th Hussars, on 40 Daimler armored cars with 75 mm guns, 30 mm armor, and a top speed of 50 kph. It included the HQ squadron, comprising an AA section with five 20 mm guns, a liaison section with 13 scoutcars, and an administration section of 4 squadrons each with an HQ, 5 platoons with 2 Daimler cars and 2 scoutcars, a heavy platoon, 2 AEC, and an infantry platoon transported by half-track. This regiment was detached from XXX Corps for Operation Perch (1), although at the time of the African and Italian campaigns it belonged to the 7th Armoured Division. Raised in 1939, its distinguishing feature was having its men wear a ceremonial uniform with cherry-colored trousers, hence the nickname "Cherry pickers".

The 7th Armoured Division was equipped with mostly Cromwell IV and Firefly tanks.

The Cromwell IV was a medium tank brought into service specially for D-Day. Its chief advantages were its speed and maneuverability. It had a crew of 5, a weight of 28 tons, speeds of 30 to 64 kph (20-40 mph) depending on the nature of the terrain, and a range of 270 km. Its armor was 76 mm thick down to just 8 in the least vulnerable spots, and it was armed with a 75 mm gun and 2 heavy machine-guns.

The Cromwell VI was armed with a 95 mm short-barrelled howitzer.

The Firefly was a Sherman tank mounting a long-barrelled 76.2 mm diameter gun which fired 17-pdr (7.7 kilo) shells capable of piercing 19 cm of armor at a range of 1,000 meters.

History

The 7th British Armoured Division, the "Desert Rats" division, which had fought with distinction in North Africa and Italy, was brought home to England to take part in the Normandy landing.

After leaving England for Egypt in December 1940, it had mixed fortunes as part of the British Eighth Army, initially against the Italians and later against the Germans in 1941 and 1942. In May 1940, it was in the Nile delta. This was when the German armies were invading France, with Italy declaring war. In September, from Libya, 200,000 Italians, commanded by Field-Marshal Graziani, launched an attack towards Egypt and the Suez Canal, then crossed the border, heading for the Nile delta. On December 8th, General Wavel's tiny army of 50,000 men responded, led by 7th A.D. which passed through Cyrenaica, reaching the Egyptian border on February 9th 1941 and coming to a halt in May against the defenses in the Halfaya Pass. However it failed to capture the fortified town of Tobruk, and had to outflank it.

General Wavel then launched a counter-attack in which the Desert Rats took part, but the Germans had the edge with their superior 88 mm AA gun used as anti-tank gun, inflicting heavy losses on the British and driving them back to their start line.

After a rest period to recuperate, the division, incorporated in November into the newly formed Eighth Army, was launched into the attack to relieve the garrison encircled at Tobruk and reconquer Cyrenaica. In December, Rommel was driven back into Tripolitania, from where he had started out in May 1941.

des revers au sein de la 8ᵉ Armée britannique. En mai 1940, elle est dans le delta du Nil. A ce moment, les armées allemandes envahissent la France, l'Italie va déclarer la guerre. En septembre, depuis la Libye, 200 000 Italiens, sous les ordres du maréchal Graziani, lancent une attaque vers l'Egypte et le canal de Suez et franchissent la frontière avec, comme objectif, le delta du Nil. Le 8 décembre, la petite armée du général Wavel, forte de 50 000 hommes, riposte avec à sa tête la *7th AD* qui, traversant la Cyrénaïque, arrive le 9 février 1941 à la frontière égyptienne et en mai butte sur la défense du col d'Halfaya. Toutefois elle n'a pas réussi à s'emparer de la place forte de Tobrouk et a dû la contourner.

Le général Wavel lance alors une contre-attaque à laquelle participent les « rats du désert » mais les Allemands, profitant de la supériorité de leur canon de DCA de 88 mm utilisé comme canon antichar infligent de lourdes pertes aux Britanniques et les obligent à revenir sur leur base de départ.

Après une période de repos pour récupérer, la division incorporée en novembre à la 8ᵉ Armée qui vient d'être formée, est lancée à l'attaque pour soulager la garnison encerclée de Tobrouk et reconquérir la Cyrénaïque. En décembre, Rommel sera repoussé jusqu'en Tripolitaine d'où il était parti en mai 1941. Le 21 janvier 1942, Rommel contre-attaque. Le 10 février, il atteint la ligne Gazala Bir-Hakeim et la franchit le 27 reprenant ainsi la Cyrénaïque. Il fait subir de telles pertes aux « rats du désert » qu'ils sont obligés de faire retraite jusqu'en Egypte en livrant des combats d'arrière-garde. Il prend Tobrouk le 21 juin et est élevé à la dignité de maréchal le lendemain. En juillet, il parvient à El Alamein, à 100 km d'Alexandrie. Ces succès, il les doit surtout aux blindés dont il dispose, les *Panzer IV* à canon de 75 court, alors que les Britanniques ne peuvent que lui opposer des *Crusader* à canon de 40 mm et des chars Général Grant à canon de 75 mm seuls capables de rivaliser avec le char allemand. Pendant la retraite sur El Alamein, la *7th AD* faisait partie du XXXᵉ Corps aux côtés de la 1ʳᵉ brigade des Forces Françaises Libres qui, sous le commandement du général Koenig, résistera pendant 16 jours à Bir-Hakeim (du 27 mai au 11 juin 1942).

C'est à ce moment qu'arrive à El Alamein un général anglais peu connu pour prendre le commandement de la 8ᵉ Armée, le général Montgomery. Il doit prendre en mains une armée pour laquelle les mouvements de conquête et de repli ne sont pas faits pour maintenir un moral élevé. La division des « rats du désert » quant à elle, est constituée d'officiers et de soldats qui, depuis longtemps, vivent, combattent, souffrent ensemble et se connaissent tous. Il y règne une camaraderie certaine. C'est ce modèle que Montgomery veut essayer d'imposer à la 8ᵉ Armée en vivant continuellement avec elle. Il abandonne sa tenue de général, remplace sa casquette par un béret, sa vareuse par un pull-over, rien ne le distingue des autres officiers.

Le 28 octobre, après une formidable préparation d'artillerie effectuée à l'aide d'une concentration de 1 000 canons, Montgomery prend l'offensive à El Alamein. Ce sera un des tournants de la guerre. La *7th AD* est maintenant équipée de chars *Sherman* à canon de 75 depuis septembre. Le président Roosevelt les a offerts à l'Armée britannique après la chute de Tobrouk. Avec la 8ᵉ armée, elle va aller de succès en succès. Fonçant le long de la côte libyenne, elle entre dans Tripoli le 23 janvier 1943 puis butte sur la ligne Mareth en Tunisie, la contourne pour atteindre Gabès et pousser ensuite vers le nord sous les ordres, depuis le 22 février, du général Erskine.

On January 21st 1942, Rommel counter-attacked. On February 10th, he reached the Gazala Bir-Hacheim line, which he crossed on the 27th, retaking Cyrenaica in the process. He inflicted heavy losses on the Desert Rats, forcing them to withdraw back to Egypt, fighting a rearguard action all the way. He captured Tobruk on June 21st and was promoted to field-marshal the following day. In July, he reached El Alamein, 100 km from Alexandria. He owed these successes mostly to his tanks, the Panzer IVs with their short 75 mm gun, against which all the British had to offer was the Crusader's 40 mm gun and the Grant's 75 mm gun, the Grant being the only tank anything like a match for the German tank. During the retreat to El Alamein, the 7th AD came under XXX Corps, alongside the Free French Army's 1st Brigade commanded by General Koenig, which held out for 16 days at Bir-Hacheim (from May 27th to June 11th 1942).

It was at this time that a little known British general arrived at El Alamein to take over command of the Eighth Army: General Montgomery. He had to take in hand an army whose movements of conquest and withdrawal were not good for morale. Meanwhile, the Desert Rats division was made up of officers and men who had lived, fought and suffered together for a long time, where everyone knew everyone else. There was a definite spirit of camaraderie. This was the model that Montgomery sought to impose on the Eighth Army by living with it all the time. He gave up wearing a general's uniform, replacing his cap with a beret, his jacket with a sweater, and so he looked just like any other officer.

On October 28th, following a terrific preliminary artillery bombardment carried out by a concentration of 1,000 guns, Montgomery went onto the offensive at El Alamein. This was to be one of the turning points of the war. Since September, the 7th AD had Sherman tanks with 75 mm guns, a gift from President Roosevelt to the British Army after the fall of Tobruk. With the Eighth Army, it ran up a string of successes. Racing along the Libyan coast, it entered Tripoli on January 23rd 1943, then came up against the Mareth Line in Tunisia, which it outflanked to reach Gabes and then drove north, General Erskine having taken over command on February 22nd.

On May 7th, the 11th Hussars Regiment in the vanguard entered Tunis. On the 8th, the remaining Germans decided to abandon North Africa, one of them being General Bayerlein.

For the 7th AD the African campaign was over. The division was then detached from the British Eighth Army and withdrawn to Homs, near Tripoli, to train for a different kind of warfare, on the continent in Italy.

The Italian campaign

The Eighth Army landed at Messina in Sicily on July 10th. On September 3rd, Italy sued for an armistice. The 7th AD was then attached to the U.S. Fifth Army commanded by General Clark, and landed at Salerno on September 15th. The Germans put up some stout resistance, and for several days the outcome of the operation hung in the balance. This was because the Desert Rats had to confront the enemy on very different terrain from what they had been used to in North Africa. No more wide open spaces; they now had to fight on narrower, more uneven terrain, and so improvise fresh techniques. The British ended up adjusting to the situation.

---

(1) L'opération Perch est le nom de code donné à la tentative de percée des lignes allemandes vers Villers-Bocage.
(1) Operation Perch was the codename of the attempted breakout through the German lines in the Villers-Bocage sector.

1. La gerboise, premier emblème de la division.
2. L'emblème porté par les hommes de la division en Normandie.
3. L'insigne personnel du *Brigadier* Hinde montrant une version primitive et artisanale de l'emblème de la division. (Coll. Musée Mémorial de Bayeux.)
4. Casque peint couleur sable avec l'emblème de la division. (Coll. part.)
5. Casque d'agent de liaison motocycliste avec l'emblème de la Gerboise, 4th Armoured Brigade. (Coll. part.)

1. The gerboa, the division's first badge.
2. Badge worn by the men of the division in Normandy.
3. Brigadier Hinde's personal insignia showing a primitive homemade version of the divisional badge. (coll. Musée Mémorial de Bayeux.)
4. Helmet painted in a sandy color with the divisional badge. (Private coll.)
5. Motorcycle liaison officer's helmet with the gerboa badge, 4th Armoured Brigade. (Private coll.)

Le 7 mai, le 11ᵉ régiment de hussards qui forme son avant-garde, pénètre dans Tunis. Le 8, les Allemands rescapés décident d'abandonner l'Afrique du Nord, parmi eux, le général Bayerlein.

La campagne d'Afrique est terminée pour la 7th AD. Elle va alors être séparée de la 8ᵉ Armée britannique et se retirer à Homs, près de Tripoli pour s'entraîner à un autre genre de guerre, la guerre sur le continent, en Italie.

La campagne d'Italie

La 8ᵉ Armée débarque à Messine en Sicile le 10 juillet. Le 3 septembre, l'Italie demande l'armistice. La *7th AD* est alors jointe à la 5ᵉ Armée américaine, sous les ordres du général Clark, et débarque à Salerne le 15 septembre. La résistance allemande est âpre et, pendant plusieurs jours, le résultat de l'opération est assez douteux. C'est que les « rats du désert » affrontent l'ennemi sur un autre terrain que celui rencontré en Afrique du Nord. Finis les grands espaces, il faut se battre sur des terrains plus étroits, plus tourmentés, utiliser des techniques différentes. Les Britanniques finissent par s'adapter.

Dans la nuit du 28-29 septembre, ils réussissent à avancer vers le nord et, dans la journée, le *6th Queen's* prend Scafati. Naples tombe le 1ᵉʳ octobre et la *1st Rifle Brigade* prend Cardite plus au nord. Le 13, la 131ᵉ Brigade d'infanterie franchit la Volturne, fonce vers Garigliano, atteint le 1ᵉʳ novembre, et le *11th Hussars*, la *1st Rifle Brigade* et un escadron du *5th RTR* y prennent position. Le reste de la division est mis en réserve. Le 7 novembre, toute la division est rassemblée à Aversa où elle abandonne tous ses véhicules et ses chars à la 5ᵉ Division Blindée canadienne qui va continuer le combat, puis, elle gagne la péninsule de Sorrente, au nord du golfe de Salerne avant le départ vers l'Angleterre.

L'insigne de la Division, le rat du désert

Jusqu'en 1940, l'insigne de la division consistait en un cercle blanc dans un carré rouge. Le général O. More Creach décida de faire figurer dans le cercle blanc un animal qui rappelait l'action de la division dans le désert, la grande gerboise égyptienne dont on trouvait un exemplaire au zoo du Caire. Le général avait proposé une prime à qui aurait la meilleure idée pour représenter la division. Selon le capitaine Warwick Charlton, c'est un caporal du 1ᵉʳ régiment de chars qui proposa le rat du désert après avoir découpé l'emblème dans un vieux bidon à essence. Son projet fut retenu et il gagna ainsi la prime promise (in cadran).

L'animal, grossièrement dessiné était rouge sur fond kaki. Lors du débarquement, l'animal mieux dessiné ressemblait plutôt à un kangourou. Il était rouge sur fond noir et demeura l'insigne de la division jusqu'à sa dissolution en 1958.

La 7th Armoured Division arrive en Normandie

La division n'était pas prévue pour faire partie de la 1ʳᵉ vague d'assaut le 6 juin, elle ne devait commencer à intervenir qu'à partir du 7, derrière la 50ᵉ Division d'Infanterie dans le secteur anglais, lorsque les plages ne seraient plus sous le feu des défenses allemandes. Toutefois le *Brigadier* Looney Hinde qui commandait la 22ᵉ Brigade Blindée devait, avec la section de reconnaissance, débarquer en même temps que le quartier général de la *50th ID* pour

During the night of September 28-29th, they managed to advance northwards, and in the course of the day, the 6th Queen's took Scafati. Naples fell on October 1st and the 1st Rifle Brigade took Cardite further north. On the 13th, the 131st Infantry Brigade crossed the Volturno River and raced towards Garigliano, arriving on November 1st, and the 11th Hussars, 1st Rifle Brigade and a squadron of the 5th RTR took up position there. The rest of the division was placed in reserve. On November 7th, the whole division was brought together at Aversa where it gave up all its vehicles and tanks to the Canadian 5th Armoured Division, which carried on the battle, and then it moved on to the Sorrento peninsula in the north of the Gulf of Salerno, before sailing home to England.

The Division's insignia, the desert rat

Until 1940, the divisional insignia consisted of a white circle in a red square. Then General O. More Creach decided to put in the white circle an animal recalling the division's action in the desert, the large Egyptian jerboa of which there was a specimen at Cairo Zoo. The General had offered a prize for whoever came up with the best idea for an emblem for the division. According to Captain Warwick Charlton, it was a corporal of the 1st Tank Regiment who suggested the desert rat after cutting one out from an old jerrican as an emblem. His suggestion was adopted and so he was awarded the prize (a dial).

The animal was roughly drawn in red on a khaki background. By D-Day, the design had been improved but now rather resembled a kangaroo. It was in red on a black ground and remained unchanged until the division was disbanded in 1958.

rechercher un lieu de rassemblement pour la division.

Au mois de mai, la 22ᵉ brigade avait été regroupée près de Ipswich, au nord-est de Londres à proximité du port de Felixstowe où elle devait embarquer. Le reste de la division était à l'est de Londres, à Brentwood et West Ham, à proximité des docks de la Tamise. Pour atteindre la Normandie, il lui faudra donc franchir le détroit du Pas-de-Calais, passer au large des défenses côtières du mur de l'Atlantique et se mettre à l'abri de la vue des Allemands en utilisant des nuages de fumée et en brouillant les radars de l'ennemi.

Le soir du 6 juin, le *4th CLY* et le *1st RTR* arrivaient au large des côtes normandes. Le *4 CLY* avait pris le départ à l'aube du 5 juin à Felixtowe après avoir embarqué ses chars sur des *L.S.T. (Landing Ship Tanks)* protégés des attaques en piqué des avions ennemis par des ballons. Il se forma au fur et à mesure que d'autres bateaux se joignaient à eux, un énorme convoi qui se glissa vers le sud. C'est sous une petite pluie fine et une houle légère que les bâtiments franchirent le Pas-de-Calais et se présentèrent devant les plages d'Arromanches et Le Hamel, vers 22 heures dans la soirée du 6 alors que la *50th ID* avait déjà pris pied à terre. Quelques bateaux avaient dérivé un peu vers l'est à la lisière de *Sword Beach* là où avait débarqué le *3rd CLY*, le régiment frère, qui faisait partie de la 4ᵉ brigade blindée, soutien de la *51st ID* écossaise.

Ayant manqué la marée, la *7th AD* doit attendre l'aube du 7 juin, à 400 mètres du bord pour commencer les opérations de débarquement au cours desquelles elle va perdre 3 chars « noyés ». A terre, les tankistes commencent par enlever les adhésifs qui avaient protégé l'intérieur de leurs chars de l'eau de mer puis, en file indienne, chaque char suivant les traces du char qui le précède pour éviter les mines et sous les sifflements des obus de marine au-dessus de leurs têtes, se dirigent vers le lieu de rassemblement dans les champs de blé, du côté de Sommervieu au nord de Bayeux.

Les 8 et 9 juin, la section de reconnaissance et l'escadron C du *4th CLY* supportent une attaque du *6th DLI (Durham Light Infantry)* de la *151st Brigade* de la *50th ID* sans rencontrer véritablement d'ennemis, à part quelques tireurs isolés perchés dans les hauts arbres. Le *4th CLY* traverse Bayeux en partie désert et arrive au Douet-de-Chouain le soir du 9 juin. Le 7, le *5th RTR* qui avait suivi le même chemin en mer que le *4th CLY* arrivait le matin et se rassemblait lui aussi près de Sommervieu ; après avoir perdu 2 chars. Le même jour, l'après-midi, c'était au tour du *Norfolk Yeomanry* et au 1ᵉʳ bataillon de la *Rifle Brigade* de débarquer à Arromanches.

La brigade des *Queen's*, qui était partie des docks de Tilbury à l'embouchure de la Tamise le 4 juin, passait au large de Southend-on-sea, franchissait le Pas-de-Calais, longeait la côte sud de l'Angleterre et s'attardait dans la rade de Spithead pour ne débarquer qu'à partir du 8 juin en un point situé dans *Gold Beach*, dans les dunes de sable à l'est d'Arromanches et à l'ouest de la Rivière, pour se rassembler à Pouligny au nord-est de Bayeux. Le *7th Queen's* n'arrivait que le 10. Le 11ᵉ Hussard, parti de Millwall Docks arrivait le 9, à Courseulles et était dirigé aussitôt vers Ryes. Le 8ᵉ Hussard parti de la côte sud, Bampert Quay à Gosport près de Portsmouth, se rassemblait à Sommervieu.

Lorsqu'ils prennent pied sur le sol français, les soldats de la *7th AD* ne connaissent pas les difficultés qui les attendent. Pour eux qui n'ont pas subi les pertes des troupes d'assaut à quelques exceptions près, les opérations de débarquement se sont pas-

The 7th Armored Division arrives in Normandy

The Division was not scheduled to arrive with the 1st assault wave on June 6th, it was only supposed to enter the fray on the 7th, coming in behind the 50th Infantry Division in the British sector, when the beaches were no longer under fire from the German defenses. However Brigadier Looney Hinde, commander of the 2nd Armored Brigade, had to land with the scout section at the same time as 50th ID headquarters in order to locate a place for the division to assemble.

In May, the 2nd Brigade had assembled near Ipswich, north-east of London, not far from the port of Felixstowe where it was to embark. The remainder of the division was at the Thames docks, in eastern London at Brentwood and West Ham. This meant that to get to Normandy, it had to pass through the Straits of Dover, sail past the coastal defenses of the Atlantic Wall, and keep out of the Germans' sight behind smoke screens and by scrambling the enemy radar.

6. Un char anti-aérien, un Crusader AA MK VI du groupe de QG de la *22nd Armoured Brigade* débarque d'un LST dans l'après-midi du 7 juin sur *Gold Beach*. On voit peint sur la tourelle le texte « Allakee fek » qu'on retrouvera aussi peint sur la tourelle d'un firefly.

7. Peu après, nous voyons une colonne de véhicules blindés, peu après le débarquement, précédés d'un autre Crusader AA MK III, le « Skyraker ». (Photos IWM.)

6. *A Crusader AA MK III of 22nd Armoured Brigade's HQ Group disembarks from au LST on the afternoon of June 7. The name painted on the side of the turret (« Allakee-fek ») will be seen again on one of the Fireflies.*

7. *Short after, we see a column of tanks.*

Des hommes de la *22nd Armoured Brigade* s'approchent de la côte normande le 7 juin 1944 et vont débarquer sur *Gold Beach*. (IWM.)

Men of the 22nd Armoured Brigade approach the Normandy coast on June 7th, 1944 before landing on Gold Beach. (IWM.)

sées relativement bien. Certains toutefois ont été particulièrement touchés par les premiers spectacles qu'ils ont rencontrés, sur la mer, sur les plages, sur terre. Les combattants sont tout de même inquiets, surtout les vétérans. La *7th AD* a construit sa réputation en Afrique du Nord et en Italie avec un matériel différent de celui qui est maintenant le sien. Les chars *Sherman* laissés en Italie semblaient plus sûrs que les *Cromwell* qui ont été livrés en Angleterre et avec lesquels ils estiment être insuffisamment familiarisés.

Ils connaissent le canon de 88 mm allemand pour sa redoutable efficacité. Ils ont affronté en Tunisie les *Tiger* du Bataillon Blindé lourd allemand, le 503, et les redoutent. Au moment où ils débarquent en Normandie, ils ne connaissent pas le bocage normand. Si le bocage est un lieu agréable pour se promener et profiter de la nature, il en est autrement lorsqu'il faut combattre dans un tel milieu, il devient alors un terrain particulièrement difficile. Les « rats du désert » ont été habitués à combattre en terrain découvert, dans de grands espaces, tourelle découverte. Ils vont devoir maintenant affronter un adversaire qu'ils ne vont pas voir mais qu'ils vont soupçonner être dans chaque haie, dans chaque maison, derrière chaque mur, dans chaque village, dans des chemins si étroits qu'ils ne pourront pas y manœuvrer à leur guise, comme dans un couloir. Ils vont peut-être affronter corps à corps des adversaires rompus à ce genre de combat surtout s'ils ont servi sur le front russe. Ils ont déjà connu cette situation. En Italie, les terrains plus accidentés que les espaces désertiques d'Afrique les avaient obligés à changer de technique de combat et ils s'y étaient adaptés rapidement.

Premiers combats

La *50th ID* a pris Bayeux mais a peu progressé vers le sud vers Tilly-sur-Seulles. Le *5th RTR* est la première unité de la division à entrer en action. Il seconde une brigade d'infanterie indépendante, la 56e, composée du *2nd Essex*, du *1st Borderer* (Sud Galles), et du *2nd Gloucester*, qui neutralise des poches de résistance allemande à Sully et Port-en-Bessin. Au cours de cette action, la division subit ses premières pertes, 2 chars et 6 hommes mais détruit

On the evening of June 6th, the 4th CLY and 1st RTR arrived off the Normandy coast. The 4th CLY had sailed out of Felixstowe at dawn on June 5th after embarking its tanks on L.S.T.s (Landing Ship Tanks) protected by barrage balloons from enemy dive bombers. Gradually, as other boats joined up, it was made to join a huge convoy which slowly headed off south. The ships crossed the Channel in slight drizzle and relatively calm seas, arriving off the beaches of Arromanches and Le Hamel at c. 22.00 hours on the evening of the 6th, by which time the 50th ID had already set foot on shore. A few boats drifted a little to the east to the edge of Sword Beach where the sister regiment had landed - the 3rd CLY, which came under the 4th Armoured Brigade, supporting the Scottish 51st ID.

Having missed the tide, the 7th AD had to wait 400 metres off shore until dawn on June 7th to begin landing operations in the course of which it lost 3 "drowned" tanks. Once ashore, the tank crews began by removing the adhesive tape keeping the seawater out of their tanks then file off, each tank following in the tracks of the tank in front in order to steer clear of mines, with naval shells whistling overhead, on their way to their mustering point in the cornfields, somewhere near Sommervieu, north of Bayeux.

On June 8th and 9th, the scout section and "C" Squadron 4th CLY lent support to an attack by the 6th DLI (Durham Light Infantry), 151st Brigade, 50th ID without really encountering the enemy apart from the odd sniper perched in the tall trees. The 4th CLY passed through a partly deserted Bayeux and came to Douet-de-Chouain on the evening of June 9th. On the 7th, the 5th RTR, having sailed by the same route as the 4th CLY, arrived during the morning and also mustered near Sommervieu after losing 2 tanks. That afternoon, it was the turn of the Norfolk Yeomanry and the Rifle Brigade's 1st Battalion to land at Arromanches.

The Queen's Brigade sailed out of Tilbury docks in the Thames estuary on June 4th, passed out to sea off Southend-on-Sea, passed through the Straits of Dover, followed the southern coast of England and waited in the Spithead roads, not to land until June 8th at a location in the Gold Beach sector, in the sand dunes east of Arromanches and west of La Rivière, before mustering at Pouligny, north-east of Bayeux. The Queen's did not arrive until the 10th. The 11th Hussars sailed from Millwall Docks, arriving at Courseulles in the Juno Beach sector on the 9th, and were at once directed towards Ryes. The 8th Hussars started out from Bampert Quay on the southern coast, at Gosport near Portsmouth, and were concentrated at Sommervieu.

On setting foot on French soil, the men of the 7th AD had no idea of the difficulties that lay ahead. They did not suffer anything like the losses of the assault troops, and with a few exceptions, landing operations passed off pretty smoothly. Some however were devastated by what they first saw in the sea, on the beaches, then on land. The fighting troops were certainly on edge, particularly the veterans. The 7th AD had built up its reputation in North Africa and Italy with different equipment from what it now had. The Sherman tanks left behind in Italy seemed more reliable than the Cromwells issued in England with which they felt unfamiliar.

They knew just how lethally effective the German 88 mm gun was. In Tunisia they had faced the Tigers of the German heavy tank battalion, the 503rd, and learned to fear them. At the time of the Normandy landings, the hedgerow country was for them a whole new experience. While the Bocage was nice country for walking in and enjoying nature, it made a parti-

8 canons de 88 mm. C'est aussi pour les Britanniques la première prise de contact avec le bocage normand et il leur faudra souvent dépêcher devant leurs chars des coéquipiers chargés de scruter les alentours avant de s'engager dans un carrefour au risque de les exposer aux tirs de snipers ou de mitrailleuses à l'affût dans les haies et les fourrés. Du 10 au 12 juin, la 22ᵉ brigade blindée est placée sous le même commandement que la 56ᵉ brigade d'infanterie pendant l'attaque sur Tilly. La tactique de cette attaque est la suivante :

Le *4th CLY* avancerait sur la gauche, vers Tilly et Juvigny, le *5th RTR* marcherait sur la droite vers le château de Verrières à travers le village d'Ellon et le hameau de Folliot (village de Juaye-Mondaye), le *1st RTR* resterait en réserve à Rucqueville pour garder les ponts sur la Seulles autour de la RN 13. Ces régiments de chars sont accompagnés de l'infanterie motorisée du 1ᵉʳ bataillon de la *Rifle Brigade* et soutenus par l'artillerie des *3rd* et *5th RHA*, du *86th Field Regiment RA* et du *65th AT Norfolk Yeomanry*.

En face, se trouvent la 352ᵉ Division d'Infanterie allemande et la *Panzer-Lehr-Division*. Cette dernière, arrivée récemment, est solidement installée dans les hameaux fortifiés entre Tilly et le château de Verrières, dans Bucéels, Bernières, le hameau de Jerusalem, avec des chars, *Panzer IV,* quelques *Panther* et une infanterie nombreuse et expérimentée.

La *7th Armoured Division* commence à subir des pertes sérieuses. Le 1ᵉʳ bataillon de la *Rifle Brigade* perd une section, le *5th RTR* déplore la destruction de 5 chars et compte 16 tués.

Le **9**, le QG du *4th CLY* a pris position près de Jerusalem et essaie de faire progresser ses escadrons mais se heurte à une violente résistance de l'adversaire. Dans la nuit cependant, il réussit à glisser ses chars et les rassembler dans les champs près du village de Bucéels après en avoir perdu quelques-uns.

cularly tricky battlefield. The Desert Rats were used to fighting in vast open spaces, with the turret hatch open. They now had to face an enemy who was nowhere to be seen but might be lurking behind every hedge, every house, every wall in every village, and the lanes were so narrow, like going down a corridor, that they could not maneuver as they pleased. They would possibly have to take on at close quarters an enemy that was an old hand at the game, especially the men who had served on the Russian front. They had already coped with a similar situation. In Italy, the terrain was hillier than the African desert, and being forced to change their combat technique, they had adapted quickly.

Early engagements

The 50th ID took Bayeux but made little headway south towards Tilly-sur-Seulles. The 5th RTR was the division's first unit into action. It gave support to an independent infantry brigade, the 56th, comprising 2nd Essex, 1st Borderers (South Wales) and 2nd Gloucesters, which mopped up pockets of German resistance at Sully and Port-en-Bessin. During this action, the division sustained its first casualties, 2 tanks and 6 men, but destroyed eight 88 mm guns. This was also for the British their first contact with the Normandy Bocage and they often had to dispatch crew members ahead of their tanks to scout the area before coming to a crossroads where they might expose themselves to fire from snipers or machine-gunners lurking in the hedgerows and thickets. From June 10th to 12th, during the attack on Tilly, the 2nd Armoured Brigade was brought under joint command with the 56th Infantry Brigade. The tactics of this attack were as follows:

The 4th CLY was to advance on the left, towards Tilly and Juvigny; the 5th RTR was to march on the right flank towards the château de Verrières, passing through the village of Ellon and the hamlet of Folliot (in the parish of Juaye-Mondaye); the 1st RTR was

Un Panzer IV *(II./Panzer-Lehr-Regiment 130)* détruit près de Bayeux par un canon antichar du *Durham light Infantry (50th Infantry Division)* le 10 juin 1944 est examiné par des hommes de la *7th Amoured Division*. Photo prise par le *Sergeant* Laing le 11 juin 1944.

A Panzer IV (II./Panzer-Lehr-Regiment 130) destroyed near Bayeux by an antitank gun of the Durham Light Infantry (50th Infantry Division) on 10 June 1944 is examined by men of the 7th Armoured Division. Photo taken by Sergeant Laing on 11 June 1944. (IWM.)

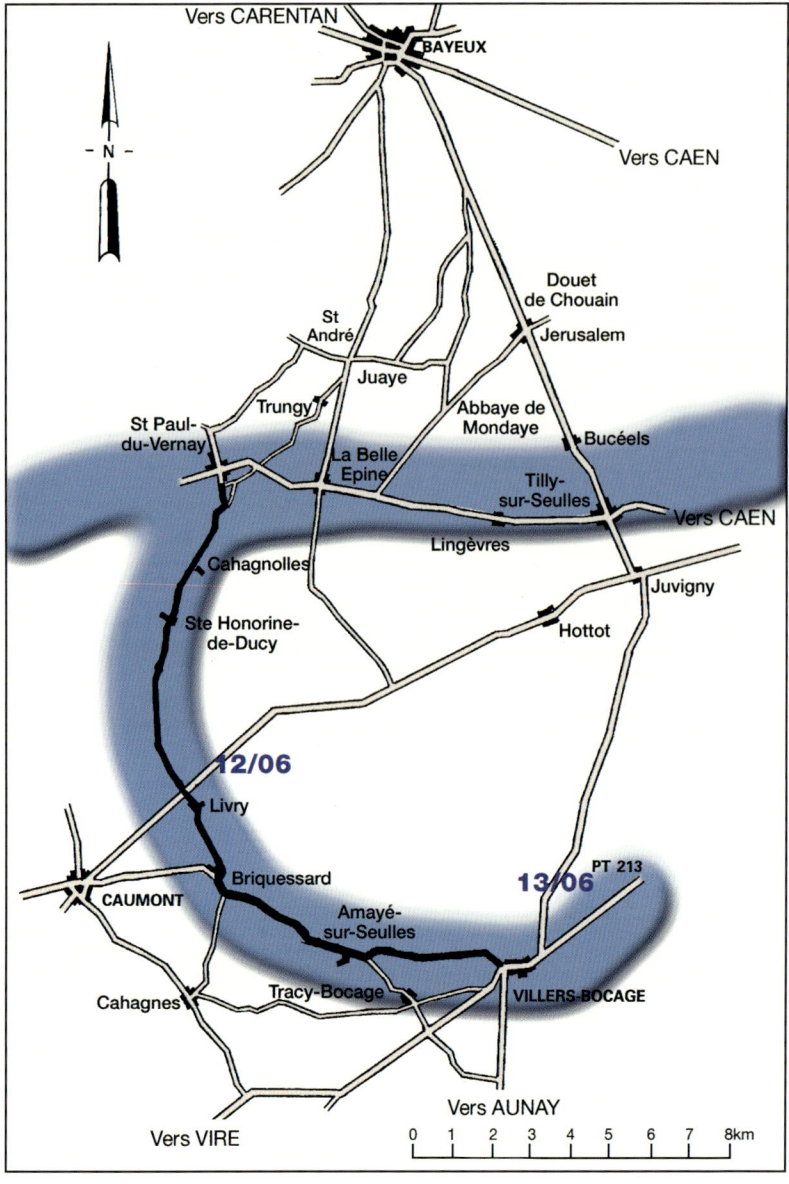

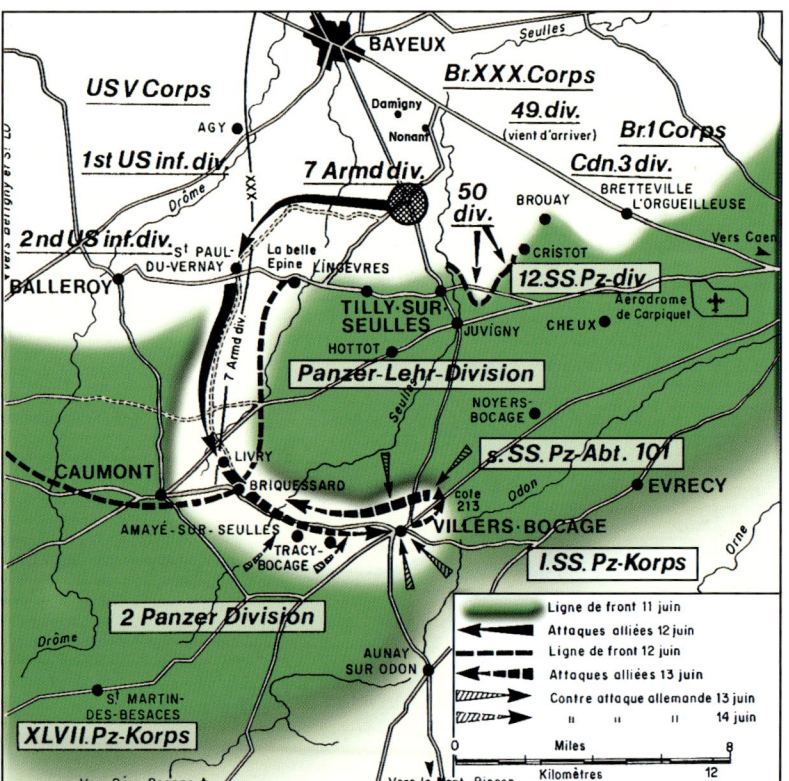

to be held in reserve at Rucqueville to guard the bridges over the Seulles in the vicinity of Highway N13. These tank regiments were to be accompanied by the motorized infantry of the 1st Battalion, the Rifle Brigade and supported by the artillery of the 3rd and 5th RHA, 86th Field Regiment RA and 65th AT Norfolk Yeomanry.

Facing them were the German 352nd Infantry Division and the Panzer-Lehr-Division. The latter, a recent arrival, was firmly entrenched in the fortified hamlets between Tilly and the château de Verrières, and at Bucéels, Bernières and the hamlet of Jerusalem, with tanks - Panzer IVs and a few Panthers - and large numbers of hardened infantry.

The 7th Armored Division began to sustain serious casualties. The 1st Battalion, Rifle Brigade lost a platoon, while the 5th RTR had 5 tanks knocked out with 16 killed.

On the **9th**, the HQ of 4th CLY took up position near Jerusalem and tried to move its squadrons forward, only to encounter fierce enemy resistance. During the night however, it contrived to slip its tanks through and assemble them in the fields near the village of Bucéels, although losing a few on the way.

On the **10th**, the 4th CLY was in the sector of the 6th DLI (50th ID), the scout section and "A" Squadron were on the left, "B" Squadron on the right, and "C" Squadron in the center. On the left, at a distance of 2.5 to 3 km on the right bank of the Seulles, was the 8th Armoured Brigade. By evening, "A" Squadron was in Bucéels, "B" Squadron, which lost one Cromwell, was to the east of the hamlet Marcel, "C" Squadron was north of Sainte-Bazile, north-east of Bucéels. But during the night, the enemy managed to infiltrate and take control of the town.

On the **11th**, the Queen's, who had taken up a defensive position around Nonant, where General Erskine had his headquarters since June 8th, was ordered to skirt Tilly via the west to reach Juvigny. The 5th and 6th Queen's, with the 1st RTR in support, attacked at dawn south of the Tilly-Verrières line but, caught in heavy artillery fire, they had to withdraw and regroup in the Douet de Chouain area. There was a battle for the bridges over the Seulles, and two bridges held by the Germans were captured by two companies of the 2nd Gloucesters, enabling the road to Tilly to be taken and held. 4th CLY's "A" and "B" Squadrons were kept back in reserve at Bucéels while "C" Squadron, placed under the 2nd Gloucesters, destroyed 2 Panthers and reaching Tilly, enabling the infantry to enter the town. The tanks then tried to outflank the town on the right, but a troop of 3 Cromwells and 1 Firefly was wiped out after going down a narrow lane with no room to maneuver. A German counter-attack forced "C" Squadron and the Gloucester companies back to Bucéels.

On the **12th**, General Montgomery now took the decision to withdraw the 7th Armoured Division and launch it to the west in Operation Perch, and to place the front at Tilly in the hands of the 50th Infantry Division (Northumbrian) and 56th Infantry Brigade. **The battle was on for possession of Villers-Bocage.**

June 12th, the 7th Armoured Division advances

The 7th Armoured Division set off at 15.00 hours, making rapid progress. It formed a long column of tanks and other vehicles behind the 8th Hussars whose "C" Squadron had not yet arrived. The regiment was followed by the scout section of the 4th CLY; "A" Company, Rifle Brigade, 4th CLY; a Sexton battery of 5th RHA, 7th Queen's; 5th RTR and "I"

Le **10**, le *4th CLY* est dans le secteur du *6th DLI (Durham Light Infantry de la 50th ID)*, la section de reconnaissance et l'escadron A sont sur la gauche, l'escadron B sur la droite, l'escadron C au centre. Plus à gauche, à une distance de 2,5 à 3 km, sur la rive droite de la Seulles, se trouve la *8th Armoured Brigade*. Dans la soirée, l'escadron A est à Bucéels, l'escadron B, qui a perdu un *Cromwell*, est à l'est du hameau Marcel, l'escadron C est au nord de Sainte-Bazile soit au nord-est de Bucéels. Mais, la nuit, l'ennemi parvient à s'infiltrer dans le bourg et à le contrôler.

Le **11**, la Brigade des *Queen's* qui a pris une position défensive autour de Nonant où le général Erskine a installé son poste de commandement dès le 8 juin reçoit l'ordre de contourner Tilly par l'ouest pour rejoindre Juvigny. Le *5th* et le *6th Queen's* attaquent à l'aube, soutenus par le *1st RTR*, au sud de la ligne Tilly-Verrières mais, pris sous un tir massif d'artillerie, ils doivent battre en retraite et se rassembler autour du Douet de Chouain. On se bat pour les ponts sur la Seulles, deux ponts tenus par les Allemands sont pris par deux compagnies du *2nd Gloucester*, les « *Glosters* », ce qui permet de conquérir et conserver la route menant à Tilly. Les *A* et *B Squadrons* du *4th CLY* restent en réserve à Bucéels tandis que le *C Squadron* placé sous la direction du *2nd Glocester* détruit 2 *Panther* jusqu'à Tilly où l'infanterie peut pénétrer. Les chars tentent alors de déborder le bourg par la droite mais une « *troop* » de 3 *Cromwell* et 1 *Firefly*, qui s'était engagée dans un chemin étroit où il n'y avait pas de place à manœuvrer, est entièrement détruite. Une contre-attaque allemande oblige l'escadron C et les compagnies de *Glosters* à se replier sur Bucéels.

Le **12**, le général Montgomery prend alors la décision de retirer la 7ᵉ Division Blindée pour la lancer vers l'ouest, dans l'opération *Perch* et de confier le front de Tilly à la 50ᵉ Division d'Infanterie *(Northumbrian)* et à la 56ᵉ Brigade d'Infanterie. **La bataille pour la possession de Villers-Bocage est lancée.**

Le 12 juin, la 7th Armoured Division s'élance

A 15 heures, la *7th Armoured Division* démarre et avance rapidement. Elle forme une longue colonne de blindés et de véhicules divers précédés par le *8th Hussars* dont l'escadron C n'est pas encore arrivé. Le régiment est suivi par la section de reconnaissance du *4th CLY*, de la compagnie A de la *Rifle Brigade*, du *4th CLY*, d'une batterie de Sexton du *5th RHA*, du *7th Queen's*, du *5th RTR* et de la compagnie I du 1ᵉʳ bataillon de la *Rifle Brigade*. La compagnie A des fusiliers, qui n'avait participé qu'à quelques escarmouches depuis le 7 juin, avait dû intervenir le matin vers 9 heures autour de quelques fermes de Jérusalem à l'ouest du Douet de Chouain pour repousser une attaque de chars allemands.

La *131st Infantry Brigade*, qui avait reçu comme objectif Tilly pour le 11, est toute surprise de continuer dès le 12 dans une autre direction. Le *5th*, le *6th Queen's* et le *1st RTR* reçoivent alors l'ordre d'occuper la position de la 22ᵉ brigade en attendant du renfort. L'attaque se développe vers l'ouest et traverse les villages et hameaux du Bas-Mougard, de Trungy, de Saint-Paul-du-Vernay, de Cahagnolles, de Sainte-Honorine-de-Ducy. A Cahagnolles, le lieutenant-colonel Goulburn, commandant du *8th Hussars* écrit dans son journal que « *ses hommes ont rencontré une patrouille américaine en voitures blindées et jeeps dont les hommes étaient couverts de poussière.* »

Company, 1st Battalion, Rifle Brigade. The Rifles' "A" Company, which had taken part in just a few skirmishes since June 7th, was brought into action that morning at c. 09.00 hours to counter a German tank attack around some farms at Jerusalem west of Douet de Chouain.

The 131st Infantry Brigade, whose assigned objective was Tilly by the 11th, was very surprised to be promptly sent off in a different direction on the 12th. The 5th, 6th Queen's and 1st RTR were then ordered to occupy the 2nd Brigade's position awaiting reinforcements. The attack was built up in the west, passing through the villages and hamlets of Bas-Mougard, Trungy, Saint-Paul-du-Vernay, Cahagnolles and Sainte-Honorine-de-Ducy. At Cahagnolles, 8th Hussars commander Lieutenant-Colonel Goulburn recorded in his diary that "we meet a U.S. patrol in armored cars and jeeps looking very dusty and excited."

Some Queen's reached the Belle Epine crossroads, but failed to penetrate further behind the German lines and gave up. The British continued south as far as Livry, where a "panzerfaust" charge stopped the 8th Hussars' leading tank in its tracks. This halted the breakthrough. Paul Lepoil saw "a tank brew up, heard ammunition exploding inside, and saw a soldier desperately trying to put out the fire with an extinguisher". Two tank crew members were killed.

Brigadier Hinde then asked Lieutenant-Colonel Goulburn to try and break out eastwards to avoid and if necessary bypass such obstacles. Lieutenant Talbot Hervey's troop moved towards the hamlet of La Croix des Landes and within two or three minutes encountered elements of the left flank of Panzer-Lehr-Division which knocked out the two leading British tanks with three rounds from an anti-tank gun. It took the Rifle Brigade, under Lieutenant Bruce Campbell, with support from the tanks of "B" Squadron of the 4th CLY, to release Livry. There were just three hours left until nightfall, so it was too late to carry on to Villers-Bocage. Brigadier Hinde gave orders to halt so as to leave the enemy no clue as to which direction the offensive was to take: south with the Americans or east to outflank the Panzer-Lehr-Division.

A small detachment reached Briquessard. Allied soldiers entered a farmyard where the Normans served them cider and calvados apple-brandy. Late that evening, they were ordered to withdraw north of the Caen-Caumont road, which they did after promising to return the next day and attempting to take their hosts along with them into the liberated sector. This forced halt was put to good use by the supply vehicles to regroup around the leading units and bring up food and ammunition. For the column was so huge and the lanes so narrow in the Bocage that the division's normal progress was slowed up by the ensuing congestion. The 7th Queen's infantrymen had to wait till one in the morning to get their promised meal, by which time they were pretty ravenous.

June 13th, the British offensive

On June 13th, the men rose at 04.30 to start out from Livry at 05.30. The 7th Armoured Division headed off towards Villers-Bocage, advancing without encountering much opposition. On the roadside, the country folk chatted with the men, offering them cider and butter, and strewing flowers over their vehicles. The tanks did not particularly keep to the roads but took to the fields - without opening the gates, complained one young farmer. They soon came to highway N175, 2 or 3 km west of Villers. In the vanguard were the 4th CLY, "A" Company, Rifle Brigade and "B" Battery, 5th RHA. Armored cars of the reconnaissance

Des *Queen's* atteignent le carrefour de la Belle Epine, mais ne peuvent s'enfoncer davantage dans les lignes allemandes et renoncent. Les Anglais continuent vers le sud jusqu'à Livry, où une charge de « *panzerfaust* » arrête brutalement le char de tête du *8th Hussars*. La percée est stoppée. Paul Lepoil voit « *le char brûler, entend des munitions exploser à l'intérieur, et voit un soldat essayer d'éteindre en vain l'incendie avec un extincteur* ». Deux tankistes sont tués.

Le *Brigadier* Hinde demande alors au lieutenant-colonel Goulburn de tenter une percée vers l'est pour éviter de rencontrer de tels obstacles et le cas échéant de les contourner. La *troop* du lieutenant Talbot Hervey se dirige vers le hameau de la Croix des Landes et, au bout de deux à trois minutes, rencontre des éléments de l'aile gauche de la *Panzer-Lehr-Division* qui, de trois coups de canon antichar mettent deux chars anglais de tête hors de combat. Il faut l'intervention de fusiliers de la *Rifle Brigade*, commandés par le lieutenant Bruce Campbell, soutenus par des chars de l'escadron B du *4th CLY* pour dégager Livry. Il ne reste que trois heures avant la tombée de la nuit, il est trop tard pour continuer vers Villers-Bocage. Le *Brigadier* Hinde donne l'ordre de s'arrêter ce qui empêchera l'ennemi de connaître la direction que prendra l'offensive, direction du sud avec les Américains ou bien direction est pour contourner la *Panzer-Lehr-Division*.

Un petit détachement atteint Briquessard. Des soldats alliés pénètrent dans une cour de ferme où des Normands leur offrent cidre et calvados. En fin de soirée, ils reçoivent l'ordre de se replier au nord de la route de Caen-Caumont et se retirent après avoir promis de revenir le lendemain et tenté d'entraîner leurs hôtes avec eux vers le secteur libéré. Cette halte forcée va être utilisée par les véhicules de ravitaillement pour se regrouper près des unités de tête et leur apporter nourriture et munitions. En effet, l'importance de la colonne est telle et les chemins sont si étroits dans le bocage, que l'encombrement qui en résulte ralentit la bonne marche de la division. C'est ainsi que les fantassins du *7th Queen's* attendront jusqu'à une heure du matin le repas qui leur était promis et qui commençait à se faire désirer.

Le 13 juin, offensive britannique

Le 13 juin, réveil à 4 h 30, départ de Livry à 5 h 30. La 7e Division Blindée prend la direction de Villers-Bocage et avance sans grande opposition. Sur le bord de la route, les ruraux parlent avec les soldats, leur offrant du cidre, du beurre, lançant des fleurs sur les véhicules. Les chars ne suivent pas particulièrement les routes et passent par les champs, *sans ouvrir les barrières* déplore un jeune cultivateur. La RN 175 est atteinte rapidement, 2 à 3 km à l'ouest de Villers. L'avant-garde est constituée du *4th CLY*, de la compagnie A de la *Rifle Brigade* et de la batterie B du *5th RHA*. Des voitures blindées du régiment de reconnaissance qui protège le flanc droit de la Division, le *11th Hussars*, signalent qu'elles ont vu à Cahagnes des blindés allemands se diriger vers Villers et qu'elles se sont accrochées avec des véhicules blindés à huit roues. Le lieutenant Charles Pearce, officier de liaison du QG du *4th CLY*, rapporte que les Britanniques inquiets, ont pris pour un Tigre, au bruit qu'il faisait, un gros canon blindé autotracté allemand qui n'a pas résisté aux tirs qui lui ont été adressés. Les chars allemands semblent éviter le combat et préfèrent rester camouflés. C'est ainsi qu'à Amayé-sur-Seulles, route d'Orval, un char allemand, en se repliant, a abandonné un poste d'observation qu'un char anglais a occupé presqu'aussitôt. Le lendemain, c'est le contraire qui se produira.

regiment guarding the Division's right flank, 11th Hussars, reported that they had seen German tanks in Cahagnes heading towards Villers and that they had engaged some eight-wheeled armored vehicles. 4th CLY HQ liaison officer Lieutenant Charles Pearce reported that, judging by the noise it made, the anxious British had mistaken for a Tiger a big German self-propelled armored gun which failed to withstand their fire. The German tanks seemed to avoid any fighting, preferring to remain camouflaged. At Amayé-sur-Seulles for instance, on the Orval road, a British tank occupied an observation post just vacated by a retreating German tank. The next day, it would be the other way round.

North of the Caumont-Villers road, in a lane called La Bresserie, one Panzer - possibly one of the Panzer-Lehr-Division's Panthers – swung round in such a hurry to get onto the Anctoville road that it shed its protective side plates.

The advance progressed so quickly that the leading units got the impression that the rest of the division was no longer following in support. However, orders to advance continued to reach them. 8th Hussars "B" Squadron advanced to Craham after passing through Cahagnes. It was ordered to stay north of Tracy-Bocage to protect 22nd Armoured Brigade's Tactical HQ (Tac Brigade) based at Amayé-sur-Seulles.

The route taken by the leading elements was as follows: the Caumont road, Vierge Noire hill, Place Jeanne d'Arc, where they arrived at around 08.00 hours, and where Monsieur Filion, who worked for the electricity board, was able to film them. Their objective being Point 213 (2), they passed quickly through Villers to set up a strongpoint at the top of the Côte des Landes hill, leaving the main body of the regiment to stop down in the town.

The vanguard reaches Villers

"A" Squadron, 4th CLY reached the Côte des Landes and dug in. The Rifle Brigade's "A" Company stopped at the bottom of the hill after the Bayeux road. The Regimental HQ (RHQ), with its 4 Cromwell tanks, preceded by 3 light Honey reconnaissance tanks, a few scout and intelligence vehicles and 2 of the 5th RHA's observation tanks, took up positions at the top of the Rue G Clemenceau, leaving "B" Squadron in the Place Jeanne d'Arc. "C" Squadron stayed behind in reserve. Armored cars belonging to the 11th Hussars guarding the division north and south of the road came under 88 mm shellfire near Tracy and disposed of two German reconnaissance vehicles. This regiment was under-strength, being made up of just the HQ and "C" Squadron, the other squadrons detached from XXX Corps not having joined up yet.

The 5th RTR, under Lieutenant-Colonel Joe Lever, remained in the rear, south of the D71 Caumont to Villers road. From Point 142 its tanks covered Villers at the bottom of the valley. All that day it came under German artillery and tank fire. Lieutenant Richard Haywood of "C" Squadron had his tank knocked out at around 14.00 hours.

Meanwhile, the 5th RHA set up its batteries of Sextons north of the D71 road, in the small valley where the Coudray and Pont Chouquet Rivers flow. The officers commanding these batteries were Captain Tim Lanyon, who ended his military career as a brigadier, and Captain Andrew Burn. On entering the Rue Pasteur, the leading tank commander asked Marcel Ozenne, "Where's Jerry?" Ozenne confessed his ignorance as to the whereabouts of the German tanks, and on seeing the forces being deployed went back down to his cellar, where he spent the day with his

Au nord de la route Caumont-Villers, dans le petit chemin de la Bresserie, un *Panzer* fait marche arrière si rapidement pour gagner la route d'Anctoville qu'il perd les jupes qui protègent ses flancs. Il s'agit peut-être d'un *Panther* de la *Panzer-Lehr-Division*.

La rapidité de l'avance est telle que les unités qui constituent l'avant-garde ont l'impression qu'elles ne sont plus sous le support du reste de la division. Mais les ordres d'avancer continuent de leur parvenir. L'escadron B du *8th Hussars* s'avance jusqu'à Craham après être passé par Cahagnes. Il lui est ordonné de rester au nord de Tracy-Bocage pour protéger le *22nd Armoured Brigade's Tactical HQ (Tac Brigade)*, installé à Amayé-sur-Seulles.

La route suivie par les premiers éléments est la suivante, route de Caumont, côte de la Vierge Noire, place Jeanne d'Arc où ils arrivent vers 8 heures et où Monsieur Filion, employé EDF, peut les filmer. Leur objectif étant la cote 213 (2), ils traversent rapidement Villers pour aller installer une position forte au sommet de la côte des Landes alors que le gros du régiment s'arrête dans le bas du bourg.

L'avant-garde arrive à Villers

L'escadron A du *4th CLY* parvient à la côte des Landes et s'y installe. La compagnie A de la *Rifle Brigade* s'arrête au bas de la côte après la route de Bayeux. L'Etat-Major du régiment, le *RHQ*, avec ses 4 chars *Cromwell* précédés de 3 chars légers *Honey* de reconnaissance, quelques véhicules de reconnaissance et de renseignement ainsi que 2 chars d'observation du *5th RHA* prennent position dans le haut de la rue G. Clemenceau. L'escadron B reste place Jeanne d'Arc. L'escadron C est plus en arrière, en réserve. Des voitures blindées du *11th Hussars* qui protégeaient la division au nord et au sud de la route ont subi des tirs d'obus de 88 mm vers Tracy et disposé de deux véhicules allemands de reconnaissance. Ce régiment est incomplet, il est constitué seulement de l'Etat-Major et de l'escadron C, les autres escadrons détachés du XXXe Corps ne l'ayant pas encore rejoint.

Le *5th RTR*, commandé par le lieutenant-colonel Joe Lever, est resté en arrière, au sud de la D 71, la route de Caumont à Villers. Ses chars peuvent, depuis la cote 142 surveiller Villers au fond de la vallée. Il subira toute la journée les tirs de l'artillerie et des blindés allemands. C'est ainsi que, vers 14 heures, le char du lieutenant Richard Haywood, de l'escadron C sera mis hors de combat.

Le *5th RHA*, quant à lui, a installé des batteries de *Sexton* au nord de la D 71, dans le petit vallon où coulent les ruisseaux du Coudray et de pont Chouquet. Les officiers qui commandent ces batteries sont le capitaine Tim Lanyon qui terminera sa carrière militaire général de Brigade, et le capitaine Andrew Burn. A son entrée rue Pasteur, le chef de char de tête demande à Marcel Ozenne, « *Où sont les Boches* » ? Ce dernier avoue son ignorance sur la présence des blindés allemands et, voyant le déploiement de forces, rentre chez lui dans sa cave où il passera la journée avec sa famille. D'autres Anglais demanderont quelquefois des renseignements aux habitants qui ne pourront répondre qu'évasivement.

Les Villérois, surpris d'abord, sont sortis sur les trottoirs et, par petits groupes, accueillent les troupes britanniques en libérateurs, « *avec plus d'intérêt que d'enthousiasme* » écrira le capitaine Christopher Milner, de la *Rifle Brigade*. « *Tout était tranquille et serein en cette matinée de juin* » constata le sergent Walter Allen de l'escadron B du *4th CLY* lorsqu'il arriva place Jeanne d'Arc. Il remarqua *une délicieuse odeur* family. Other British soldiers occasionally asked the inhabitants for information, but got only guarded answers.

The people of Villers, although surprised at first, came out into the streets and in small groups welcomed the British troops as liberators, "with interested rather than enthusiasm", Captain Christopher Milner of the Rifle Brigade later wrote. "All was quiet and serene. [...] Sgt. Walter Allen, a Cromwell commander in no. 1 Troop, strolled over to talk to Lockwood. He remarked upon the delicious smell of newly-baked bread coming from some unknown location, and the two of them did a 'Bisto Kids' act as, with noses in the air, they tried to detect the origin of the appetizing aroma. Any thoughts of fresh loaves were speedily dashed as the peace was shattered by the noise of machine-gun fire plus the sharp crack of an 88 as a German Tiger tank lumbered into the town". The smell of fresh bread was soon mixed with the stench of powder and smoke. This was the smell of Villers-Bocage that Walter Allen would never forget.

Soon the local people, who had been waiting to see this for four years, began to clap, throw flowers on the tanks and bring out bottles of wine. Monsieur Doublet cracked open his last bottle of champagne which "was feeling lonely" in his cellar. "Bread and wine, it was all very exciting", recalled Corporal Watt, a veteran of the North African desert and Italy, and a half-track commander with the Rifle Brigade.

Two German soldiers were captured, one in the Rue Montebello, on failing to see the British tanks coming as he was putting some groceries he had taken into his bicycle saddlebags; the second after his sidecar had broken down at the intersection of the Rue Montebello and the Rue Saint Martin. They were lifted up onto the leading tank and acted as guides. The two Germans did not stay long on the British tanks. As the front of the column came up to Mademoiselle Bazin's building in the Rue Georges Clemenceau, just past the Bras d'Or Hotel garages (now number 39), they spotted an open gate, jumped off the tank in among some civilians watching the tanks drive past, and raced off towards the meadows located along the "chemin de la fontaine pourrie" (Rotten Fountain Lane). The British could do nothing to stop them, not even shoot at them, and so they got off scot-free.

In the Place de la Liberté, close to the bandstand in front of the Hospice, the fire brigade were already hard at work, having been called out during the night, and had not yet taken back their "Lafly". Throughout the day, they fought the fires caused by tanks destroyed in the battle. Obviously they had to remove the source of the fires by trying to put out the flames coming from the brewed-up tanks in order to protect the town buildings effectively. But that afternoon most of the burning tanks were German, and the British soldiers got the wrong end of the stick, criticizing the firemen's initiative.

At what is now number 55 Rue Georges Clemenceau, Monsieur Delafontaine, a gendarme, heard the rumble of tracked vehicles approaching from the west as he finished washing and shaving. He looked out of his window to see who they were and, to his great surprise, they were British tanks. He saw almost a

---

(2) Cote 213. Sur les cartes anglaises en 1944, le sommet de la côte des Landes était de 213 m. Sur les cartes I.G.N., il est de 217 m.

(2) Point 213. On the British maps in 1944, the Côte des Landes hilltop was marked at 213 m. On the French ordnance survey map, it rises to 217 m.

W.D. Allen.

Stan Lockwood.

de pain fraîchement cuit et le nez en l'air avec son copain Stan Lockwood, il essaye de trouver l'origine de cet arôme appétissant. Nous verrons plus loin que tout espoir de trouver une miche chaude disparut au moment où le silence fut brisé par des tirs de mitrailleuse et le craquement d'un obus de 88 mm d'un Tiger entrant dans la ville. L'odeur de pain frais se mélangea bientôt à celle de la poudre et de la fumée ». C'est cette odeur de Villers-Bocage qui restera dans la mémoire de W. Allen.

Rapidement les Villérois, qui espéraient voir un tel spectacle depuis quatre ans, applaudissent, fleurissent les chars, sortent des bouteilles de vin. Monsieur Doublet débouche la dernière bouteille de champagne qui « s'ennuyait » dans sa cave, « *bread and wine, it was all very exciting* », (pain et vin, c'était très excitant) se souviendra le caporal Watt, ancien du désert d'Afrique du Nord et d'Italie, chef d'un *half-track* de la *Rifle Brigade*.

Deux soldats allemands sont faits prisonniers, le premier rue Montebello, qui n'avait pas vu les chars anglais arriver tandis qu'il mettait dans les sacoches de sa bicyclette des provisions prélevées dans une épicerie, le second en panne de *side-car* à l'intersection des rues Montebello et Saint-Martin. Ils sont hissés sur le char de tête et serviront de guides. Les deux Allemands ne vont pas rester longtemps sur les chars anglais. Au moment où la tête de la colonne atteint l'immeuble de Mademoiselle Bazin, rue Georges Clemenceau, après les garages de l'hôtel du « Bras d'or » (39 de la rue actuelle), ils voient un portail ouvert, sautent du char au milieu des civils qui regardent passer les tanks et se précipitent vers les herbages situés le long du « chemin de la fontaine pourrie ». Les Anglais qui ne peuvent intervenir et utiliser leurs armes, les laissent partir.

Place de la Liberté, près du kiosque à musique, devant l'Hospice, les pompiers, qui ont été appelés durant la nuit, sont déjà à pied d'œuvre et n'ont pas encore rentré leur « Lafly ». Dans la journée, ils vont combattre les incendies provoqués par les chars détruits au cours de la bataille. Evidemment, pour protéger efficacement les immeubles du bourg, il sera essentiel de supprimer la source des incendies en essayant d'arrêter les flammes qui s'échapperont des tanks détruits. L'après-midi, les chars qui brûleront seront surtout allemands et les soldats britanniques interprétant mal l'action des pompiers, condamneront leur initiative.

Au 55 (actuel) de la rue Georges Clemenceau, Monsieur Delafontaine, gendarme, entend le grondement de véhicules à chenilles venant de l'ouest au moment où il finit sa toilette. Il va alors à sa fenêtre pour les identifier et, à sa grande surprise, reconnaît des blindés anglais. Il voit passer une dizaine de chars et autant de chenillettes et d'autochenilles qui dépassent le carrefour de la route de Bayeux puis, 3 chars légers qui s'arrêtent avant le carrefour, enfin 4 chars moyens qui se placent sur une centaine de mètres à droite de la rue, devant son immeuble et une voiture blindée légère sous sa fenêtre. Il ne sait pas qu'il a devant lui le QG d'un régiment.

Il y avait quelques Allemands dans Villers la veille. Ce devait être une compagnie sanitaire qui a quitté la commune au petit jour et a été remplacée par de petits groupes de combattants qui, dès qu'ils ont été prévenus d'une attaque éventuelle, se sont fait ignorer dans l'agglomération. Le matin, Madame Fromont et les siens, qui étaient allés voir passer les Britanniques place Richard-Lenoir, ont entendu des coups de feu du côté de la gare et des balles siffler non loin d'eux place du marché, alors qu'ils rentraient chez eux rue Emile-Samson pour se mettre à l'abri. En fin de matinée, le Maire, Monsieur Doublet, avait envoyé

dozen tanks pass and as many bren gun carriers and half-tracks going over the Bayeux road junction, then 3 light tanks which stopped before coming to the crossroads, and finally 4 medium tanks, which were strung out along a hundred meters down the right side of the street, in front of his own building, with a light armored car under his window. Little did he know that he was looking at a Regimental HQ.

There had been some Germans in Villers the day before. It was probably a medical company which moved out at first light, to be replaced by small groups of combat troops who immediately crawled into the woodwork on hearing of a possible attack. That morning, Madame Fromont and her family, who had gone to see the British pass by the Place Richard-Lenoir, on their way back home to shelter in the Rue Emile Samson, heard shots in the vicinity of the station and bullets whistling close by in the marketplace. Late that morning, the Mayor, Monsieur Doublet, dispatched a fireman, Monsieur Choin, to the market to gather news. He reported back that he had seen some German light vehicles parked under the trees. After 09.00 hours, once the column had passed, the streets were fairly deserted, with just some firemen, members of the Red Cross emergency and civil defense teams looking around for casualties. The euphoria did not last long. Everyone soon sensed that a battle was about to begin and that they had no time to lose in seeking the safety of shelter somewhere, anywhere, preferably as far as possible out of the town center and off the main streets.

When the British column arrived, Monsieur René Roger was at the Bayeux road crossroads and driving a horse and cart belonging to the farm at Hauts Vents. He instantly dropped everything and ran across the fields to his work place. After the battle, the horse made its own way back unhurt to its stable. But all that was left of the cart were the two shafts dangling from the harnesses, the rest had been blown to smithereens.

The British take up position

The regimental commander, Lieutenant-Colonel Viscount Lord Arthur Cranley, left his tank in the hands of his second-in-command, Major Carr. He got into a scout car to take him to Point 213, at the summit of the Côte des Landes. He noted with concern that his unit had forged on blindly without sending anyone on ahead to reconnoitre. He forked off, leaving the Bayeux road to his left, and on his right he saw a roadsign marked N175, Caen 24 (km). He carried on past the Rifle Brigade company, which had stopped to his right, until he came to a halt at the top of the Côte des Landes, where he set up his forward command post from which to prepare for the arrival of another tank squadron and the leading battalion of the Infantry Brigade, the 7th Queen's.

The vehicle of the Rifle Brigade company following immediately behind stopped on a level with the farm-track to Hauts Vents. Bringing up the rear of the column were the two 6 pounder (57 mm) anti-tank guns and carriers.

The half-track drivers had stopped their vehicles nose to tail, so as to let the relief units through, although such a position goes against standard military procedure. The soldiers took this opportunity to stretch their legs and light up a cigarette while awaiting orders from their troop commanders. "I was in my half-track, wrote Captain Milner, second-in-command, from the rear to the front of the company, picking up the platoon commanders and the mortar sergeant en route, in response to an urgent 'O' group summons from my company commander, James Wright. He in turn

56

aux nouvelles un pompier, Monsieur Choin, vers le marché. Ce dernier lui a rapporté qu'il avait vu des véhicules légers allemands stationnés sous les arbres. Après 9 heures, une fois la colonne passée, il y avait peu de monde dans les rues, seulement les pompiers, des membres des équipes d'urgence de la Croix Rouge et de la défense passive à la recherche de quelque blessé. L'euphorie a duré peu de temps. Chacun a vite compris qu'une bataille était proche et qu'il fallait trouver au plus vite un endroit abrité et sûr, éloigné autant que possible du centre de la ville et des rues principales.

A l'arrivée de la colonne anglaise, Monsieur René Roger est au carrefour de la route de Bayeux et conduit un attelage de la ferme des Hauts Vents. Sans perdre de temps, il abandonne son équipage et gagne le lieu de son travail en courant à travers champs. Après la bataille, le cheval indemne rentrera seul à son écurie. De la charrette, il ne restera que les deux brancards suspendus aux harnais, le reste du véhicule aura été déchiqueté.

Les Britanniques prennent position

Le lieutenant-colonel Viscount Lord Arthur Cranley, qui commande le régiment, quitte son char et le confie au *Major* Carr, commandant en second. Il monte dans un scout-car pour gagner la cote 213, sommet de la côte des Landes. Il constate avec inquiétude que son unité s'est portée en avant sans trop savoir où elle va et que personne n'est allé en reconnaissance. Il laisse sur sa gauche la route de Bayeux et peut voir sur sa droite le panneau indiquant *N 175, Caen 24*. Il continue, passe la compagnie de la *Rifle Brigade* arrêtée sur sa droite et s'arrête au haut de la côte des Landes où va être installé son poste de commandement avancé d'où il pourra préparer l'arrivée d'un autre escadron de chars et du bataillon de tête de la Brigade d'Infanterie, le *7th Queen's*.

Le premier véhicule de la compagnie de la *Rifle Brigade,* qui suivait s'est arrêté au niveau de l'allée qui mène à la ferme des Hauts Vents. La colonne qui suit se termine par les deux canons antichars de 57 mm et leurs carriers.

Les conducteurs des *half-tracks* ont arrêté leurs véhicules serrés les uns derrière les autres, « pare-chocs à pare-chocs », *nose to tail*, pour ne pas gêner le passage de renforts, bien qu'une telle position soit contraire aux instructions militaires habituelles. Les soldats profitent de cet arrêt pour mettre pied à terre, « griller », une cigarette en attendant les ordres de leurs chefs de section. « *J'étais dans mon Half-track*, écrit le capitaine Milner, commandant en second, *allant de la tête à la queue de la compagnie pour répondre à l'ordre urgent de mon commandant de compagnie, le Major James Wright, rassemblant les chefs de section et le sergent mortier. James Wright avait reçu l'ordre de Arthur Vicomte de Cranley, commandant le 4th CLY d'atteindre la cote 213. Nous roulions rapidement le long de la route droite, passant les riflemen qui avaient mis pied à terre et leurs half-tracks espacés à intervalles égaux. Une section manquait, la section éclaireur du lieutenant Alan Matter dont les carriers avaient été embarqués dans un bateau différent du nôtre et qui n'avaient pas encore rejoint. Son rôle aurait été en la circonstance, de se porter en avant pour recueillir des renseignements sur la présence éventuelle d'un ennemi. Il aurait pu repérer et approcher les chars ennemis et les renseignements qu'il aurait pu recueillir ainsi lors de l'avance des blindés et après leur installation à la cote 213 auraient pu être précieux pour les chars de l'escadron et le reste du régiment* ».

has just been briefed by Arthur, Viscount Cranley, commanding 4 CLY, at Point 213 [...] We were motoring rapidly along the straight road, past the dismounted riflemen and their evenly-spaced half-tracks, but not past the scout platoon because Alan Mather's carriers had been shipped in a different vessel and they had not yet caught up with us." In this situation, he says, his job would have been to go on ahead and collect intelligence as to possible enemy presence. He could have located and approached the enemy tanks and any information collected in this way during the tank advance and after they had taken up position on Point 213 might well have made all the difference for the squadron's tanks and the rest of the regiment's."

Just before 09.00 hours, the **positions** of the British units were as follows:
- the tanks of "A" Squadron were spread out around Point 213, two troops in the fields south of highway N175, the others along the road,
- at Point 213, the tactical headquarters with: Lieutenant-Colonel Cranley and his adjutant, Captain W.C. Rose; "A" Squadron commander Major Peter Scott, Captain James Wright, commander of "A" Company, the Rifle Brigade, and his officers, Lieutenants Campbell, Coop and Parker; Sergeant Gale commanding the mortar detachment, and just one rifle troop, under Corporal Nicholson, detached to "A" Squadron under Captain Christopher Milner, and which joined them after the start of the German attack.

Captain Roy Dunlop, artillery observation officer with the 5th RHA, had pulled his Cromwell up on the roadside a little short of the hilltop. The tank was recognizable by the letters RD on a blue and red ground on the front. This officer was deputizing for Captain Paddy Victory, held up in the Rue Clemenceau following engine trouble with his Cromwell.
- at the bottom of the Côte des Landes, at the entrance to the road to Epinay-sur-Odon, the tanks of "A" Squadron HQ Sq.
- then the half-tracks of the Rifle Brigade, followed by 4 Lloydcarriers of the anti-tank platoon under Lieutenant Roger Butler, which after being called to Point 213, had headed off south to reconnoitre,
- at the Bayeux road junction, part of the 4th CLY's scout section, comprising 3 Honey tanks and a scout car,
- from Lemonnier Farm to the James building, the 4 RHQ Cromwells,
- on the other side of the street, Lieutenant Pearce's scout car,
- after the bend in the Rue Georges Clemenceau, two artillery observation tanks belonging to the 5th RHA, Major Dennis Wells Sherman, with a wooden gun, and Captain Paddy Victory's Cromwell,
- in the Rue Clemenceau and the Rue Pasteur, various tanks, Captain Maclean's medic's half-track and the rest of the scout section of 4th CLY,
- in the Place Jeanne d'Arc, "B" Squadron, 4th CLY,
- further on, on the Caumont road, "C" Squadron.

The 7th Battalion, Queen's Royal Regiment, under Lieutenant-Colonel Desmond Gordon, was to occupy the ground taken by the 4th CLY around Point 213, and to use it as a springboard from which to pursue the offensive. "In the meantime the Bn. had bebussed and was preparing to take up a firm base position on the high ground east of the village when orders were issued to move at once into Villers Bocage itself and cover its main exits. The carriers and A tk guns moved off at once, the leading Coy encountered a German staff car and two motor cycle combinations which suddenly appeared from a side track.

Lt.-Col. Lord Cranley.

Peu avant 9 heures, les **positions** des unités anglaises sont les suivantes :

- les chars de l'escadron A sont dispersés autour de la cote 213, deux *troops* dans les champs au sud de la RN 175, les autres le long de la route,

- à la cote 213, le poste de commandement tactique où se trouvent : le lieutenant-colonel Cranley avec son assistant le capitaine W.C. Rose, le *Major* Peter Scott commandant de l'escadron A, le capitaine James Wright, commandant de la compagnie A de la *Rifle Brigade* et ses officiers, les lieutenants Campbell, Coop et Parker, le sergent Gale qui commande le détachement de mortiers, une section seulement de *riflemen*, sous les ordres du caporal Nicholson, détachée auprès de l'escadron A, le capitaine Christopher Milner, qui les rejoindra après le début de l'attaque allemande.

Le capitaine Roy Dunlop, officier d'observation d'artillerie du *5th RHA* a arrêté son *Cromwell* sur le bord de la route un peu au-dessous du sommet de la côte. Le char est reconnaissable aux lettres RD sur fond bleu et rouge qu'il porte sur l'avant. L'officier remplace le capitaine Paddy Victory resté dans la rue Clemenceau, suite à des ennuis de moteur de son *Cromwell*.

- Au bas de la côte des Landes, à l'entrée de la route qui mène à Epinay-sur-Odon, les chars du QG de l'escadron A, le *HQ Sq.*,

- puis les *half-tracks* de la *Rifle Brigade* suivis des 4 *Loydcarriers* de la section antichar du lieutenant Roger Butler, lequel appelé cote 213, était parti en reconnaissance vers le sud,

- au carrefour de la route de Bayeux, une partie de la section de reconnaissance du *4th CLY*, composée de 3 chars *Honey* et d'un scout-car,

- de la ferme Lemonnier à l'immeuble James, les 4 *Cromwell du RHQ*,

- sur le trottoir opposé de la rue, le scout-car du lieutenant Pearce,

- après le virage de la rue Georges Clemenceau, deux chars d'observation d'artillerie, le *5th RHA*, le *Sherman* à canon en bois du *Major* Dennis Wells et le *Cromwell* du capitaine Paddy Victory,

- rue Clemenceau et rue Pasteur des véhicules blindés épars, le *half-track* médical du capitaine Maclean et le reste de la section de reconnaissance du *4th CLY*,

- place Jeanne d'Arc, l'escadron B du *4th CLY*,

- plus loin, route de Caumont, l'escadron C.

Le 7ᵉ bataillon du *Queen's Royal Regiment*, commandé par le lieutenant-colonel Desmond Gordon devait occuper le terrain conquis par le *4th CLY* autour de la cote 213, et partir de la base ainsi créée pour poursuivre l'offensive. « Vers 9 heures, il se préparait à investir la ville et en couvrir les entrées prin-

Les hommes de la compagnie de Wittmann, vont vivre ainsi la nuit du 12 au 13 juin à l'est de Villers-Bocage : violents tirs d'artillerie, ils ne leur permettront pas de se reposer avant l'engagement sur le front. (BA.)

The men of Wittmann's company would have spent the night of June 12-13th east of Villers-Bocage like this: violent artillery fire, allowing them no rest before being committed in the front line. (BA.)

The leading platoon suffered some casualties when the car and combinations had been ditched and the crews taken to the fields, where they commenced sniping. These had to be dealt with before the Coy could move on. Three prisoners were taken and the remainder dispersed." This was doubtless a leadin element of 2. Panzer-Division.

Lieutenant-Colonel Gordon (3) positioned his companies, "A" Company in the station sector, "B" Company in the town center, Major French's "C" Company along the Rue Pasteur and the Rue Clemenceau, and his reserve, "D" Company, near the cemetery. He then proceeded to set up his HQ in the "down the hill" part of town, where he came under fire from some German 88 or 105 mm guns positioned southwest of Villers, maybe the ones at Maisoncelles Pelvey, at La Hogue or on the bank of the Seuline. Later, elements of Major R.C. Freeman's "D" Company skirted Villers from the north by the cemetery, entering the town along the Rue Curie.

General Montgomery, monitoring the division's progress from his HQ at Creully, was able to wire with satisfaction the provisional outcome of the offensive on Caen to 21st Army Group Chief-of-Staff Major General Francis de Guingand, in Portsmouth. Little did he know that just one Tiger of 2nd Company, SS-Panzer-Abteilung 101, under its commander, SS-Obersturmführer (Lieutenant) Michael Wittmann, was about to put a spanner in the entire works.

The "heavy tanks" of SS-Panzer-Abteilung 101

The heavy tank battalion of I. SS-Panzer-Korps was raised in July 1943 from the 3rd Heavy Tank Company of the Leibstandarte SS Adolf Hitler tank regiment. It belonged to and was directly dependent on I. SS-Pz-Korps. During the early months of 1944, it was in Belgium and northern France, where the young crews trained in Germany and Italy completed their military and tactical training. The heavy tank battalion was issued with PzKpfw VI tanks, better known as Tiger tanks. When the Tiger appeared on the battlefields of Russia in September 1942, it was the most powerful tank in the world. It had an 88 mm gun, two machine-guns and a grenade thrower. No enemy shell could penetrate its frontal armor except from very close range. On the down side, its weight made it difficult to operate and it tended to break down as well. Also, it had a limited range on account of its huge fuel consumption.

Its specifications were as follows:

- crew, 5,

- weight, 57 tons, maximum speeds: 38 kph road speed, 19 kph all terrain,

cipales. Immédiatement, les carriers et les canons antichars de la compagnie D s'étaient mis en route pour prendre position. La compagnie de tête avait rencontré alors une voiture d'état-major allemande escortée de deux side-cars qui débouchaient de la route d'Aunay. La voiture et les side-cars étaient renversés et leurs équipages s'étaient enfuis tout en faisant usage de leurs armes. La section de tête avait eu quelques pertes mais les Anglais étaient venus à bout des Allemands avant que la compagnie ne se mette en mouvement. Trois Allemands étaient faits prisonniers et les autres dispersés ». Il s'agissait sans doute d'une avant-garde de la 2. Panzer-Division.

Le lieutenant-colonel Gordon (3) plaça ses compagnies, la compagnie A dans le secteur de la gare, la compagnie B dans le centre ville, la compagnie C du Major French le long des rues Pasteur et Clemenceau, la compagnie D, de support, près du cimetière. Il installe ensuite son QG dans le bas du bourg où il subira les tirs de canons allemands (88 ou 105) en position au sud-ouest de Villers, peut-être de ceux installés à Maisoncelles Pelvey, à la Hogue, sur le bord de la Seuline. Plus tard, des éléments de la compagnie D du Major R.C. Freeman contourneront Villers par le nord en longeant le cimetière et pénétreront dans la ville par la rue Curie.

Le général Montgomery, qui suivait l'avance de la division depuis son QG de Creully, pouvait transmettre avec satisfaction à son chef d'état-major du QG du 21ᵉ Groupe d'Armée à Portsmouth, le *Major General* Francis de Guingand, le résultat provisoire de l'offensive vers Caen. Il ignorait tout simplement qu'un seul *Tiger* de la 2ᵉ compagnie de la *SS-Panzer-Abteilung 101*, sous les ordres de son chef, le *SS-Obersturmführer* (lieutenant) Michael Wittmann, allait contrarier tous ses projets.

Les « chars lourds » de la *SS-Panzer-Abteilung 101*

Le bataillon de chars lourds du *I. SS-Panzer-Korps*, formé en juillet 1943, était issu de la 3ᵉ compagnie de chars lourds du régiment de chars, de la *Leibstandarte SS Adolf Hitler*. Il faisait partie du *I. SS-Pz-Korps* et en dépendait directement. Dans les premiers mois de 1944, il se trouve en Belgique et dans le nord de la France où les jeunes équipages, qui avaient été formés en Allemagne et en Italie, complètent leur instruction militaire et tactique. Le bataillon de chars lourds est équipé de chars *PzKpfw VI*, plus connus sous le nom de Tigre ou *Tiger*. Lorsque le Tigre est apparu sur les champs de bataille de Russie en septembre 1942, il était le plus puissant char de combat du monde. Il était armé d'un canon de 88 mm, de deux mitrailleuses et d'un lance-grenades. Il possédait un blindage frontal qu'aucun obus adverse ne pouvait transpercer sauf à très faible

- consumption, 570 to 850 liters per 100 km, hence a range of 60 to 100 km depending on the terrain,
- armor, front 100 mm thick, sides 80 mm, rear 82 mm,
- armament, 88 mm gun firing shells able to pierce 120 mm of armor at 100 m, 112 at 500 m, 102 at 1,000, two heavy machine-guns, one hull-mounted, one on the turret, and a grenade thrower on the turret roof for short-range defense. The Tiger, designed by Professor Ferdinand Porsche, was especially effective in defensive action and generally fought with no supporting infantry. It was called upon by the troop, sometimes even just a tank at a time, to support infantry actions and to serve in a way as a "firefighter" when the enemy attacked in force.

On June 6th 1944, the 101st Battalion's theoretical structure was as follows:

- a command and supply company under the battalion commander, his deputy and a signalling officer, each of whom had a Tiger with extensive radio equipment and an unfurlable antenna. This company included a supply section, an ammunition and fuel section, a medical section and a heavy vehicle repair section,

- three combat companies of three troops with four Tigers each. The commander and his adjutant each had a Tiger,

- a so-called light company, detached from the command company, with one reconnaissance, scout, engineers and AA section. Only Battalion 101st had this independent company,

- a workshop company.

The commander (Kommandeur) was battalion commander Hein von Westernhagen, 1st Company was under Captain Möbius, 2nd under Lieutenant Michael Wittmann, 3rd under Lieutenant Raasch, 4th under Lieutenant W. Spitz, and the workshop company was commanded by Lieutenant Klein. The 101st Battalion was virtually wiped out in September 1944. The survivors joined the 501st Battalion when it was refitted in October.

The battalion on its way to Normandy

On June 6th, the battalion was north of Beauvais, some 70 km north of Paris. In the time needed to assemble the scattered companies, the battalion was ready to set off for the Normandy front during the night of June 6-7th. It was recommended anyway as far as possible to move only under cover of night, to avoid being attacked by Allied fighter-bombers. The convoy passed through Gournay-en-Bray, Lyons Forest, Morgny where it came under aerial attack on the 7th, and Les Andelys where the bridge across the Seine had been bombed and was no longer capable of taking the weight of a 57 ton Tiger tank. This entai-

distance. Par contre, sa masse le rendait difficile à manœuvrer et sa mécanique n'était pas très sûre. D'autre part, du fait de son énorme consommation, son autonomie était faible.

Ses caractéristiques étaient les suivantes :

- équipage, 5 hommes,
- poids, 57 tonnes, vitesse maximum, 38 km/h sur route, 19 en tout terrain,
- consommation, 570 à 850 litres aux 100 km, ce qui lui donnait une autonomie de 60 à 100 km selon le terrain,
- blindage, 100 mm d'épaisseur à l'avant, 80 sur les flancs, 82 à l'arrière,
- armement, canon de 88 mm dont les obus pouvaient pénétrer 120 mm de blindage à 100 m, 112 à 500 m, 102 à 1 000, deux mitrailleuses lourdes, une dans la caisse, une sur la tourelle, un lance-grenades sur le toit de la tourelle pour la défense rapprochée. Le Tigre, conçu par le professeur Ferdinand Porsche avait surtout une action défensive et combattait le plus souvent sans infanterie d'accompagnement. Il était appelé par section, voire même à l'unité pour soutenir des actions d'infanterie et servir en quelque sorte de « pompier » lorsque l'ennemi attaquait en force.

Le 6 juin 1944, le bataillon 101 est composé théoriquement :

- d'une compagnie de commandement et de ravitaillement ayant à sa tête le commandant du bataillon, son adjoint et un officier de transmissions qui ont à leur disposition chacun un Tigre pourvu d'installations radio importantes et d'une antenne « parapluie ». On trouve dans cette compagnie, une section de ravitaillement, une section de munitions et de carburant, une section sanitaire et une section de réparation de véhicules lourds,
- de trois compagnies de combat composées de trois sections chacune de quatre Tigres. Le commandant et son adjoint disposent chacun d'un Tigre,
- d'une compagnie dite légère, détachée de la compagnie de commandement, composée d'une section de reconnaissance, d'une section d'éclairage, d'une section de génie et d'une section de D.C.A. Cette compagnie existe seulement dans le bataillon 101, elle est indépendante,
- d'une compagnie atelier.

Le commandant *(Kommandeur)* est le chef de bataillon Hein von Westernhagen, la 1ʳᵉ compagnie est sous les ordres du capitaine Möbius, la 2ᵉ du lieutenant Michael Wittmann, la 3ᵉ du lieutenant Raasch, la 4ᵉ du lieutenant W. Spitz, la compagnie atelier est sous la responsabilité du lieutenant Klein. Le bataillon 101 va pratiquement disparaître en septembre 1944. Ses rescapés seront regroupés au sein du bataillon 501 lors de sa remise sur pied en octobre.

Le bataillon en route vers la Normandie

Le 6 juin, le bataillon se trouve au nord de Beauvais, à environ 70 km au nord de Paris. Le temps de rassembler les compagnies dispersées et le bataillon peut prendre la route du front de Normandie dans la nuit du 6 au 7. Il lui est d'ailleurs recommandé de circuler le plus possible la nuit pour éviter les attaques des chasseurs bombardiers alliés. Le convoi va passer par Gournay-en-Bray, la forêt de Nyons, Morgny où, le 7, il subit une attaque aérienne, les Andelys où le pont qui franchit la Seine, bombardé, n'est plus capable de supporter le poids d'un char Tigre de 57 tonnes. Il doit alors faire un large détour vers le sud, vers Paris où il défilera sur les Champs-Elysées dans un but évident de propagande avant de regagner le

led a broad detour southwards to Paris where the battalion paraded down the Champs-Elysées, obviously for propaganda purposes, before moving up to the front. In the woods near Versailles, the battalion was the target of a major night attack from the air, with the 3rd Company and the workshop company sustaining heavy losses.

During the long journey, the companies' remaining tanks encountered mechanical problems, forcing their commanders to leave them where they stood until they could be repaired, with the attendant risk of leaving them at the mercy of a marauding Allied plane. The road followed passed through Dreux, Verneuil, Argentan and Falaise. The companies sustained further casualties, with 9 killed and 21 wounded on the Creil road, at Versailles, at Argentan where Sergeant Kleber was killed, and at Falaise.

The theoretical strength of a company was 14 tanks, but on June 12th, when Michael Wittmann's 2nd Company reached Villers after passing through Epinay-sur-Odon, it was down to 6, and 4 of those required attention the following day. The 1st Company under Captain Möbius was some 10 km further east in the Noyers area, and was down to 8 tanks. Meanwhile, the 3rd Company was in Falaise on June 13th. Its forwardmost elements, 4 Tigers and supply column vehicles, reached Evrecy on June 14th where during the night of the 14-15th they suffered a devastating bombing raid that killed 130 townspeople. The Germans lost 3 Tigers, 18 men killed and 11 wounded. That same night, Aunay-sur-Odon was bombarded for the second time after the raid on the 12th, and the whole town was reduced to rubble, leaving 56 dead. The 101st Battalion had its headquarters at Baron-sur-Odon.

In the afternoon of June 12th, Wittmann's company reached Epinay-sur-Odon and moved to the foot of the Côte des Landes where, under the cover of sunken lanes, it found shelter south of highway N175. This afforded a chance to get some rest while waiting till the supply vehicles following behind caught up.

That evening, Monsieur Henri Robine saw two tanks. One had started to move up his farm track leading to the highway, before turning back and taking shelter in the Hauts Vents farmyard. The commander parked his tank in a barn, then lay down in the grass where, plainly dog-tired, he fell fast asleep. The other tank stayed under the trees of the country lane known as "the old Caen road" to the east of the farm. According to civilians who were sheltering at the ciderhouse on the Epinay road, these two tanks were accompanied by three other tanks, some light vehicles and German soldiers.

During the night, the Tigers moved on at least three times. They must have been located by spotter planes because they were hounded by naval shells, which however caused no damage.

I.SS-Panzer-Korps liaison officer Lieutenant Jürgen Wessel made contact with the Panzer-Lehr-Division which informed him of the skirmishes during the afternoon of the 12th between German units and British Hussars guarding the 22nd Armoured Brigade's left flank. The officer alerted the Panzer-Korps HQ, which in turn alerted 1 and 2 Companies of the 101st Heavy Battalion and two companies of 12. SS-Pz-Div. Hitlerjugend as well.

Thus, on June 12th, the Germans knew that a British attack was on the way but not the exact direction it was to take, and the few tanks on the lookout in various places sidestepped the battle to let the Allied column advance, the better to neutralize it later on. So General Montgomery's surprise attack turned out to be nothing of the sort.

front. Dans les bois, près de Versailles, le bataillon subit, de nuit, une importante attaque aérienne dont sont surtout victimes la 3e compagnie et la compagnie atelier.

Au cours du long trajet, les chars rescapés des compagnies doivent faire face à des incidents techniques qui obligent les commandants à les laisser sur place, en attendant un dépannage, au risque de les mettre à la merci d'un avion allié en maraude. La route suivie est la suivante : Dreux, Verneuil, Argentan, Falaise. Les compagnies subiront des pertes, 9 tués et 21 blessés sur la route de Creil, à Versailles, à Argentan où est tué le sergent Kleber, à Falaise.

La dotation théorique d'une compagnie est de 14 chars mais, le 12 juin, quand la 2e compagnie, celle de M. Wittmann arrive à Villers après être passée par Epinay-sur-Odon, elle n'en compte plus que 6 et même 4 d'entre-eux doivent être mis en révision le lendemain. La 1re compagnie, celle du capitaine Möbius, qui se trouve du côté de Noyers, 10 km environ plus à l'est, est réduite à 8. Quant à la 3e, elle est à Falaise le 13 juin. Les éléments les plus avancés de cette dernière, 4 Tigres et des véhicules du train et de ravitaillement, arriveront le 14 juin à Evrecy où ils subiront dans la nuit du 14 au 15 le terrible bombardement qui fera 130 morts parmi les habitants du bourg. Les Allemands perdront 3 Tigres et auront 18 tués et 11 blessés. Cette même nuit, Aunay-sur-Odon subira son 2e bombardement après celui du 12 et disparaîtra complètement sous les ruines. Il y aura 56 morts. La poste de commandement du bataillon 101 est à Baron-sur-Odon.

1. Les Hauts Vents, l'ancienne route de Caen, section subsistante de ce chemin où la compagnie de Wittmann se trouvait en position à l'aube du 13 juin 1944. (EG/Heimdal.)
2. Photo prise probablement le 14 juin d'un Tiger camouflé dans le même chemin ou à proximité. (BA.)

1. Les Hauts Vents, on the old Caen road, the remaining section of this lane where Wittmann's company was in position at dawn on June 13, 1944. (EG/Heimdal.)
2. Photo probably taken on June 14th of a Tiger camouflaged in the same lane or thereabouts. (BA.)

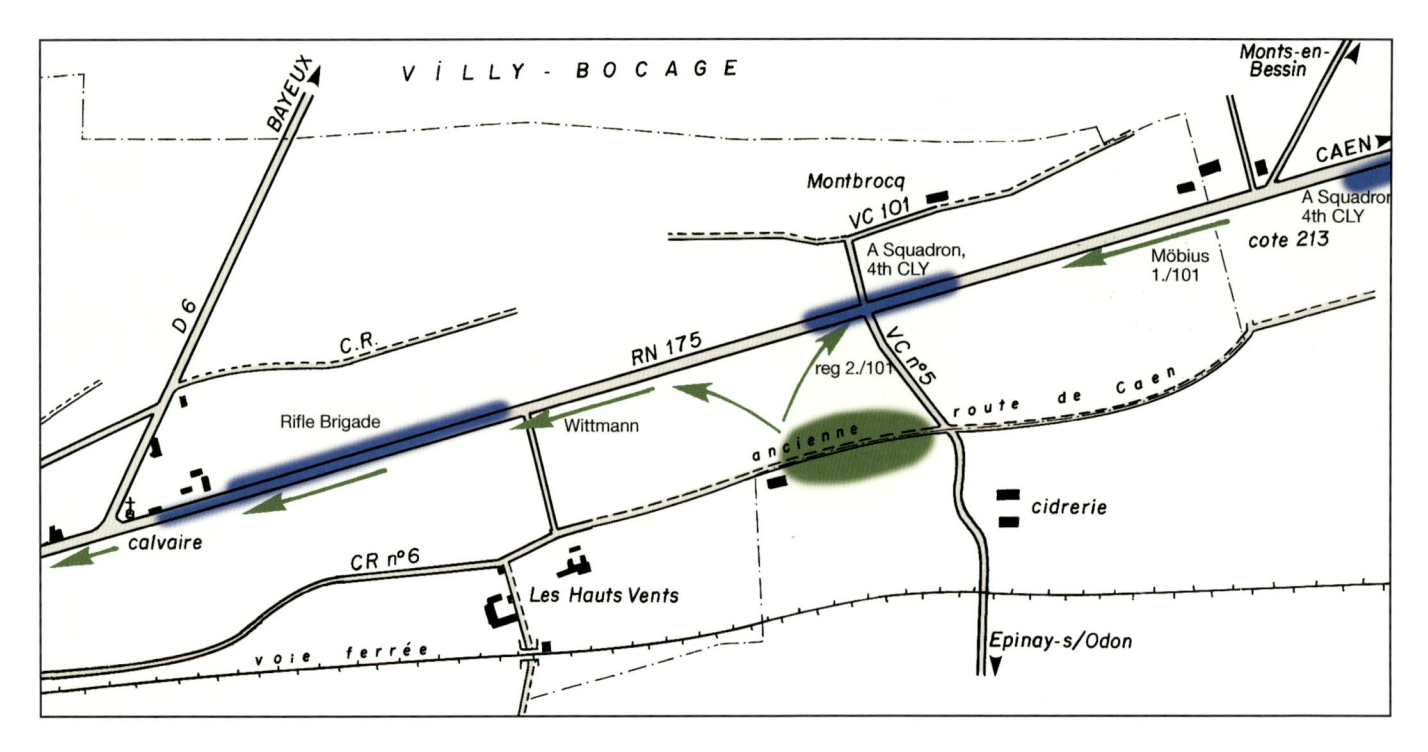

Tandis que Wittmann attaque en dévalant sur Villers-Bocage en ouvrant le feu sur les véhicules de la *Rifle Brigade*, les autres chars de sa compagnie remontent vers la cote 213. (Heimdal.)

While Wittmann attacked by swooping down on Villers-Bocage and opening fire on the Rifle Brigade vehicles, his company's other tanks moved up towards Point 213. (Heimdal.)

Dans l'après-midi du 12 juin, la compagnie de Wittmann arrive à Epinay-sur-Odon et se dirige vers le bas de la côte des Landes où, sous les couverts et dans les chemins creux, elle va trouver un abri au sud de la RN 175. Là, elle va pouvoir se reposer en attendant que ses véhicules de ravitaillement qui la suivent, la rejoignent.

Ce soir-là, Monsieur Henri Robine a vu deux chars. L'un s'est avancé dans l'allée de sa ferme qui mène à la route nationale puis a fait demi-tour et s'est réfugié dans la cour de la ferme des Hauts Vents. Le chef de char a alors rangé son engin sous un hangar puis s'est allongé dans l'herbe où, sûrement fatigué, il s'est endormi rapidement. L'autre est resté sous les arbres du chemin rural dit de « l'ancienne route de Caen » à l'est de la ferme. Selon des civils qui s'étaient réfugiés à la cidrerie, route d'Epinay, ces deux chars étaient accompagnés de trois autres tanks, de véhicules légers et de soldats allemands.

Au cours de la nuit, les Tigres vont changer au moins trois fois de place. Ils ont dû être repérés par des avions d'observation car ils vont être harcelés par des tirs d'obus de marine mais ne subiront aucune perte.

Le lieutenant Jürgen Wessel, officier de liaison auprès du *I.SS-Panzer-Korps*. avait pris contact avec la *Panzer-Lehr-Division* qui lui avait fait part des escarmouches de l'après-midi du 12, opposant unités allemandes et Hussards britanniques qui protégeaient le flanc gauche de la 22ᵉ Brigade Blindée. L'officier avait alerté le QG du *Panzer-Korps*, lequel avait mis en alerte les compagnies 1 et 2 du bataillon lourd 101 et même deux compagnies de la *12. SS-Pz-Div. Hitlerjugend*.

Ainsi, le 12 juin, les Allemands savent qu'une attaque britannique se prépare mais ils en ignorent la direction exacte et les quelques chars en observation ici et là se sont écartés pour éviter le combat et laisser s'avancer la colonne alliée pour mieux la neutraliser par la suite. En fait, il semble que l'attaque surprise du général Montgomery n'en soit plus une.

L'attaque de Michael Wittmann

Les Allemands ont, semble-t-il, évacué Villers, mais ils ont laissé une unité autour de la ville. Le **13 juin,**

Michael Wittmann's attack

The Germans seemed to have pulled out of Villers, but they left a unit around the town. **On June 13th, before 09.00,** *Wittmann at his command post was informed by a warrant officer he had sent to reconnoitre that British tanks were advancing along the highway. Through his binoculars, he could see a column of vehicles passing on highway N175 just 200 yards away from him. First tanks climbing up to the top of the hill and some more stopping at the bottom, then some half-tracks which also stopped, tightly packed together behind each other unguarded along the roadside. Wittmann, who had built up a considerable reputation as a tank commander on the Russian front, had soon sized things up. Before him he had a squadron of Cromwells looking for a favorable position on Point 213, and behind them, a motorized infantry company about to take a break. He thought to himself that these two British units had to be particularly vulnerable in such a formation.*

This was a situation that Montgomery had not catered for. It upset his whole planned outflanking of the Panzer-Lehr-Division from the south, and delayed the fall of Caen and the success of the Allied offensive in Normandy.

According to the 101st Battalion historian, Patrick Agte (in Tiger), on the morning of June 13th, Wittmann was not in his Tiger but at his CP. He had 5 Tigers in addition to his own, those of Second Lieutenant Hantusch (n° 221), Adjudants Brandt (223) and Lötzsch (233), Sergeants Stief (234) and Sowa (222). Having made his decision on the spur of the moment to attack the British, he began heading for his tank a little way off, but then, to save time he jumped into the nearest tank, Stief's n° 234 which he quickly abandoned on hearing the engine making a funny noise, jumping instead into n° 222, Sowa's tank, which was already out of the sunken lane. His crew were Corporal Boldt, the loader, Sergeant W Müller the driver, and Corporal Günther Jonas, the radio operator. Stief, now being surplus to requirements, was ordered to tell the others.

At 09.05, Wittmann set off in his Tiger. With two more tanks behind him, along the roadside, he took the same direction as the Rifle Brigade's half-tracks. A British liaison officer, Sergeant O'Connor, spotted the tanks

avant 9 heures, Wittmann, à son poste de commandement, est informé par un sous-officier qu'il a envoyé se renseigner sur la situation, que des blindés britanniques avancent sur la route nationale. Dans ses jumelles, il voit alors une colonne de véhicules passer sur la RN 175 à 200 mètres de lui. Des chars d'abord qui gagnent le sommet de la côte et quelques autres qui s'arrêtent au bas de la montée, puis des *half-tracks* qui s'arrêtent également, serrés les uns derrière les autres sur le bas-côté de la route et sans protection. Wittmann, qui s'est forgé une forte réputation de chef de char sur le front russe, juge rapidement la situation. Il a devant lui, un escadron de *Cromwell* qui est en train de chercher une position favorable cote 213 et, derrière, une compagnie d'infanterie portée qui se prépare à se mettre au repos. Il pense alors que ces deux unités britanniques ainsi disposées doivent être particulièrement vulnérables.

Cette situation, Montgomery ne l'avait pas prévue. Elle va bouleverser ses plans, faire échouer le projet de contournement de la *Panzer-Lehr-Division* par le sud, retarder la prise de Caen et la réussite de l'offensive alliée en Normandie.

Selon Patrick Agte *(in Tiger)*, historien du Bataillon 101, le matin du 13 juin, Wittmann n'est pas dans son Tigre mais à son PC. Il dispose de 5 Tigres en plus du sien, les chars du sous-lieutenant Hantusch (n° 221), des adjudants Brandt (223) et Lötzsch (233), des sergents Stief (234) et Sowa (222). Sa décision prise rapidement d'attaquer les Anglais, il se dirige vers son char, arrêté à quelque distance mais, désirant perdre le moins de temps possible, il bondit dans le char le plus proche, le 234 de Stief qu'il abandonne rapidement en entendant un bruit suspect dans le moteur et saute dans le 222 qui est déjà sorti du chemin creux, le char de Sowa. Ses équipiers sont le caporal Boldt, pourvoyeur, le sergent W. Müller, conducteur, le caporal Günther Jonas, radio. Stief, « mis à pied », est chargé de prévenir les autres.

A 9 h 05, Wittmann lance son Tigre. Suivi de deux autres chars, longeant le chemin, il prend la même direction que les *half-tracks* de la *Rifle Brigade*. Un agent de liaison anglais, le sergent O'Connor voit les chars sortir de leur abri et vient en informer le lieutenant A.P. de Pass qui, incrédule, répond qu'il s'agit de chars alliés. Le sergent, sûr de lui, arrivé à la cote 213, avertit le *Major* Wright commandant de la compagnie A de la *R.B.* Ce dernier lui répond qu'effectivement des chars allemands sont partout autour, trois à droite et aussi deux à gauche.

Soudain, les Tigres changent de direction et se dirigent vers la RN. Le dernier char s'arrête et commence à tirer tandis que Wittmann fonce vers la colonne anglaise avec l'intention de séparer l'escadron de *Cromwell* des *half-tracks* de la *Rifle Brigade*, et de descendre vers Villers pendant que les autres Tigres attaqueront les chars anglais installés à la cote 213. Son premier tir de 88 touche au bas de la côte au moment précis où, écrit Ch. Milner, « *mon half-track dépasse le dernier char du 4th CLY qui avançait sans rien soupçonner. Le char touché explose en flamme et obstrue la route. A peu près 100 yards plus en avant, nous passons un char qui a une haute tourelle et un canon de 17 livres* » (un Firefly). Il est touché à son tour, fait un tête-à-queue et commence à attaquer vers le sud tout ce qui a pu éliminer le char qui le suivait.

Après avoir détruit 3 chars du *SqHQ*, 2 à droite et 1 à gauche, le Tigre fait un écart à gauche vers Villers et rugit vers le bas de la côte en mitraillant les *half-tracks* de la *Rifle Brigade*, section après section, en tir direct, sans avoir besoin de corriger le tir. Le lieutenant de Pass se précipite vers son *half-track* pour se procurer un *PIAT*, il est tué au moment où il grim-

breaking cover, and came up to inform Lieutenant A.P. de Pass who, in disbelief, answered that they were Allied tanks. Being quite sure he was not mistaken, on reaching Point 213, the sergeant informed the R.B.'s "A" Company commander, Major Wright, who confirmed that there were German tanks all around them, three on the right and another two on the left.

Suddenly, the Tigers changed direction and moved towards the highway. The last tank stopped and started firing as Wittmann accelerated towards the British column, intending to split the Cromwell squadron from the Rifle Brigade's half-tracks and head off down towards Villers, leaving the other Tigers to attack the British tanks at Point 213. His first 88 round hit the bottom of the hill at the very moment when, wrote C. Milner, when my truck had just passed the rearmost CLY tank (which was facing forward, suspecting nothing) [...] the tank was hit and burst into flames. As we passed the next one (it had a high tur[ret] and was a 17 pounder) perhaps a 100 yards further on [a Firefly], it swung round and began to engage whatever had shot up its neighbor from the south."

After destroying 3 SqHQ tanks, 2 on the right and 1 on the left, the Tiger swerved off left towards Villers and roared down the hill machine-gunning the Rifle Brigade's half-tracks, troop after troop, at pointblank range without bothering to adjust his aim. Lieutenant de Pass raced to his half-track to fetch a PIAT, but was killed while mounting his vehicle. Some riflemen tried to engage the Tiger with the company's two 57 mm anti-tank guns but were mown down. Others ran desperately for open country as machine-gun fire raked the ditch into which they had dived for cover.

Corporal Watt recalls, "We were then in an armored column which was nose-to-tail at the side of the road (against all army training) when suddenly a Tiger tank appeared and hit several of our vehicles, which were so close we had no room to maneuver. I ordered my crew out of my half-track A/V to the side of the road, and as I went back for my maps, my own vehicle was hit. As we could see nothing from where we were and could hear much shouting, I went to find a better vantage point, and when I got into the open was hit by machine-gun fire." Losing a great deal of blood, Watt took cover in a farm building where he passed out. He was captured and treated at a field hospital. On recovering, he escaped and was taken in by civilians to whom he expressed his gratitude, praising their courage. "I definitely remember that it was a shocking sight to see", he concludes.

At the top of the Côte des Landes, at the junction with the Bayeux road, it was just one long column of armored vehicles in flames. In this brief battle, the Rifle Brigade's "A" Company lost 9 half-tracks, 4 Loydcarriers, 2 carriers, and two 6-pdr anti-tank guns. The Battle of Villers had begun.

One may wonder whether Wittmann really destroyed the company of half-tracks single-handed or whether maybe he got help from another Tiger lying in ambush in a little apple plantation behind the cross. Two eye-witness accounts lend some credence to this possibility. The day after the battle, at a house in Villy where he had taken shelter after deserting, a young Slovenian soldier who was "volunteered" for the 130th Artillery Regiment of the Panzer-Lehr-Division overheard a version of this episode in the conversation of two German officers. In addition, according to local inhabitants, there was a Tiger on the spot which may even have run out of fuel, as rumor later had it. Without moving, this tank was indeed capable of intervening and taking part in the destruction of the British vehicles around the Calvary, from the moment Wittmann launched his attack and after his passage, while he carried his action on into the town.

Georg Hantusch Tiger « 221 ».

Jürgen Brandt, Tiger « 223 ».

Georg Lötzsch, Tiger « 233 ». (BA.)

pe dans son véhicule. Quelques *riflemen* essaient d'engager le Tigre avec les 2 canons antichars de 57 mm (6 pounds) de la compagnie. Ils sont abattus. D'autres courent désespérément en rase campagne pendant que la mitrailleuse ratisse le fossé dans lequel ils avaient pensé trouver refuge.

Le caporal Watt raconte « nous étions une colonne blindée qui était pare-chocs à pare-chocs sur le côté de la route quand soudain un Tigre apparut et atteignit plusieurs de nos véhicules arrêtés sans espace pour manœuvrer. J'ordonnai à mes hommes de sortir du half-track et, comme je retournai chercher mes cartes, mon propre véhicule était touché. Comme nous ne pouvions voir d'où nous étions et comme j'entendais de nombreux tirs, je cherchai une meilleure position. C'est à ce moment que j'étais touché par le feu d'une mitrailleuse ». Perdant son sang en abondance, Watt se réfugiait dans un bâtiment de ferme où il s'évanouissait. Fait prisonnier, il a été soigné dans un hôpital de campagne. Rétabli, il s'est évadé puis a été recueilli par des civils à qui il exprime sa reconnaissance et dont il vante le courage. « I definitely remember that it was a shocking sight to see » conclut-il (Je me souviendrai toute ma vie que c'était un spectacle terrifiant à voir).

Du haut de la côte des Landes jusqu'au carrefour de la route de Bayeux, ce n'est qu'une longue colonne de véhicules blindés qui flambent. Au cours de cette courte passe d'armes, la compagnie A de la *Rifle Brigade* a perdu 9 *half-tracks*, 4 *Loydcarriers*, 2 carriers, 2 canons antichars de 6 pounds. La bataille de Villers commence.

On peut se demander si Wittmann était vraiment seul pour anéantir la compagnie de *half-tracks* et s'il n'a pas été aidé par un autre Tigre en embuscade dans un petit plant de pommiers derrière le calvaire. Deux témoignages semblent confirmer cette hypothèse. Un jeune soldat d'origine slovène incorporé « malgré lui » dans le 130ᵉ régiment d'artillerie de la *Panzer-Lehr-Division*, a surpris le lendemain de la bataille dans une maison de Villy où il s'était réfugié, après sa désertion, le récit de cet épisode dans la conversation de deux officiers allemands. D'autre part, selon un Villérois, un Tigre se trouvait à cet emplacement, peut-être même en panne d'essence, fut-il colporté par la suite. Sans bouger, ce char en effet, pouvait intervenir et participer à la destruction des véhicules anglais autour du calvaire, depuis le moment où Wittmann a lancé son attaque et après son passage, pendant qu'il poursuivait son action dans le bourg. Ce peut être aussi, vers ce deuxième char qu'a été dirigé le canon antichar mis en place à la queue de la colonne par des *riflemen*.

Le *Brigadier* Hinde qui avait décidé de quitter son QG de la 22ᵉ Brigade Blindée, pour se rendre à la cote 213 examiner la situation avec Lord Cranley,

It may also be at this second tank that riflemen aimed the anti-tank gun at the rear of the column.

Brigadier Hinde, who had decided to leave his 22nd Armoured Brigade's HQ and take his radio operator, Sergeant George Kay, along in his scout car to Point 213 to take stock of the situation with Lord Cranley, doubled back on hearing approaching gun and machine-gunfire. Following behind came another scout car, perhaps that of Lieutenant Charles Pearce, a reconnaissance troop liaison officer, and which had stopped on the left-hand side of the Rue Georges Clemenceau on the sidewalk in front of n° 55, the home of Mr Delafontaine, who was a gendarme.

Lieutenant Pearce reported that when he arrived, he saw a German tank some 600 yards ahead of the RHQ tanks. Being powerless to do anything in his scout car, he raced up to the leading Cromwell to warn the tank commander and advise him to find someplace where he would not be a sitting target. He saw the Panzer break cover, spring towards the stationary column and later expressed his astonishment that no British tank opened fire. He felt afterwards that had the German tank been taken on rightaway by the 4th CLY's Cromwells and Fireflies, it could have been damaged and the outcome of the encounter would have been very different.

At the locality called **le Calvaire**, at the intersection of the highway and the Bayeux road, half of the regiment's reconnaissance troop was ahead of the HQ tanks, with the other half to the rear, actually in Villers. With 3 Honey light tanks armed with 37 mm guns, this unit followed behind Lieutenant Ingram's tank. Wittmann knocked out the 3 tanks and also a half-track following behind.

The driver of Lieutenant Simmonds' scout car, Lance-Corporal Donald Hammacort, recalls the episode as follows. "On 13th June Lt. Simmons used his recce tank and I had Corporal Cooke as my commander. [...] Just as I got to the corner in the main street, all hell broke loose ahead of me when German tanks ambushed our column at very close range. In the confusion which followed, Lt. Simmonds moved for the troop to follow him back the way we had come. I was the last of four vehicles. Unfortunately his tank was hit and went up in flames, and a half-track following him was also hit. Luckily I had stopped opposite an entrance to a field so I was able to reverse and head back the way we had come. The Germans mortared our positions and there was intensive sniper fire. Unfortunately I was caught outside and away from my scout car and I took refuge under a tank. After what seemed an eternity there was a lull in the battle and the remnants of the troop managed to break out, and we eventually met up with other units of the regiment."

Lieutenant Simmonds was taken in by civilians who tended his wounds, gave him shelter and saw him off him once they felt the danger had passed. Meanwhile, the commander of the leading Honey "Calamity", Lieutenant Ingram, had been killed in the action.

Wittmann swept on down the **Rue Georges Clemenceau**. There, parked in line over a distance of some 100 yards, on the right side of the road in front of the Lemonnier farm and the Laurent meadow, were the 4th CLY's 4 HQ Cromwells.

The first tank was that of the regiment's second-in-command, Major Carr, Colonel Cranley having left for Point 213 in a scout car. The commander did not know what had caused the explosions he heard on Highway N175. He thought maybe it was German battery fire until he suddenly saw the Tiger emerging from the smoke produced by the burning vehicles and he fired two 75 mm shells which just harmlessly bounced off the German tank. Wittmann stopped

dans son scout-car en compagnie de son radio, le sergent George Kay, fait demi-tour lorsqu'il entend les tirs de canons et de mitrailleuses se rapprocher. Un autre scout-car le suit, celui peut-être, du lieutenant Charles Pearce de la section de reconnaissance, officier de liaison, qui était arrêté sur le côté gauche de la rue Georges Clemenceau sur le trottoir de l'immeuble n° 55, là où habitait Monsieur Delafontaine, gendarme.

Le lieutenant Pearce a rapporté que lorsqu'il est arrivé, il a vu un véhicule blindé allemand à 600 yards environ devant les chars du *RHQ*. Ne pouvant rien faire dans son scout-car, il s'est précipité vers le *Cromwell* de tête pour prévenir le chef de char et lui conseiller de se déplacer vers un endroit où il ne serait pas une cible immobile. Il a vu le *Panzer* sortir d'un couvert, s'élancer vers la colonne arrêtée et s'est étonné plus tard qu'aucun char anglais n'ait ouvert le feu. Il pensa par la suite que si le char allemand avait été attaqué dès le départ par les *Cromwell* et les *Fireflies* du *4th CLY*, il aurait pu être endommagé et le résultat de l'accrochage aurait été bien différent.

Au lieu-dit **le calvaire**, à l'intersection de la route nationale et de la route de Bayeux, se trouve la moitié de la section de reconnaissance du régiment devant les chars de l'Etat-Major, l'autre moitié est derrière, dans Villers même. Cette unité forte de 3 chars légers *Honey* armés de canons de 37 mm est précédée du char du lieutenant Ingram. Wittmann détruit les 3 chars et un *half-track* qui les suivait.

Le *Lance corporal* (soldat de 1re classe) Donald Hammacort, chauffeur du scout-car du lieutenant Simmonds raconte ainsi cet épisode. *« Le 13 juin, le lieutenant Simmonds utilisait son char de reconnaissance et le scout-car que je conduisais était commandé par le caporal Cooke. Au moment où nous arrivons au carrefour de la rue principale, un feu d'enfer éclatait devant nous, à l'instant où un char allemand engageait notre colonne à très courte portée. Dans la confusion qui suivait le lieutenant Simmonds manœuvrait pour que la section recule vers la route que nous avions prise pour venir. Mon scout-car était le dernier des quatre véhicules. Malheureusement son char était touché, prenait feu et un* half-track *qui le suivait était atteint aussi. Heureusement, en reculant, j'ai pu me réfugier dans un verger à la lisière de la ville. Nous étions sous le feu d'obus de mortier et de tirs de snipers très actifs. J'étais alors projeté hors de mon scout-car et je me réfugiai sous un char. Après un temps qui me parut une éternité, il se produisit une accalmie et ce qui restait de la section se mit en demeure de retrouver les autres unités ».*

Le lieutenant Simmonds, blessé, a été recueilli par des civils qui l'ont soigné, mis à l'abri, et libéré lorsqu'ils ont estimé qu'il n'y avait plus de danger. Par contre, le lieutenant Ingram qui commandait le *Honey* de tête « Calamity », avait été tué lors de l'action.

Wittmann continue sa route et s'engage dans la **rue Georges Clemenceau**. Là, stationnent, l'un derrière l'autre, les 4 *Cromwell* du QG du *4e CLY* sur une distance de 100 mètres environ, du côté droit de la route devant la ferme Lemonnier et l'herbage Laurent.

Le premier char est celui du *Major* Carr, commandant en second du régiment que le colonel Cranley a quitté pour se rendre en scout-car à la cote 213. Le commandant ignore l'origine des explosions qu'il entend sur la RN 175. Il pense à des tirs de batterie allemande, quand, soudain, il voit le Tigre sortir de la fumée produite par les véhicules incendiés et lui décoche 2 obus de 75 qui ricochent sur le char allemand sans lui provoquer de dégâts. Wittmann s'arrête et d'un seul obus de 88 mm bien ajusté, met le *Cromwell* en flammes. Bien que blessé, le *Major* Carr peut se sauver. Il sera fait prisonnier. Son pointeur qui était descendu du char est indemne. les autres

Le calvaire : cette vue **(1)** prise après les combats montre la route de Caen avec le panneau de la N175 et, en regardant vers l'est, les véhicules détruits de la *Rifle Brigade*. On voit le même endroit **(2)** actuellement très modernisé avec le château d'eau sur la gauche. On aperçoit à gauche le carrefour du calvaire qu'on voit ici **(3)** au début du xxe siècle avec la route de Bayeux. (Heimdal - H. Marie : 3.)

*The roadside cross: this picture **(1)** taken after the battle shows the Caen road with the N175 signpost and, looking eastwards, the destroyed Rifle Brigade vehicles. We see the same place **(2)** as it is today, considerably modernized with the water tower on the left. The crossroads with the cross can be seen on the left and again here **(3)** as it was in the early 20th century, with the Bayeux road. (Heimdal - H. Marie: 3.)*

1. Le deuxième char Cromwell est celui du lieutenant John L.C. Cloudsley Thompson qu'on voit ici en janvier 1944 en tant que *Gunnery-Instructor* à Sandhurst après avoir été grièvement blessé dans le désert en mai 1942. Il sera volontaire pour le Jour J et rejoindra ici le *4th CLY*. (Coll. H. Marie.)

2. Le Cromwell (T 189592) du lieutenant Cloudsely Thompson a été détruit après avoir reculé contre ce mur à l'entrée de Villers-Bocage. (BA.)

Les lieux ont maintenant bien changé et, le mur de la grange contre lequel s'appuyait ce char membres d'équipage, J. Pumphrey, pourvoyeur et H. Ramsbottom, radio, assez brûlés, sont faits prisonniers.

Mis en alerte, les autres chars du *RHQ* commencent à faire marche arrière, mais leur faible vitesse dans cette position, 2 miles à l'heure, ne leur permet pas de s'opposer par la fuite à l'attaque de l'Allemand.

Le deuxième char est celui du lieutenant John L.C. Cloudsley-Thompson qui commande les chars du *RHQ*. Il a déjà reçu un obus antichar mais sans trop de dommage. Le lieutenant commande à son conducteur de reculer et de passer à travers la haie qui borde l'herbage Laurent, entre les bâtiments de la ferme Lemonnier et le mur de la propriété James. Lorsque le Tigre est à 30 mètres environ, il lui décoche des obus de 75 qui ricochent sur le blindage. Un obus de mortier de 38 mm ne fait pas plus de dégâts. L'Allemand réplique aussitôt, « *l'obus de 88 pénètre dans la coque du char, passe près de l'épaule du sergent mitrailleur, puis entre les jambes du lieutenant avant d'atteindre le moteur où il explose, projetant un jet de flammes, de la peinture brûlée et quelques éclats* ». Indemnes, les 5 occupants, sautent du char et, sous les tirs de mitrailleuses, se cachent dans les buissons. Après avoir repris leurs esprits, les Anglais s'éloignent du lieu des combats. Ils longent des murs, traversent l'herbage de Monsieur Bernouis puis la route d'Epinay, longent un talus, franchissent une clôture de barbelés, plongent dans une tranchée et rampent dans l'herbe jusqu'à la porte d'un bâtiment dans lequel ils pénètrent. Cet immeuble est une cave à demi enterrée où ils trouvent des rondins de bois, deux chaises cassées, une caisse de bouteilles de vin et ... un petit fût de calvados. Cette cave devait être située entre la rue Foch et la rue Auguste Briard actuelle.

and with a single well-aimed 88 mm shell, brewed up the Cromwell. Although wounded, Major Carr managed to bale out but was captured. His gun-layer, who was not in the tank, was unhurt. The other crew members, ammunition server J. Pumphrey and radio operator H. Ramsbottom were badly burned and were taken prisoner.

Thus alerted, the other leading RHQ tanks started to back off, but their low speed in reverse gear, 2 mph, prevented them from fending off the German attack by retreating.

The second tank belonged to Lieutenant John L.C. Cloudsley-Thompson, commanding the RHQ tanks. It had already received an anti-tank shell, sustaining little damage however. The lieutenant ordered his driver to reverse and pass through the hedge bordering the Laurents' meadow, between the Lemonnier farm buildings and the wall of the James' property. When the Tiger was about 30 yards away, it fired one of its 75 mm shells at him, which just bounced off the armor plating. A 38 mm mortar shell proved equally harmless. The Germans replied immediately, then an 88mm shell whizzed between the wireless-operator's head and Cloudsley-Thomson's, near the gunner sergeant's shoulder, then between the lieutenant's legs before hitting the engine, where it exploded, spraying out flames, burnt paint and shrapnel. The 5 occupants baled out unhurt and hid in the bushes under machine-gun fire. After regaining consciousness, the British moved away from the scene of the battle. They stayed close to the walls, crossed Monsieur Bernouis' meadow then the Epinay road, down an embankment, across a barbed wire fence, dived into a trench and crawled through the grass to the door of a building which they entered. This building was a semi-underground cellar where they found some logs, two broken chairs, a case of wine and… a small cask of calvados apple-brandy. This cellar must have been somewhere between the Rue Foch and what is now the Rue Auguste Briard.

The 3rd tank belonged to Colonel Cranley's adjutant, intelligence officer Captain Bernard W.C. Rose, who

a été abaissé **(photo 3)**, on voit ces lieux **(photo 4)** sous l'angle de la photo d'époque. (EG/Heimdal.)

1. The second Cromwell tank was that of Lieutenant John L.C. Cloudsley-Thompson, seen here in January 1944 as a gunnery-instructor at Sandhurst after being seriously wounded in the desert in May 1942. He volunteered for D-Day and joined the 4th CLY. (Coll. H. H. Marie)

2. Lieutenant Cloudsley-Thompson's Cromwell (T 189592). It was destroyed after reversing up against this wall on the road into Villers-Bocage. (BA.)

The place has now changed a lot, and the barn wall against which the tank came to rest has been lowered **(photo 3)**; here we see the spot **(photo 4)** as viewed in a contemporary photograph. (EG/Heimdal.)

Le 3ᵉ char est celui du capitaine Bernard W.C. Rose, officier de renseignement, adjoint du colonel Cranley parti lui aussi à la cote 213. Le chef de char est le capitaine Patrick Dyas, assistant du capitaine Rose. Avant d'amorcer sa manœuvre de recul, il était devant le char de Cloudsley-Thomson, et il avait été légèrement blessé à la tête par un éclat de métal provenant de projectiles de mitrailleuses, expédiés depuis les fenêtres d'un immeuble proche et qui avait ricoché sur le bord de la tourelle.

En septembre 1993, Pat Dyas raconte son histoire en français. « *L'équipage de mon char était composé de mon conducteur, d'un radio-opérateur en avant et dans la tourelle, moi, un canonnier et d'un autre radio, aussi canonnier. Le canonnier m'a dit : "il me faut faire pipi". Tout était calme, et j'ai dit O.K. mais vite et de demander des nouvelles au char du colonel parce qu'en approchant de la ville il y avait eu ordre de silence radio. Lorsque le canonnier était en dehors du char, la bataille a commencé. Le char du commandant était en flammes. J'ai donné l'ordre à mon conducteur de renverser vivement dans un virage à droite. Nous étions dans un jardin. Tout d'un coup, un char allemand Tigre, était devant moi, passant à l'ouest, seulement à 10 mètres de moi. Mais c'était impossible de tirer sans un canonnier. J'avais mis le radio-opérateur à la place du canonnier et demandé qu'il vérifie les munitions* ». Dyas se trouvait alors dans le jardinet de la maison James, au n° 70 actuel de la rue Clemenceau.

Le 4ᵉ char est le char du sergent-major régimentaire Gerald Holloway. Le conducteur J.C. Trevor-Roper fait reculer le tank mais, au moment où il arrive devant le n° 64 actuel, un obus de 88 mm envoyé par Wittmann immobilise le *Cromwell*. Le conducteur est tué, Holloway sera fait prisonnier un peu plus tard.

Pat Dyas sort alors de son refuge et se lance à la poursuite du Tigre pour le frapper à l'arrière, là où le blindage est moins épais donc plus vulnérable. Quatre-vingts mètres plus loin, devant l'entrée de la ferme Guéroult, il voit le char allemand arrêté au milieu de la fumée à environ cinquante mètres devant lui. Il stoppe et a le temps d'envoyer deux obus de 75 mm sans causer de dommage au *Panzer*. Wittmann manœuvre alors pour faire pivoter son canon de 88 de 180° et envoie vers le *Cromwell* un obus qui pénètre dans le char par le côté gauche de la tourelle. Dyas et son conducteur Mike Lindrea peuvent s'extraire de leur engin en feu mais les deux radio-opérateurs, Buck Taylor et Steve Stevenson sont tués. Légèrement brûlé, le capitaine traverse la rue et se précipite vers le char proche du sergent-major détruit précédemment où il constate que la radio est encore en état de marche. Il s'empare de l'appareil et essaie d'entrer en relation avec le poste de commandement à la cote 213 pour tenir Cranley au courant de la situation. Ce dernier lui répond que les chars de l'escadron A et son état-major sont encerclés par des Allemands soutenus par des Tigres et que sa position inconfortable devient désespérée.

« *Après cela* continue Dyas, *quelques fantassins allemands ont commencé à tirer contre moi d'une fenêtre. J'ai sauté une haie, je suis entré dans un jardin. Une jeune femme était là, elle m'a dit "vous êtes blessé, venez avec moi" et nous nous sommes cachés dans une porcherie. Elle m'a expliqué qu'elle connaissait toutes les petites rues d'à côté et que c'était possible de trouver mes amis. Elle a pris ma main, m'a dit "vite" et en courant nous avons réussi* ».

Après avoir éliminé les quatre *Cromwell* du *RHQ*, Wittmann continue son passage meurtrier et détruit, rue Clemenceau, le *Sherman* d'observation de la batterie K du *5th Royal Horse Artillery* du *Major* Dennis

Captain Patrick Dyas.

Le char Cromwell du Captain Pat Dyas a été détruit côté nord de la route (voir p. 99, see p. 99).

5. Le char Cromwell (T187608) du RSM, Gerald Holloway, a été détruit derrière celui du lieutenant Cloudsley-Thompson. Ce char a dû être atteint au niveau de sa suspension. (BA.)

6. Actuellement, les lieux ont été modifiés sauf la petite maison. (EG/Heimdal.)

5. RSM Gerald Holloway's Cromwell tank (T187608) was destroyed behind Lieutenant Cloudsley-Thompson's. The tank must have been hit on the level of the suspension. (BA.)

6. Today, the spot has changed except for the small house. (EG/Heimdal.)

(4) Les Villérois qui ont vu le canon en bois gisant devant le Sherman se sont interrogés sur l'utilité d'un tel subterfuge. Ce char utilisé par l'artillerie qui était un véritable P.C. blindé mobile destiné à abriter des observateurs et du matériel radio ne devait pas être encombré par des munitions et un affût de canon. Son rôle était de transmettre des renseignements aux batteries sur les positions de l'ennemi et devait approcher le plus près possible du front. Un canon factice le faisant ressembler aux autres chars lui évitait de se faire repérer et de devenir une cible privilégiée pour l'ennemi qui connaissait l'importance d'un tel poste. Le Sherman portait sur l'avant la lettre X qui signifiait qu'il s'agissait d'un char de commandement. Il faisait partie de la batterie K du *5th RHA*.

Wells muni d'un canon factice en bois (4), devant l'Hôtel du Bras d'Or, le scout-car de l'officier de renseignement et rue Curie, le *Cromwell OP (Operation Post)* du capitaine Paddy Victory du *5th RHA* qui avait cherché un endroit de repli, mais n'avait pu faire demi-tour dans cette rue étroite.

Le Tigre s'élance alors dans la **rue Pasteur**. Il se dirige sans se douter que, **place Jeanne d'Arc**, se trouvent l'escadron B du *4th CLY* et les premiers éléments de la compagnie D du 7ᵉ Bataillon d'Infanterie du *Queen's Royal Regiment*, la compagnie de renfort qui compte dans ses rangs une section de canons antichars et une section de mortiers de 60 mm.

Place Jeanne d'Arc, le sergent Stan Lockwood qui commande la *troop* 2 en tête de l'escadron B, arrête son *Firefly*, écrit-il, contre la maison située « à l'angle de la vieille place, là où la route d'Amayé-sur-Seulles rejoint la rue principale », la rue Pasteur. Les *Cromwell* et les 3 autres *Fireflies* sont derrière lui, le long de la route de Caumont. Il y a peu de monde dans la rue. Les Villérois doivent être retournés chez

had also set off for Point 213. The tank commander was Captain Rose's adjutant, Captain Patrick Dyas. Before beginning his withdrawal maneuver, he was in front of Cloudsley-Thomson's tank and had received a slight shrapnel wound to the head after machine-gunfire from the windows of a nearby building had ricocheted off the hatch rim.

In September 1993, Pat Dyas told his story. "My own tank had the driver, and a specialist radio-operator in front. In the turret, apart from myself, was the gunner and another radio-operator, who was also trained as a gunner. My gunner suddenly said he needed to use a toilet urgently. Seeing no sign of trouble I agreed, but said "be quick – and while you are out of the tank, ask for any news from the Colonel's tank" because we had been ordered to keep radio silence. .... therefore no information. While the gunner was out everything happened at once! Several shots of 88mm came down the main street; I saw the 2nd Command's tank go up in flames [...] and I told our driver to reverse, as quickly as possible, and then right hand down to swing off the road, and into somebody's front garden, where we stopped. The next

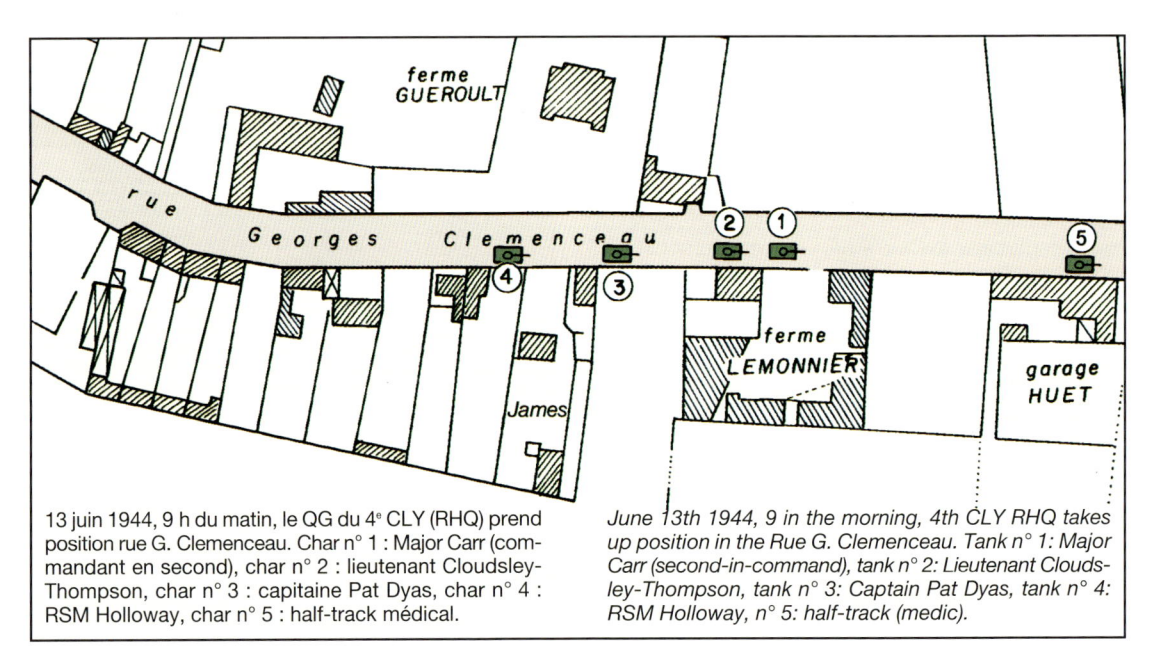

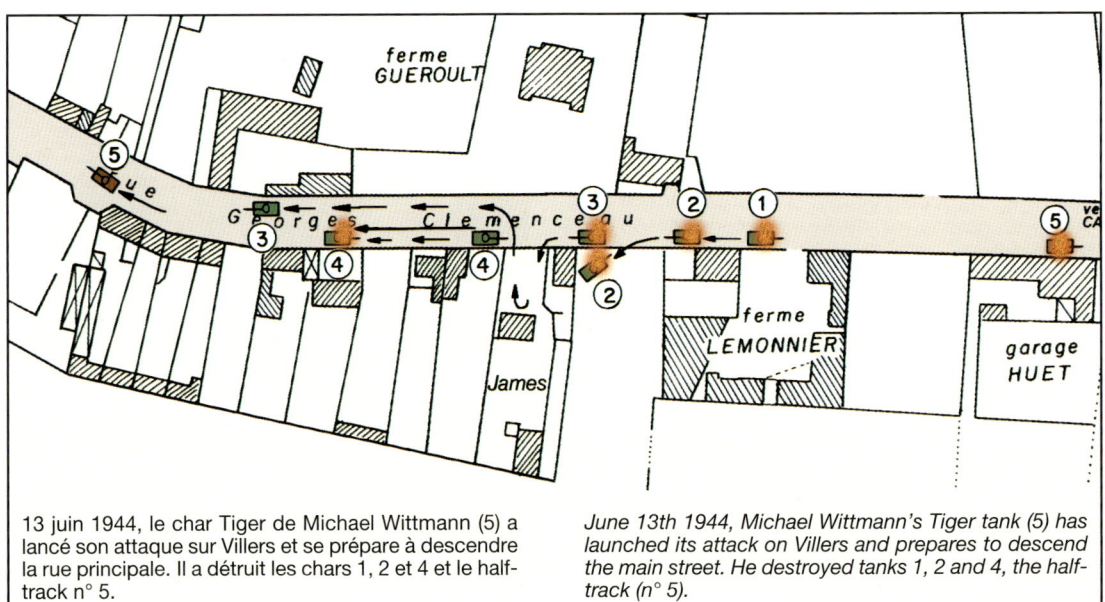

13 juin 1944, 9 h du matin, le QG du 4ᵉ CLY (RHQ) prend position rue G. Clemenceau. Char n° 1 : Major Carr (commandant en second), char n° 2 : lieutenant Cloudsley-Thompson, char n° 3 : capitaine Pat Dyas, char n° 4 : RSM Holloway, char n° 5 : half-track médical.

June 13th 1944, 9 in the morning, 4th CLY RHQ takes up position in the Rue G. Clemenceau. Tank n° 1: Major Carr (second-in-command), tank n° 2: Lieutenant Cloudsley-Thompson, tank n° 3: Captain Pat Dyas, tank n° 4: RSM Holloway, n° 5: half-track (medic).

13 juin 1944, le char Tiger de Michael Wittmann (5) a lancé son attaque sur Villers et se prépare à descendre la rue principale. Il a détruit les chars 1, 2 et 4 et le half-track n° 5.

June 13th 1944, Michael Wittmann's Tiger tank (5) has launched its attack on Villers and prepares to descend the main street. He destroyed tanks 1, 2 and 4, the half-track (n° 5).

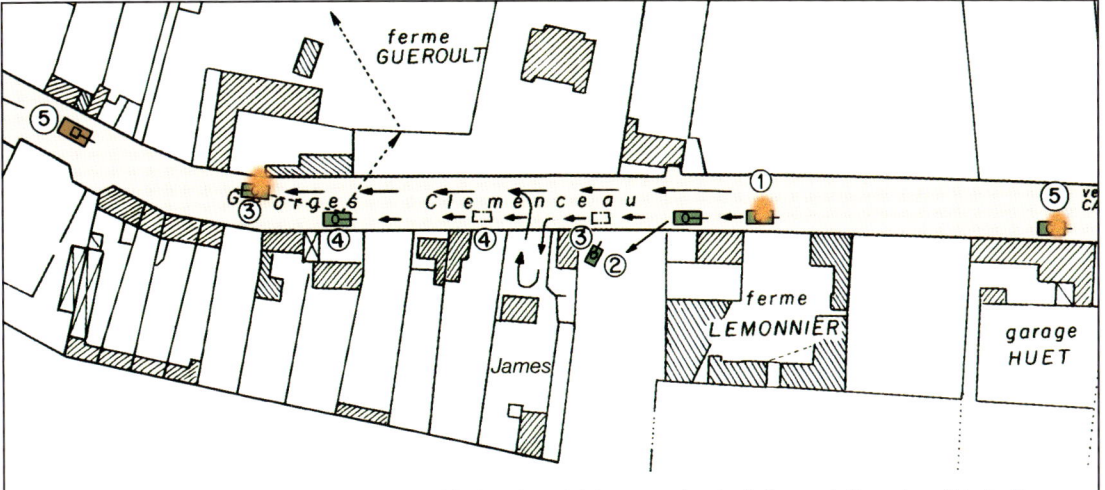

June 13th 1944. Dotted lines, Captain Dyas (3) set off in pursuit of the German. His tank destroyed, Dyas hid on the Guéroult farm and joined up with the British in the Place Jeanne d'Arc, on foot, with the help of a young Frenchwoman. Wittmann stopped his tank, traversed its gun 180° to the rear ("to six o'clock") and destroyed Dyas's n° 3 Cromwell.

13 juin 1944. En pointillés, le capitaine Dyas (3) s'est lancé à la poursuite de l'allemand. Son char détruit, Dyas se cache dans la ferme Guéroult et rejoint les Anglais place Jeanne d'Arc, à pied, grâce à l'aide d'une jeune femme française. Wittmann arrête son char, dirige son canon de 180° vers l'arrière (« à 6 heures ») et détruit le Cromwell de Dyas n° 3. Le n°4 est celui de d'Holloway.

eux après le passage du *RHQ*. Tout est tranquille. Les équipages descendent de leur char pour se dégourdir les jambes jusqu'au moment où ils entendent des crépitements de mitrailleuses et le craquement d'un obus de 88 mm. Tous regagnent leur char précipitamment au moment où, ils voient un scout-car dont un des occupants agite les bras pour les prévenir d'un danger, descendre la rue Pasteur à vive allure. Il s'agit du lieutenant Charles Pearce, officier de liaison qui a quité le RHQ dès qu'il a vu le Tigre commencer à engager les Cromwell pour aller prévenir l'escadron B du danger qui se présentait. Aussitôt qu'il a eu franchi le virage du haut de la rue Clemenceau, il a aperçu le capitaine John Philip-Smith à la tête de la demi-section de reconnaissance et lui a conseillé de replier ses véhicules. Il est surpris de ne pas voir aussitôt derrière les chars des éclaireurs, les Cromwell de l'escadron B. Il descend la rue Pasteur complètement déserte et le premier char qu'il rencontre est le Firefly de Stan Lockwood qu'il fait se déplacer de quelques dizaines de mètres vers l'avant pour affronter le Tigre.

Le conducteur fait avancer le *Firefly*, Lockwood peut alors voir à environ 200 mètres, le Tigre, tourelle de travers, en train de tirer dans la rue latérale. Il a le temps d'expédier deux obus de 17 livres avant que le Tigre ne dirige son canon dans sa direction. La déflagration du premier tir provoque l'effondrement de la façade de l'immeuble près duquel il se trouve. Ardoises, morceaux de bois, pierres s'écroulent provoquant la chute d'un tireur allemand caché dans la maison et entraîné sous les débris. Un des deux tirs touche le Tigre, provoquant flammes et fumée sans pénétrer le blindage. Une fois son canon placé dans la bonne direction, Wittmann réplique deux fois sans toucher le *Firefly*, ses obus se perdant dans l'immeuble. L'Anglais tire encore deux fois avant que, l'Allemand, soupçonnant la présence de nombreux blindés, ne fasse demi-tour.

Selon Lockwood, c'est son premier tir qui a provoqué la chute de la façade alors qu'il est le plus souvent admis que c'est le tir de Wittmann qui a abattu la maison. Une anecdote surprenante pendant ou aussitôt après cet échange d'obus est racontée par Lockwood. Selon lui, « *a little old lady* (une petite vieille dame) *tenant un sac à provisions à la main est sortie précipitamment d'une maison pour entrer dans une autre un peu plus loin* ».

moment the Tiger tank came down the street, and passed right in front of us, about 10 meters away. But I could not shoot without a gunner. I put the radio-operator into the gunner's position – and checked the correct ammunition was loaded".

Dyas was then in the small garden of the James house, at what is now n° 70 Rue Clemenceau.

The 4th tank belonged to the regimental sergeant-major, Gerald Holloway. The driver, J.C. Trevor-Roper, forced the tank back but, just as he passed in front of what is now n° 64, the Cromwell was stopped by an 88 mm shell from Wittmann. The driver was killed and Holloway was taken prisoner a little later.

Pat Dyas then broke cover and set off after the Tiger to strike it in the rear, where the armor was not so thick and hence more vulnerable. Eighty yards further on, on a level with the Guéroult farm track, he saw the German tank motionless amid a cloud of smoke some fifty yards ahead of him. He pulled up and had time to fire two 75 mm shells, causing no damage to the Panzer. Wittmann then traversed his 88 mm gun through 180° and fired a shell at the Cromwell, penetrating the tank on the left side of the turret. Dyas and his driver Mike Lindrea managed to bale out of their burning tank, but the two radio-operators, Buck Taylor and Steve Stevenson, were killed. Slightly burned, the captain crossed the street and raced towards the tank near the sergeant-major that had been knocked out earlier but whose radio he saw was still working. He grabbed the wireless set and tried to get through to his command post at Point 213 to keep Cranley informed of the situation. Cranley told him that the tanks of "A" Squadron and its HQ were encircled by Germans, with Tigers in support, and that its position from being uncomfortable was now getting desperate.

Dyas goes on, "Then some German infantry started shooting at me from windows, and I jumped a hedge into a garden, and landed beside a French girl, who said "You are wounded – come quickly with me" and we hid in a pigsty. She explained she knew all the side roads, and could lead me to my friends. This she did, and we ran together".

After knocking out the four RHQ Cromwells, Wittmann continued on his deadly way and, in the Rue Clemenceau, destroyed the observation Sherman of Battery K, 5th Royal Horse Artillery, commander Major Dennis Wells, armed with a dummy gun made of

(4) The people of Villers who saw the wooden gun lying in front of the Sherman wondered how this subterfuge was supposed to work. The tank, used by the artillery, was actually a veritable mobile armored CP used to conceal observers and wireless equipment and had no room for ammunition or a gun carriage. Its purpose was to pass on intelligence to the batteries regarding the enemy's positions and it had to get up as close as possible to the front line. A dummy gun made it look like an ordinary tank instead of standing out as a prime target for the enemy, who knew just how important it was. The Sherman carried a letter X on the front, indicating that it was a command tank. It belonged to "K" Battery 5th RHA.

Wittmann rompt le combat

Charles Pearce a raconté :

«

J'ai réalisé alors en approchant du Quartier Général de l'escadron B qu'il n'y avait, à mon avis, aucune marque de commandement dans le régiment ni aucun lien avec la Brigade, vers l'arrière. Il me semblait que le Tigre, mis à part le sergent Lockwood, n'aurait rencontré aucune opposition tout le long de la rue Pasteur jusqu'au bas de la ville.

Quand je suis arrivé au Quartier Général de l'escadron B, j'ai vu la major I.B. (Ibby) Aird assis dans la tourelle de son Crommwell et je l'ai salué. Je lui ai parlé du sort du RHQ, du Tigre, de l'action du sergent Lockwood. Il dit qu'il ne savait rien. Je lui ai raconté l'attaque contre l'escadron A mais il devait déjà être au courant. Alors je dis qu'il devait alerter son escadron, qu'il n'y avait rien entre lui et l'escadron de tête, quelque un mile et demi au loin. J'ai offert d'alerter son escadron pour lui avec ma radio mais il ne m'a absolument rien dit, ne me donnant aucune réponse. Je ne savais plus que faire.

Alors, tombé du ciel, le Major Peter McColl (O.C. CSq) apparut et voulut savoir ce que signifiait l'arrêt de l'arrière de l'escadron B et aussi de l'escadron C qui bloquaient la route de Villers-Bocage, ainsi le bataillon des Queens ne pouvait pas entrer dans la ville. Je racontai à Peter McColl ce que j'avais dit à Aird et que je ne pouvais pas obtenir de réponse de lui. Peter décida Aird à alerter l'escadron et lui faire prendre position loin de la route. A ma surprise, I.B. Aird commanda l'action.

Un Quartier Général provisoire du régiment était immédiatement constitué pour les escadrons B et C avec le reste de la section de reconnaissance sous le commandement, à l'ancienneté, de Ibby Aird. McColl était commandant en second. En fin d'après-midi Pat Dyas arriva à pied et donna des détails complets sur l'attaque du RHQ à Ibby Aird qui le nomma adjoint au RHQ provisoire.

Ayant perçu qu'il avait devant lui un adversaire armé d'un canon dont la puissance approchait celle du sien, Wittmann décide de rejoindre les autres chars de sa compagnie pour trouver du renfort, se ravitailler en carburant et en munitions et aider ses hommes à éliminer le reste du *A-Squadron* autour de la cote 213.

C'est à ce moment que, selon l'historien des bataillons de chars Tigre, Peter Agte *(in Tiger)*, le Tigre que commande Wittmann est touché par un obus de canon antichar au niveau d'une chenille et arrêté devant le magasin Huet-Godefroy, rue Pasteur. L'Allemand range son char sur le trottoir pour laisser le chemin libre en vue d'une prochaine attaque, le canon dirigé vers Caen. Espérant récupérer plus tard le char accidenté, il l'abandonne sans y mettre le feu et s'enfuit avec son équipage après avoir « *détruit tous les véhicules qui se trouvaient à sa portée* » peut-on lire dans le rapport du général Dietrich, commandant du *I. SS-Panzer-Korps*.

Il aurait alors gagné à pied le QG de la *Panzer-Lehr* situé au château d'Orbois près du carrefour du Lionvert soit à 7 kilomètres au nord de Villers, pour mettre au courant des opérations le général Bayerlein et lui demander du renfort. Ainsi, selon l'auteur allemand, le Tigre de Wittmann aurait été mis hors de combat le matin du 13 juin.

L'action éclair de Wittmann a fait l'objet de nombreuses narrations différentes à commencer par celle d'un auteur anglais, Daniel Taylor dans « *Villers-Bocage through the lens* ». D'après lui, le Tigre aurait été arrêté au carrefour de la route de Bayeux, au moment

wood (4), in front of the Hôtel du Bras d'Or, also the intelligence officer's scout car and in the Rue Curie, the OP (Operation Post) Cromwell of Captain Paddy Victory of 5th RHA, who had been looking for somewhere to withdraw to, but had been unable to turn round in this narrow street.

The Tiger then sprang forward in the Rue Pasteur, little knowing that in the Place Jeanne d'Arc were "B" Squadron 4th CLY and leading elements of "D" Company 7th Infantry Battalion, Queen's Royal Regiment, the reserve company which included an anti-tank gun troop and a 60 mm mortar troop.

In the Place Jeanne d'Arc, Sergeant Stan Lockwood, in command of no. 2 Troop leading "B" Squadron, stopped his Firefly, he writes, up against the house located "in a corner of the old Square when the road from Amayé-sur-Seulles joined the main street", the Rue Pasteur. The Cromwell and 3 other Fireflies were behind him, along the Caumont road. The street was deserted. The townsfolk must have gone back home after the RHQ had passed. Everything was quiet. The crews got out of their tanks to stretch their legs until they heard bursts of machine-gunfire and the whistle of an 88 mm shell. They all rushed back to their tanks, and just then they saw a scout car speeding down the Rue Pasteur with one of its occupants waving his arms to warn them of some danger. This was the liaison officer, Lieutenant Charles Pearce, who as soon as he saw the Tiger starting to engage the Cromwells left RHQ to warn "B" Squadron of approaching danger. Rounding the bend in the Rue Clemenceau, he saw Captain John Philip-Smith leading half of the reconnaissance troop, and advised him to pull back his vehicles. He was surprised not to see "B" Squadron's Cromwells right behind the scout tanks. He went down the deserted Rue Pasteur and the first tank he came across was Stan Lockwood's Firefly, which he had move up a couple of dozen yards to take on the Tiger.

The driver moved his Firefly forward, and then Lockwood saw the Tiger some 200 meters away, with its turret turned away as it fired down a side street. He had time to dispatch two 17-pdr shells before the Tiger aimed its gun in his direction. The first shell on exploding demolished the front of the building just next to him. Slates, pieces of wood and stonework collapsed on top of a German sniper who had been hiding in the house. One of the two shots hit the Tiger, producing flames and smoke but without penetrating the armor. Once he had lined up his gun, Wittmann fired, twice missing the Firefly, his shells being lost in the building. The Britisher fired twice more until the German, suspecting the presence of numerous tanks, turned round and made off.

According to Lockwood, it was his first shot that brought down the front of the house, although it is generally accepted that the shot that demolished the house was fired by Wittmann. An odd thing happened during or just after this exchange of shellfire, according to Lockwood: "a little old lady came out with a shopping bag in her hand and scuttled off to enter another house further on".

Wittmann breaks off the battle

Charles Pearce recalls:

"I realized now as approached B Sqn HQ that to my view, there was no adequate Regimental control nor any rear link to Brigade. It seemed to me that the Tiger, apart from Sergeant Lockwood, would have had no opposition all the way to the bottom of the town.

When arrived at B Sqn I saw Major I.B. (Ibby) Aird (OCB Sqn) sitting in the turret of his Cromwell and

où il revient, touché aux chenilles par un canon antichar de la *Rifle Brigade,* mis rapidement en batterie au début de l'attaque allemande et situé au niveau du château d'eau. Ainsi, Pat Dyas aurait attaqué le char allemand à son retour de la rue Pasteur et non au moment où il descend vers le centre du bourg. Cette version est d'ailleurs celle écrite par Dyas lui-même en anglais et en français, 49 ans après la bataille.

« J'avais décidé de suivre le Tigre parce que le blindage en arrière est moins épais. Malheureusement pour moi, il a rencontré deux autres chars anglais (Stan Lockwood et Bob Moore) qui l'avaient attaqué et il avait fait un virage de 180° et je l'avais rencontré face à face. J'étais, sûr, il était là et mon canonnier était prêt. Nous avons tiré contre lui, il était frappé mais pas gravement. Puis c'était son tour et avec un coup de 88 mon char était détruit ».

Il faut remarquer qu'entre l'endroit où Dyas s'est replié et l'endroit où son char a été touché, la distance relativement courte, une centaine de mètres, peut-être parcourue en quelques instants, alors qu'il a dû falloir un temps plus long à Wittmann pour aller jusqu'à la rue Pasteur, s'arrêter plusieurs fois pour ajuster 3 chars anglais et deux autres véhicules et revenir jusqu'au virage de la rue Clemenceau. De plus, il ne faut pas oublier que cette bataille de chars s'est déroulée en peu de temps dans la confusion au milieu de la fumée des explosions et sous la pluie.

D'autres auteurs pensent que Wittmann, contournant les « riflemen » arrêtés, serait descendu seul vers Villers, laissant aux autres Tigres de sa compagnie le soin d'attaquer l'escadron de *Cromwell* et la compagnie de la *Rifle Brigade.* Il aurait alors détruit les chars du *RHQ* et les chars d'observation d'artillerie avant de faire demi-tour rue Pasteur. A son retour, il aurait rencontré Dyas de face et aurait pu attaquer la compagnie d'infanterie en venant de l'ouest, ce qui expliquerait la position du canon antichar dirigé vers Villers.

Mais dans cette hypothèse, ce serait ignorer le témoignage du caporal Watt qui a vu le Tigre venant de la direction de Caen sortir d'un nuage de fumée, en mitraillant les *half-tracks* de la *Rifle Brigade* les uns après les autres, et le croquis réalisé par le capitaine Milner pour raconter par le dessin l'attaque du matin des chars allemands.

Pour conforter le récit de Peter Agte, on peut rappeler que le journal de la 22ᵉ Brigade Blindée note qu'entre 9 h et 11 h, des Tigres rodaient dans Villers et que l'un d'eux avait été mis hors de combat. D'autre part, Monsieur Stanislas de Clermont-Tonnerre était dans Villers avec les équipes d'urgence en fin de matinée et a vu le char allemand. Il précise même que ce char, un Tigre, a été remorqué quelques jours plus tard jusqu'à la lisière du petit bois de Maisoncelles-Pelvey, après le passage à niveau, route d'Aunay où, volontairement mal camouflé et avec d'autres épaves, il a servi de leurre pour les avions alliés.

Le 19 juin, un reporter américain, Tom Treaner, écrivait dans le *New's Chronicle* que Wittmann, lorsqu'il franchit le carrefour de la route de Bayeux, « se tient debout dans la tourelle de son Tigre, salue en soulevant son casque et sourit ». Que Wittmann soit debout dans sa tourelle, scrutant le terrain à la recherche d'un adversaire éventuel, cela peut se concevoir car il en avait l'habitude. Mais qu'il salue ironiquement en soulevant son casque, cela est plus difficile à admettre venant d'un officier SS, véritable professionnel de la guerre.

Il semble que Wittmann ait parcouru les rues Clemenceau et Pasteur, tourelle ouverte, pour mieux

saluted him. I told him about RHQ, the Tiger tank and that I had seen Sergeant Lockwood. He did not acknowledge me nor did he say anything.

I told him about the attack on A Sqn but he must already have known about that. I then said that he must alert his Sqn, as there was nothing between him and A Sqn, some 1 1/2 miles away. I also offered to alert his Sqn for him on my wireless, but said absolutely nothing, giving me no response in any way. I was absolutely at my wit's end.

Then, out of the blue, Major Peter McColl (OC C Sqn) appeared and wanted to know what the hold up was as the back of B Sqn, and also C Sqn, were blocking the road into Villers-Bocage so the Queen's battalion could not get through into the town.

I told Peter McColl what I had said to Ibby Aird and explained that I could get no response from Ibby. Peter, in no uncertain terms, told Ibby Aird to alert and deploy B Sqn and move them off the road.

To my surprise, Ibby Aird did take action.

A temporary RHQ was immediately set up for B and C Sqns, together with the remaining half of the recce, under Major Ibby Aird as acting CO (he was senior to Peter) and Major Peter McColl as acting 2 1/C. In the late afternoon, Capt Pat Dyas appeared on his own on foot and gave a complete rundown to Ibby Aird about the attack on RHQ. Ibby installed Pat as Adjutant at the temporary RHQ.

On seeing that he was facing an enemy armed with a weapon nearly as powerful as his own, Wittmann decided to join the rest of his company's tanks to seek reinforcement, to refuel and rearm and help his men see off the remainder of "A" Squadron from around Point 213.

It was at this time, according to the historian of the Tiger tank battalions, Peter Agte (in Tiger), that the Tiger commanded by Wittmann was shelled by an anti-tank gun, with damage to a track, and parked in front of Huet-Godefroy's store in the Rue Pasteur. The German left his tank on the sidewalk with its gun aimed towards Caen, to leave the way free for a later attack. Hoping to recover the damaged tank later on, he just left it without setting fire to it and escaped with his crew after first "destroying any vehicles within range", we read in the report by I. SS-Panzer-Korps commander General Dietrich.

He is then thought to have proceeded on foot to Panzer-Lehr HQ located at the Château d'Orbois near the Lion-vert crossroads 7 kilometers north of Villers, to report to General Bayerlein on his operations and request reinforcements. Thus, according to this German author, Wittmann's Tiger was knocked out on the morning of June 13th.

Wittmann's lightning action has been the subject of many different accounts, starting with that of a British author, Daniel Taylor, in "Villers-Bocage through the Lens". In his opinion, the Tiger was stopped at the Bayeux road junction, on the way back after its tracks had been hit by one of the Rifle Brigade's anti-tank guns, quickly brought into position on a level with the water tower at the start of the German attack. This would mean that Pat Dyas attacked the German tank on his way back from the Rue Pasteur and not as he was heading down towards the town center. This is in fact the version given by Dyas himself in English and French, 49 years after the battle.

"I then decided to follow the Tiger, as its thinnest armor was at the back, and that was the best chance of killing it. Unfortunately, he met two other English tanks further on (Stan Lockwood & Bod Moore), who fired at him, and he turned around, and we met face-to-face at the street corner. I knew he was there, and my gunner ready, so I fired two shots at him, but

Walter Lau.

observer le champ de bataille et diriger les tirs de son canonnier, mais dès qu'il avait repéré une cible, il plongeait à l'intérieur de son char. Bien lui en prit d'ailleurs, lorsque l'obus que lui décocha le *Major* Carr toucha la plaque frontale du Tigre. Des soldats anglais ont raconté cette anecdote, d'autres l'ont rapportée dans des écrits, mais sans avoir vécu la scène. « *C'était une rumeur que j'ai rapportée* » dira Cloudsley-Thomson et son épouse ajoutera « *c'était un mythe* ».

L'escadron A est encerclé

Pendant que Wittmann effectue sa rapide incursion dans Villers, les autres Tigres de sa compagnie prennent position autour de la côte des Landes aidés par les hommes de la 4ᵉ compagnie de grenadiers SS. qui, bien camouflés dans les haies et les arbres, empêchent les Britanniques de revenir vers Villers par des tirs de mitrailleuses et d'armes individuelles. Son Tigre ayant été emprunté par Wittmann, Sowa est « à pied ». Le *Rottenführer* Lau, pointeur du char de Stief lui signale que le Tigre est sans commandant et l'invite à remplacer son chef de char. Ils sont au bas de la côte des Landes, sur la gauche de la colonne de véhicules en flammes. A droite, 2 *Cromwell* cherchent à revenir vers Villers, ils sont neutralisés tous les deux.

A la cote 213, les officiers rassemblent autour du QG avancé installé dans la maison Ruelle, les 10 *riflemen* détachés près de l'escadron A, les chars en état de combattre et les quelques véhicules restants, 3 *half-tracks* et 2 scout cars. Chacun se met en place, le caporal Nicholson avec 6 hommes au sud de la RN 175, vers l'ancienne route de Caen, et le lieutenant Campbell avec 4 *riflemen* là où la route de Monts-en-Bessin rejoint la route nationale, le sergent Gale couvre le chemin qui mène au bois de Villy, le lieutenant Butler est à l'affût le long d'une haie au sud de la route nationale, le lieutenant Coop surveille un petit chemin situé en avant du lieutenant Campbell, le capitaine Milner a précisé sa position : « *je restai moi-même pour couvrir le chemin au nord-est de la cote 213 et Gale, le sergent mortier, couvrait la voie qui mène au petit bois situé vers le nord (vers Villy). J'étais derrière une petite maison, le long d'un chemin sur lequel une femme d'allure imposante avançait vers nous (route de Monts). Nous l'avons prise pour une espionne, mais après une heure de captivité, elle s'échappait. Etait-ce une espionne ?* »

Cette femme qui semblait venir d'un endroit d'où une mitrailleuse avait tiré sur les Anglais de la cote 213 a été soupçonnée par Campbell d'avoir été envoyée par les Allemands pour rendre compte de la force et des positions du détachement anglais. Le lieutenant l'interpella et la confia à Milner qui, parlant le français, a pu l'interroger.

Milner continue : « *peu de temps après, un char du CLY qui s'engageait en avant dans le chemin qui longe la maison était touché et l'équipage blessé sautait du char sauf le conducteur qui était mort. Les soldats semblaient cloués au sol, aussi je rampai le long de la route et les aidai à accepter la sûreté relative de la maison* ». Pendant plus de deux heures, en fin de matinée, les rescapés subissent le feu de l'artillerie allemande.

Parce que sa mitrailleuse s'est enrayée, Milner ne peut participer à la bataille. Il se cache dans un champ de foin, près du jardin de Monsieur Ruelle et va attendre la nuit, menacé par les tirs des canons du *RHA* qui pilonnent la cote 213, avant de regagner les lignes anglaises le matin du 14 juin au Mesnil de Tracy.

only damaged him slightly – and then he shot at my tank, and destroyed it".

It should be noted that the relatively short distance from where Dyas withdrew to where his tank was hit, some hundred yards, may have been covered in a matter of seconds, whereas it must have taken Wittmann longer to go to the Rue Pasteur, stop several times to knock out 3 British tanks and two other vehicles and get back to the bend in the Rue Clemenceau. It should also be remembered how this tank battle took place amid great confusion with the smoke from the explosions and in the rain, and it was all over very quickly too.

Other writers think that Wittmann, bypassing the riflemen who had stopped, went down to Villers alone, leaving his company's other Tigers to attack the squadron of Cromwells and the Rifle Brigade company. He would then have destroyed the RHQ tanks and the artillery observation tanks before turning round in the Rue Pasteur. On his return, he would have met Dyas head on and could have attacked the infantry company coming from the west, which would explain the position of the anti-tank gun trained on Villers.

But if this were how it actually happened, we would have to ignore the testimony of Corporal Watt, who saw the Tiger coming from the Caen direction emerge from a cloud of smoke machine-gunning the Rifle Brigade's half-tracks one after the other, and the sketch drawn by Captain Milner to give a pictorial account of the morning attack by the German tanks.

To back up Peter Agte's version, one may recall that the 22nd Armoured Brigade's log notes that between 09.00 and 11.00, there were Tigers lurking in Villers and one of them had been knocked out. Also, Monsieur Stanislas de Clermont-Tonnerre was in Villers with the emergency teams late that morning and he saw the German tank. He even states that this tank, a Tiger, was towed a few days later to the edge of the little wood at Maisoncelles-Pelvey, after the level crossing on the Aunay road, where, along with other wrecks it was deliberately poorly camouflaged and used as a decoy for Allied aircraft.

On June 19th, an American reporter, Tom Treanor, wrote in the News Chronicle that, as he crossed the junction with the Bayeux road, Wittmann "*was there standing in the turret, waving his hat and bowing*". For Wittmann to be standing up in his hatch, scouring the area in search of a possible enemy, is conceivable because that is something he was in the habit of doing. But for him to raise his helmet in ironic greeting is hardly possible on the part of a professional soldier like this SS officer.

It seems that Wittmann passed through the Rues Clemenceau and Pasteur with the hatch open, to get a better view of the battlefield and direct his gunner's aim, but once he had located a target, he would dive back inside his tank. Which was just as well in fact, as one shell fired on him by Major Carr hit the Tiger's front armor. British soldiers told the story, which found its way into the writings of others who had not witnessed the scene. "It was a rumor I reported" said Cloudsley-Thomson, and his wife added, "it was a myth".

"A" Squadron is encircled

While Wittmann was making his lightning raid in Villers, the other Tigers of his company took up position around the Côte des Landes, helped by the men of the 4th SS Grenadier Company who were well camouflaged among the trees and hedges, and prevented the British from heading back to Villers with their machine-guns and personal weapons. Sowa was "on foot" since Wittmann had borrowed his Tiger. Rottenführer Lau, the gunlayer in Stief's tank, told him

L'officier qui commande l'escadron A, le *Major* Peter Scott, est tué par un obus.

Le colonel Cranley prend alors la douloureuse décision de se rendre. Il ordonne aux tankistes de mettre le feu à leurs chars, à l'exception de ceux qui abritent des blessés. Quant à lui, il part dans son scout-car vers le sud, vers la voie de chemin de fer avec l'espoir de trouver une route de sortie. Lors de cette tentative, il est fait prisonnier.

Ce 13 juin, des civils ont passé la journée au milieu de la bataille. Ils s'étaient réfugiés près des bâtiments des fermes des Hauts Vents et de la cidrerie, route d'Epinay, d'où ils pouvaient gagner rapidement les abris qu'ils avaient confectionnés. Ces abris rudimentaires consistaient en des tranchées profondes recouvertes de tôles, de fagots, de terre et de toutes sortes de matériaux capables de les protéger d'éclats de bombes ou d'obus et de balles, mais insuffisants pour les protéger de projectiles explosifs arrivant « de plein fouet ».

Auguste et André Loir et leurs parents habitaient aux Hauts Vents. Aux premiers coups de feu, ils avaient quitté les bâtiments de leur ferme pour se réfugier dans un chemin creux proche. Ils y sont restés 24 heures sans pouvoir en sortir, des soldats allemands les empêchant d'aller prendre des nouvelles d'amis terrés dans un fossé à quelques dizaines de mètres d'eux.

Dans cette partie Est de la commune, des civils ont été tués. Madame Louise Jouanne, cultivatrice, habitait cote 213. Sa maison, proche de la route de Caen, aux Landes de Monbrocq, avait été détruite dès le 6 juin par une bombe. Elle s'était réfugiée avec sa famille à la cidrerie Robergé, au haut de la côte qui monte depuis Epinay-sur-Odon. L'endroit leur paraissait calme, pourtant il semblerait que ce soit par cette route que les Tigres de la 2ᵉ compagnie sont arrivés, venant d'Epinay où ils étaient la veille en fin d'après-midi. Elle avait laissé ses animaux à la ferme et, le matin du 13, elle avait voulu aller les soigner et traire ses vaches pour rapporter un peu de lait à son petit-fils âgé de 14 mois. Avant de revenir à son refuge, elle avait persuadé ses voisins M. et Mme Lerouilly restés véritablement au milieu des chars de se joindre à elle. Ils ont été tués tous les trois par balles, ainsi que le petit chien qui les accompagnait, au retour vers leur refuge. Peut-être ont-ils été exécutés. Leurs corps retrouvés le lendemain seront inhumés sommairement dans un trou d'obus par les secouristes du château conduits par M. Henri de Clermont-Tonnerre près de leurs habitations. Une balle avait transpercé le portefeuille de M. Lerouilly, resté dans sa poche.

Arrivée des renforts allemands

Arrivé au QG de la *Panzer-Lehr-Division* à Orbois, Wittmann aurait mis au courant des événements de Villers, le lieutenant-colonel Kauffmann, officier d'état-major, puis le QG du *I. SS-Panzer-Korps*.

La *Panzer-Lehr-Division,* bloquée par les combats dans le secteur de Tilly, ne peut dégager que 15 *Panzer IV* du IIᵉ bataillon du régiment 130 et les met sous le commandement du capitaine Helmut Ritgen qui reçoit mission de bloquer les sorties nord de Villers. Depuis Orbois, il se dirige vers Villers mais, lorsqu'il approche de la ville, au nord d'une petite vallée, il tombe sous le feu de violents tirs de canons antichars. Un de ses chars, celui du *Feldwebel* Dobrowski, reçoit un coup au but et prend feu. Sans appui d'infanterie et d'artillerie, il ne peut avancer davantage. Bayerlein, prévenu, lui ordonne de revenir vers Villy. Il replie alors ses chars vers le château de Par-

that the Tiger had no commander, and invited him to take over command. They were at the bottom of the Côte des Landes, to the left of the column of burning vehicles. On the right, 2 Cromwells were taken out as they tried to make their way back to Villers.

At Point 213, the officers brought together at the forward HQ in the Ruelle home the 10 riflemen detached to "A" Squadron, the battle-worthy tanks and the few remaining vehicles: 3 half-tracks and 2 scout cars. Each went into action, Corporal Nicholson with 6 men south of highway N175 towards the old Caen road, and Lieutenant Campbell with 4 riflemen where the Monts-en-Bessin road joins the highway, with Sergeant Gale covering the lane leading to Villy Wood, Lieutenant Butler keeping watch along a hedge south of the highway, and Lieutenant Coop guarding a lane ahead of Lieutenant Campbell. Captain Milner describes his position: "I set myself to cover the lane running east north east from Point 213 and the mortar sergeant covered the track leading into and out of the wood directly to our north. I was behind a small cottage. It was along this track that a peasant woman walked up to us, was detained as a possible spy, but after an hour or so she slipped away. Was she a spy?"

This woman who seemed to come from a spot on Point 213 from where a machine-gun had been fired at the British was suspected by Campbell of having been sent by the Germans to report on the British detachment's strength and positions. The lieutenant challenged her and handed her over to Milner who spoke French and was able to interrogate her.

Milner goes on: "Not long after, a CLY tank, edging forward around a bend along my lane, was hit and the wounded crew baled out having suffered the driver killed. They seemed to be pinned down so I crawled along the road and helped them to get back to the comparative safety of the cottage." The survivors came under German artillery fire for over two hours late that morning.

Milner's machine-gun was jammed and he could take no part in the battle. So he lay low until nightfall in a hayfield near Monsieur Ruelle's garden, under threat from the RHA guns pounding Point 213, and got back to the British lines at Mesnil de Tracy on the morning of June 14th.

"A" Squadron's commanding officer, Major Peter Scott, was killed by a shell.

Colonel Cranley then took the painful decision to surrender. He ordered the tank crews to set fire to their tanks, except for those containing casualties. He himself headed off south towards the railroad in his scout car in the hope of finding a way through. He was captured in the attempt.

On that June 13th, a few civilians spent the day in the middle of the battle. They had taken cover close to the Hauts Vents farm buildings and the cider-house on the Epinay road, from where they could quickly get to the shelters which they had built. These rudimentary shelters consisted of deep trenches covered with metal sheeting, bundles of firewood, earth and all kinds of materials to protect them from shrapnel from bombs, shells and bullets, but were not enough to protect them from an explosive projectile scoring a direct hit.

Auguste and André Loir and their parents lived at the Hauts Vents. On hearing the first shots, they had left their farm buildings to take shelter in a nearby sunken lane. They remained there for 24 hours, unable to leave as German soldiers prevented them even from going for news of friends hiding in a ditch just a few dozen yards further on.

On this eastern side of the town, a number of civilians were killed. Madame Louise Jouanne, a farmer,

fouru-sur-Odon, afin de faire réparer les panzers endommagés avant de revenir vers Villers pour occuper les extérieurs de la ville, empêcher les Anglais de revenir vers leurs bases et neutraliser toute tentative d'attaque.

Une dizaine de chars remis en état descendent la côte des Landes, de chaque côté de la route nationale. 4 *Panzer IV* sont à gauche et approchent au sud de Villers. A un carrefour, ils entrent dans la ville, les deux premiers sont détruits. Le 3e et le 4e reçoivent alors l'ordre de se replier.

Au cours de cette manœuvre, un tankiste du *4e Panzer IV* appartenant à la 5e compagnie du régiment 130, le pointeur Hans Burkhard, a vu sur sa gauche, près d'un bouquet d'arbres, 2 Tigres sans doute de la 2e compagnie, immobilisés. Un peu plus tard, les 2 *Panzer IV* vont rencontrer deux des 8 Tigres de la 1re compagnie de la *SS-Pz.-Abt. 101*, commandée par le capitaine Möbius qui, venant de Noyers, est arrivée cote 213 vers 13 h.

Lorsqu'il est revenu cote 213 à bord d'une voiture amphibie, Wittmann a retrouvé les grenadiers qui accompagnaient sa compagnie en train de nettoyer le terrain avec ses Tigres et de faire prisonniers la plupart des Anglais survivants. Les prisonniers étaient conduits *in a palatial château* écrira K.K. Weightman où ils étaient interrogés. Il s'agit peut-être du château de Missy qui servait de lieu de repos aux Waffen-SS de la *Hitlerjugend*.

Les Britanniques prennent position dans Villers

Dès que Wittmann quitte Villers le matin, la ville devient relativement calme. Les *Queen's* recherchent les emplacements les plus favorables pour répondre à une contre-attaque allemande et défendre le bourg. Ils prennent position aux soupiraux des caves et aux fenêtres, dans les ruelles pour mettre en place leurs mitrailleuses, leurs PIAT, leurs canons antichars de 57. Les rues étroites et les ruelles sont des endroits idéaux pour permettre à des antichars de tirer efficacement sur un panzer.

Madame Julien Thorel a vu deux *Queen's* installer ce qu'elle décrit comme étant un mortier dans la salle à manger du Maire, M. Albert Doublet, à un endroit d'où ils vont pouvoir contrôler la place de l'Hôtel de Ville et le carrefour du boulevard Joffre et de la rue Pasteur.

Le calme relatif permet à quelques civils de se risquer à sortir de leur refuge pour apporter une aide éventuelle à des victimes civiles ou à des soldats blessés. C'est ainsi que vers **11 heures**, Messieurs Fromont et P. Bertou vont conduire un officier blessé à un poste de secours près de l'église place Jeanne d'Arc.

M. Jean Marie et son père, sapeurs-pompiers, vont aider et réconforter Madame Letellier dont le mari a été tué rue Georges Clemenceau par un éclat d'obus. Plus tard, ils vont prendre la direction de Tracy et passeront place du marché où ils verront une mitrailleuse anglaise bien installée dans le café Dumont, situé à l'angle de la rue Emile-Samson et de la route d'Evrecy.

En **fin de matinée**, le *Major* Aird, qui commande l'escadron B du *4th CLY* stationné place Jeanne d'Arc, décide d'organiser la défense de la ville et d'essayer de prendre contact avec l'escadron A.

Il dirige la *troop* 3 vers le nord, vers la rue Saint-Martin et les abords du cimetière, avec le *Firefly* du sergent Bob Moore, la *troop* 4 commandée par le lieutenant Bill Cotton vers le sud, la *troop* 1 de Walter Allen vers la route de Caumont, sur les virages de la

Bob Moore.

lived on Point 213. Her house, at Landes de Monbrocq just off the Caen road, had been demolished by a bomb back on June 6th. She had taken shelter with her family at the Robergé cider-house, at the top of the hill up to Epinay-sur-Odon. The place seemed quiet enough, however this apparently was the road on which the Tigers of the 2nd Company arrived from Epinay, where they had been late on the previous afternoon. She had left her stock on the farm, and on the morning of the 13th, she wanted to go and see to them, milk her cows and bring back some milk for her 14-month old grandson. Before heading back for the shelter, she persuaded her neighbors Monsieur and Madame Lerouilly, who were caught right in the middle of the tanks, to come with her. All three of them were killed by bullets on the way back to the shelter, as was the little dog they had brought along. They may have been executed. Their bodies were found the next day and were hastily buried in a shell crater near their homes by first-aid workers from the château led by Monsieur Henri de Clermont-Tonnerre. A bullet had passed straight through Monsieur Lerouilly's wallet in his pocket.

Arrival of German reinforcements

On arriving at Panzer-Lehr-Division HQ at Orbois, Wittmann will have reported the events at Villers to staff officer Lieutenant-Colonel Kauffmann and to I. SS-Panzer-Korps HQ.

Being held up by the engagements in the Tilly sector, the Panzer-Lehr-Division could only release 15 Panzer IVs of 2nd Battalion, 130th Regiment, and these were placed under the command of Captain Helmut Ritgen, who was detailed to cut off the northern exits from Villers. He moved from Orbois towards Villers, but came under heavy anti-tank gun fire as he approached the town on the north side of a small valley. One of his tanks, Feldwebel Dobrowski's, was hit and burst into flames. With no infantry or artillery support he could advance no further. Bayerlein, on being informed, ordered him to head back towards Villy. He then withdrew his tanks towards the château at Parfouru-sur-Odon, to repair the damaged panzers before returning to Villers to occupy the outskirts of the town, prevent the British from returning to their bases, and thwart any attempted attack.

About ten overhauled tanks drove down the Côte des Landes on either side of the highway. 4 Panzer IVs were on the left and approached Villers from the south. At a crossroads, they entered the town, and the first two were knocked out. The 3rd and 4th then received orders to fall back.

During this operation, a tank crew member of the 4th Panzer IV belonging to the 5th Company, 130th Regiment, gun layer Hans Burkhard, saw near a clump of trees to his left 2 Tigers doubtless belonging to the 2nd Company, at a standstill. A little later, 2 Panzer IVs joined two of the 8 Tigers of the 1st Company, SS-Pz.-Abt. 101, commanded by Captain Möbius, which arrived at Point 213 from Noyers at around 13.00.

On returning to Point 213 in an amphibious car, Wittmann found the grenadiers accompanying his company busy mopping up the area with the Tigers and rounding up most of the British survivors. The prisoners were brought to "a palatial château", writes K.K. Weightman, where they were interrogated. This may have been the château at Missy which was being used as a resting place for the Waffen-SS of the Hitlerjugend.

The British take up position in Villers

As soon as Wittmann left Villers in the morning, the town became relatively quiet. The Queen's looked

route de jonction et les autres chars sur les croisements avec les routes de Vire et d'Aunay. La troupe 4 est composée de 4 chars, le *Cromwell VI* à canon de 95 mm du lieutenant, de 2 *Cromwell* à canon de 75 mm, celui du sergent Leonard Grant et celui du caporal George Horne dont le radio est le jeune *trooper* de 19 ans, Harold Currie et 1 *Firefly* commandé par le *lance-corporal* Bobby Bramall. Par la route d'Aunay, puis à travers champs, la *troop* essaie de contourner Villers par le sud, se heurte à une unité d'infanterie allemande dont elle se débarrasse à coups de mitrailleuse, touche quelques véhicules de ravitaillement et prend la direction de l'est vers la gare. Bramall, qui a vu des véhicules à hauteur du château, les a entreprises à la mitrailleuse avant de s'apercevoir que quelques-uns étaient des véhicules de la Croix Rouge.

A midi, P. Bertou a vu trois chars s'arrêter quelques instants dans la cour des établissements Rivière avant de prendre la direction de la gare par le boulevard Joffre. Là se trouve la compagnie A du *7th Queen's* que le lieutenant-colonel Desmond Gordon a installée le matin. L'objectif de Cotton est de rejoindre la cote 213 par le sud, mais il se heurte à une importante unité d'Allemands supportés par des panzers qui l'oblige à se replier en deçà de la voie ferrée. Constatant la situation, il se décide à se replier vers le centre du bourg afin de préparer une position solide autour de la place de l'Hôtel de Ville et de la rue Pasteur d'où il pourrait attaquer les chars allemands sur les flancs au lieu de les attaquer de face. Sachant que les Tigres ont du mal à se mouvoir dans les rues étroites, il pense qu'ils emprunteront plutôt la rue principale de Villers.

Vers 13 heures, Madame Thorel, voit 3 chars anglais et un canon antichar sur la petite place du monument aux morts de la guerre 14-18, derrière l'Hôtel de Ville. Trois chars seulement sur les quatre envoyés par le *Major* Aird. Il doit manquer le *Firefly* de Bramall qui a quitté le groupe et s'attarde à contrôler quelques rues adjacentes. Il rejoindra plus tard. Bill Cotton qui dispose d'un obusier de 95 mm à canon court sait que les obus à pouvoir destructif certain qu'il peut envoyer, sont sans effet sur les blindages. Il faudrait des obus perforants à couche externe en métal mou qui s'écrasent sur le blindage des chars avant de le pénétrer, ce qui permet d'éviter le ricochet. Voyant que son char ne lui sera d'aucune utilité, il l'abandonne à l'abri dans un garage de la cour de la quincaillerie Marie, place de la mairie pour le reprendre plus tard. Avec ses tankistes, les *lance-corporals* Hodgson et Payne, les *troopers* W. Jones et Humphreys, il va diriger la manœuvre à pied. Pendant ce temps, Bramall et son *Firefly* les a rejoints. De l'avis même des tankistes, les obus de 17 livres du *Firefly* étaient les plus efficaces alors que ceux des canons de 75 mm des *Cromwell* étaient « d'un petit effet » même tirés à bout portant.

En **début d'après-midi**, les Allemands décident de lancer leur attaque. Dès que la 1re compagnie du bataillon 101 arrive à la cote 213, le capitaine Möbius qui la commande prend contact avec Wittmann. Ce dernier le met au courant de la situation et ils décident tous les deux de lancer l'offensive. Les Allemands sont en force car ce sont les 8 Tigres de la 1re compagnie qui vont prêter main forte aux *Panzer IV* de la *Panzer-Lehr-Division* et attaquer le bourg. Une section est déjà à Villers. Elle est commandée par le sous-lieutenant Philipsen dont le char 111 sera touché et arrêté dans la ville. Evitant la route nationale en partie obstruée, ils descendent de la côte des Landes en traversant les herbages parallèlement à l'ancienne route de Caen et à la voie ferrée, traversant les haies et renversant les clôtures, ils pénètrent dans Villers par la route d'Evrecy.

for the most suitable sites to meet a German counter-attack and defend the town. They took up position behind basement vents and windows, and in back streets to set up their machine-guns, PIATs and 6 pounder anti-tank guns. The narrow streets and lanes were ideal places for effective anti-tank fire against a panzer.

Madame Julien Thorel saw two Queen's setting up what she describes as a mortar in the dining-room of the Mayor, Monsieur Albert Doublet, from which they covered the Town Hall square and the junction between the Boulevard Joffre and the Rue Pasteur.

*The lull caused some civilians to venture out of their shelters to help any civilian casualties or wounded soldiers. Thus at around **11.00**, Messrs Fromont and P. Bertou took a wounded officer to a first-aid post near the church in the Place Jeanne d'Arc.*

Monsieur Jean Marie and his father, both firemen, came to assist and comfort Madame Letellier, whose husband had been killed by shrapnel in the Rue Georges Clemenceau. Later they headed off towards Tracy and passed through the marketplace where they saw a well positioned British machine-gun in the Café Dumont on the corner of the Rue Emile Samson and the Evrecy road.

*In **late morning**, Major Aird, commanding "B" Squadron, 4th CLY stationed in the Place Jeanne d'Arc, decided to organize the defense of the town and try and contact "A" Squadron.*

He ushered no. 3 Troop north towards the Rue Saint Martin and the cemetery area, with Sergeant Bob Moore's Firefly, no. 4 Troop under Lieutenant Bill Cotton to the south, Walter Allen's no. 1 Troop near the Caumont road, on the bends of the link road, and the other tanks on the Vire and Aunay road junctions. No. 4 Troop was made up of 4 tanks, Lieutenant XXX's Cromwell VI with a 95 mm gun, 2 Cromwells with 75 mm guns, those of Sergeant Leonard Grant and Corporal George Horne, whose radio operator was a young 19 year-old trooper, Harold Currie, and 1 Firefly commanded by Lance-Corporal Bobby Bramall. By the Aunay road, then across the fields, the troops tried to outflank Villers via the south, encountering a German infantry unit which it fended off with machine-gunfire, hitting a few supply vehicles before heading off east towards the station. Bramall, who saw these vehicles on a level with the château, started machine-gunning them until he realized that some of them were Red Cross vehicles.

Le prince von Schönburg-Waldenburg, qui commandait le IIe Bataillon du *Panzer-Lehr-Reigment 130*, a été tué le 9 juin 1944. Il est enterré dans le cimetière de Parfouru-sur-Odon (au sud-est de Villers) avec plusieurs membres de son équipage. (EG/Heimdal.) Les armes de la famille du Prince von Schönburg étaient peintes sur les chars de son bataillon comme le « 634 » détruit à Villers (ci-dessous).

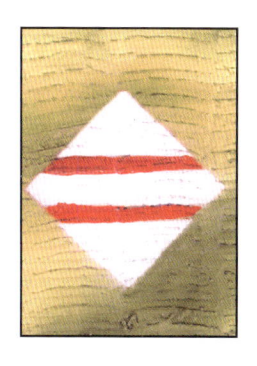

Prince von Schönburg-Waldenburg, commander of the 2nd Battalion Panzer-Lehr-Regiment 130, was killed on June 9th 1944. He was buried in the cemetery at Parfouru-sur-Odon (southeast of Villers) along with several of his crew members. (EG/Heimdal.)

Prince von Schönburg's family coat of arms were painted on his battalion's tanks as on tank "634" destroyed at Villers.

Rolf Möbius mène la contre-attaque avec les Tiger de sa compagnie.

Rolf Möbius leads the counterattack with his company's Tigers.

M. Clément Bernouis, sa famille, quelques voisins dont le Dr François voient avec inquiétude et un certain effroi par le soupirail de la cave de l'immeuble, situé au n° 50 actuel de la rue Clemenceau, les Tigres passer dans un bruit de tonnerre. Si les Allemands sont peu nombreux en ville et évitent le combat, ils sont par contre disposés à la périphérie et pénètrent avec précaution dans les bâtiments à la recherche de soldats dispersés lors des combats du matin. Une petite unité approche de la maison et pénètre au sous-sol, dans la cave à charbon, un local supposé le plus sûr pour servir d'abri à ceux qui s'y sont réfugiés. Lorsque le premier Allemand entre, il est intrigué par le parfum caractéristique des cigarettes anglaises distribuées le matin par les Anglais. Le soldat prévient un officier qui inspecte la cave et interroge les occupants sur la présence d'un Anglais parmi eux. L'officier n'écoute pas les dénégations du Dr François et, soupçonneux, fait sortir tout le monde et balaie le tas de charbon de rafales de mitraillette. Le bruit de la fusillade est entendu par un jeune homme qui, affolé, s'enfuit en courant, criant partout que la famille Bernouis avait été fusillée dans la cave. A un autre moment, une telle incursion aurait pu mal se terminer pour les réfugiés.

At noon, P. Bertou saw three tanks stop for a few moments in the Rivière factory yard before heading for the station via the Boulevard Joffre. "A" Company 7th Queen's was there, Lieutenant-Colonel Desmond Gordon having set it in place that morning. Cotton's objective was to reach Point 213 from the south, but he encountered a sizable German unit with panzers in support, forcing him to withdraw short of the railroad. Noting the situation, he decided to withdraw to the town center to prepare a firm position around the Town Hall square and the Rue Pasteur from which to attack the German tanks from the side instead of attacking them head on. Knowing that the Tigers had trouble driving through the narrow streets, he thought they would probably take the main street through Villers.

At c. 13.00, Madame Thorel saw 3 British tanks and an anti-tank gun in the small 1914-18 war memorial square, behind the Town Hall: just three of the four tanks sent by Major Aird. The missing tank must have been Bramall's Firefly, which had left the group and stayed back to check some nearby streets. It caught up later. Bill Cotton, who had a short-barrelled 95 mm howitzer, knew that he could fire shells with unquestionable destructive capacity but which would not pierce the armor. This would require armor-piercing shells with a soft metal outer layer that would be flattened on the tank's armor plating before penetrating, so as to prevent them from just bouncing off. Seeing that his tank was going to be of no use to him, he parked it in the shelter of a garage in the yard of Marie's hardware store in the Town Hall square, to pick it up later on. With his tank crew, Lance-Corporals Hodgson and Payne, Troopers W. Jones and Humphreys, he decided to lead the operation on foot. Meanwhile, Bramall had joined them in his Firefly. Even the tank crews agreed that the Firefly's 17-pdr shells were the most effective, whereas the Cromwell's 75 mm guns had "little effect" even when fired at pointblank range.

Early that afternoon, the Germans decided to launch their attack. As soon as the 1st Company, Battalion 101 reached Point 213, its commander, Captain Möbius, contacted Wittmann who reported the situation to him, and together they decided to launch the offensive. The Germans were in force because the 1st Company's 8 Tigers were there to back the Panzer-Lehr-Division's Panzer IVs and attack the town. One troop was already in Villers. It was commanded by Second Lieutenant Philipsen whose tank n° 111 was hit and stopped in the town. Avoiding the partly blocked highway, they went down from the Côte des Landes, across the fields parallel to the old Caen road and the railroad, ploughing through hedges and knocking down fences, and entered Villers via the Evrecy road.

Through the basement window of the building which is now n° 50 Rue Clemenceau, Monsieur Clément Bernouis, with concern and some fear his family and some neighbors, including Dr. François, saw the Tigers thunder past. While the few Germans in the town center avoided fighting, there were plenty standing ready on the outskirts cautiously looking in buildings for any soldiers that might have scattered during the morning's fighting. A small unit approached the house and entered the coal cellar in the basement, supposed to be the safest place to take shelter by those who went down there. When the first German entered, he was intrigued by the characteristic smell of British cigarettes distributed that morning by the British. The soldier informed an officer, who inspected the cellar and questioned the occupants as to the presence of a British soldier among them. The suspicious officer would not believe Dr. François' denials, so had everyone leave and then strafed the

Cette photo, prise peu après les combats, nous montre le monument aux morts situé entre la poste et la mairie. Les chars de Bill Cotton vont se regrouper au nord de ce monument et les hommes du 1/7th Queens vont bloquer, au sud, le Tiger qui voulait s'attaquer aux chars de Cotton. (BA.)

This photo, taken shortly after the battle, shows the war memorial, located between the post office and the Town Hall. Bill Cotton's tanks were to muster to the north of this monument and the men of 1/7th Queens to the south would cut off the Tiger trying to attack Cotton's tanks. (BA.)

Le piège anglais va fonctionner

Renforcés par des unités de la *Panzer-Lehr-Division* et confiants en la puissance de leurs Tigres, les Allemands vont attaquer Villers de deux côtés, semble-t-il, avec pour objectif principal la possession du nœud routier que constitue la localité et ne pas laisser derrière leurs lignes une menace anglaise importante. Ils vont ainsi lancer leurs chars vers le centre de la ville, par la rue Pasteur, vers l'Hôtel de Ville et la place Jeanne d'Arc et, à travers le quartier de la gare où la construction est disséminée, vers la route d'Aunay. Ils vont devoir affronter pendant deux heures des Britanniques véritables « *casseurs de chars* », en position défensive, qui vont se montrer redoutables pour les panzers.

Le capitaine Möbius a décidé de lancer deux sections de chars sur Villers, une, composée de 2 Tigres et 1 Panzer IV de la *Panzer-Lehr-Division*, rescapé d'une unité malmenée par les *Queen's*, vers le centre de la ville, et une section de Tigres vers le sud. La première section descend la rue Clemenceau, longe la place Richard-Lenoir et arrive rue Pasteur. C'est dans cette période qu'un Tigre aurait abattu une maison d'angle simplement en pénétrant dans l'immeuble pour neutraliser un canon antichar qui le menaçait. Dans la zone de l'Hôtel de Ville, le lieutenant Cotton et son équipage, à pied, Horne, Grant et Bramall, dans leurs chars, attendent que les chars allemands, dont ils entendent le bruit des moteurs, se présentent devant eux, là où ils ont tendu leur piège.

Au moment où le *Panzer IV* va atteindre la place, le caporal Horne va exécuter une manœuvre hardie que Cotton n'a encore jamais vue. Le lieutenant la raconte ainsi « *lorsque le chef de char du* Cromwell *voit apparaître à environ vingt mètres de lui, à l'angle formé par la rue Pasteur et la rue qui longe l'Hôtel de Ville, le* Panzer IV *suivi d'un Tigre, il se glisse entre les deux Allemands, envoie à bout portant un obus sur le* Panzer IV *qui s'arrête en flammes et se précipite à reculons dans la rue proche, avant que le Tigre qui suit et qui commence à diriger son canon vers lui ne puisse l'atteindre* ». Les Tigres répondent en envoyant des obus perforants vers le char anglais à travers les maisons, puis cessent leurs tirs croyant sans doute avoir eu raison du *Cromwell*.

Le premier Tigre s'engage sur la place, présentant son flanc gauche au *Firefly* de Bramall qui le touche avec un obus de 17 livres au niveau de la chenille gauche « *car le char, privé de direction, va s'écraser dans le magasin de confection Lepiètre dont la vitrine s'écroule sous le choc* (rue Pasteur, n° 64 actuel). *Des fantassins lancent alors des grenades sur le char qui prend feu. Aucun survivant ne s'échappe* ».

Trop près d'une cible éventuelle et gêné par la fumée, Bramall a été dans l'obligation d'adopter une technique empirique pour ajuster son tir. Cette façon de faire consiste à fixer un point précis à travers la lumière du tube vide, d'immobiliser le canon dans cette position et de tirer lorsque l'objectif arrive au point fixé. En l'occurrence, ce point devait être situé sur la façade d'un des immeubles de la rue Pasteur, dans une porte, une fenêtre par exemple.

Selon beaucoup d'auteurs, le Tigre mis hors de combat était celui de l'*Oscha* (adjudant) Ernst Krieg de la compagnie 1. Or ce nom ne figure pas dans la liste des tankistes de cette compagnie, ni même dans celle du bataillon. Par contre, on trouve le nom de l'*Oscha* Heinrich Ernst, char 113, dans la liste des tués à Villers le 13 juin 1944.

Deux chars étaient passés. Il devait en rester encore un entre les deux rues latérales couvertes par les chars anglais et des canons antichars. Cotton va à

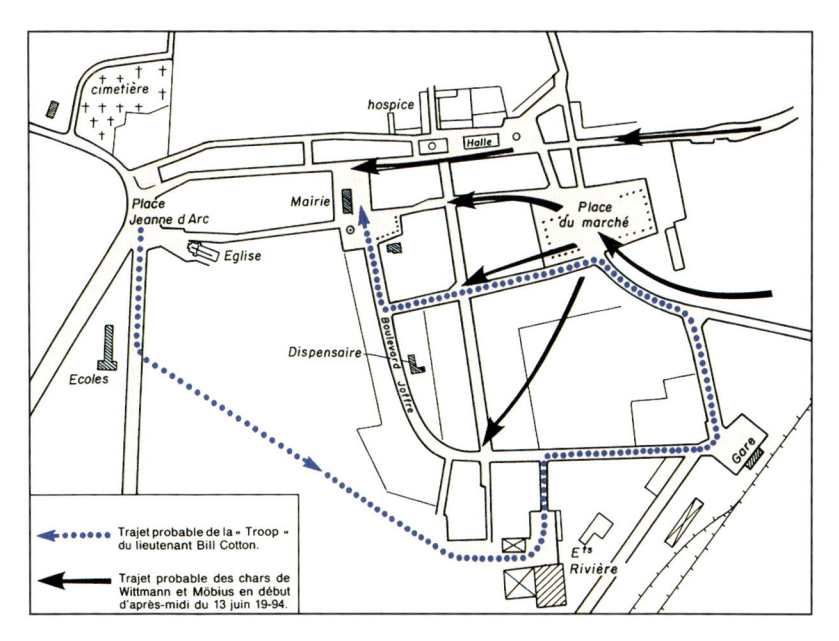

Nous voyons ici les axes probables de la contre-attaque et le trajet de la *Troop* de Bill Cotton passée devant les établissements Rivière et la gare. (Heimdal.)

coal heap with bursts of machine-gunfire. The noise was heard by a young man who ran off panicstricken, shouting all over the place that the Bernouis family had been shot in the cellar. At any other time, a raid like this would probably have ended in tragedy for the refugees.

The British trap works

Reinforced by some Panzer-Lehr-Division units and trusting in the power of their Tigers, the Germans attacked Villers from two sides, with their main objective apparently being to hold the major crossroads in the locality and to avoid having a serious British threat behind their lines. Accordingly, they launched their tanks on the town center, via the Rue Pasteur, towards the Town Hall and the Place Jeanne d'Arc and, through the station quarter with only scattered buildings, on the way to the Aunay road. For two hours they faced the formidable British tank destroyers in defensive positions, which gave the Panzers a hard time.

Captain Möbius decided to launch two tank troops on Villers, one composed of 2 Panzer-Lehr-Division Tigers and 1 Panzer IV, a survivor from a unit badly mauled by the Queen's, towards the center of the

Ci-dessous : position approximative de la *Troop* de Bill Cotton en embuscade près de la mairie avec son char **(1)**, les Cromwell du *Corporal* Horne **(2)**, du *Sergeant* Grant **(3)**, du Firefly du *Sergeant* Bramall **(4)** et du 6 pdr **(5)** des Queens.

Here we see the likely lines of the counterattack and the route taken by Bill Cotton's troop as it passed in front of the Rivière factory and the station. (Heimdal.)

Below: the approximate position of Bill Cotton's troop waiting in ambush near the Town Hall with his tank **(1)**, the Cromwells of Corporal Horne **(2)** and Sergeant Grant **(3)**, Sergeant Bramall's **(4)** Firefly and the Queens 6 pdr **(5)**.

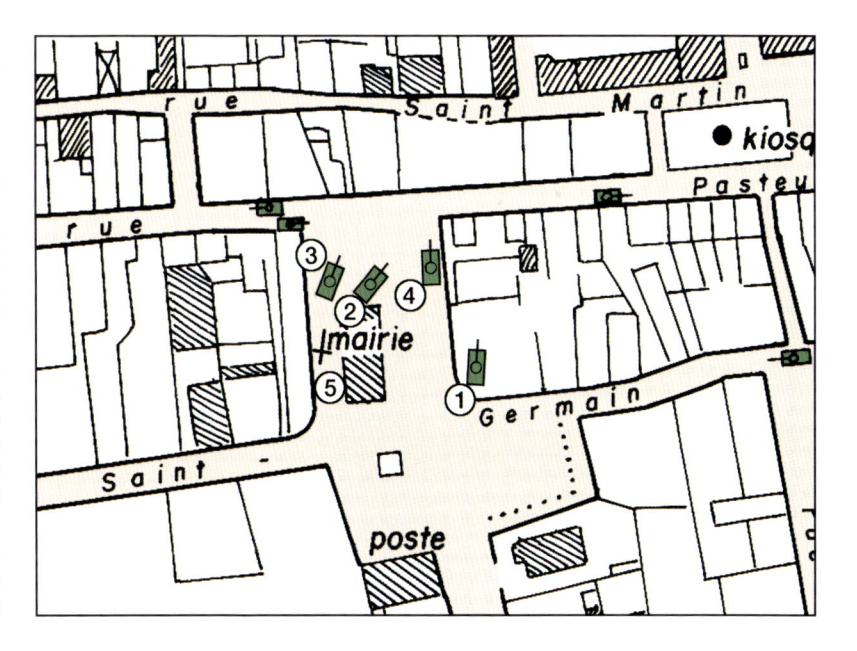

Page ci-contre : ces deux illustrations d'artistes britanniques évoquent bien la violence de la bataille de chars dans Villers-Bocage, surtout la vue en couleurs. Mais la seconde vue, extraite de l'*Illustrated London News* du 1er juillet 1944, est plus fantaisiste. Le char britannique est un Crusader, engin utilisé par la *7th Armoured Division* dans le désert mais plus en Normandie. Etant donné la position de ce char, la rue principale devrait descendre et ne pas monter, etc. (Coll. W. Theffo.)

Opposite page: these two illustrations by British artists, especially the color picture, give a good idea of the violence of the tank battle in Villers-Bocage. But the second picture, taken from the Illustrated London News dated July 1st 1944, is not so realistic. The British tank is a Crusader, a tank the 7th Armoured Division used in the desert but no more in Normandy. Given this tank's position, the main street should be going down, not up, etc. (Coll. W. Theffo.)

---

pied voir ce qui se passe dans la rue Pasteur et preuve de son sang-froid et de son flegme, comme la pluie commence à tomber, il se protège sous un parapluie récupéré dans les ruines voisines. En fait, il voit deux Tigres l'un derrière l'autre, l'un d'eux canon tourné vers Caen, et se demande comment il va les obliger à descendre vers leur objectif. Un officier de renseignement du *7th Queen's* en position dans un immeuble va utiliser une technique enseignée dans les écoles d'infanterie, qui consiste à jeter sur le char une couverture imbibée d'essence et de l'enflammer. Le char répond en chargeant vers la fenêtre d'où vient l'attaque et écroule une partie du mur de l'immeuble.

« *Les deux Tigres sont menacés par une unité anglaise armée de* PIAT *qui leur expédie des obus à charges creuses depuis les maisons mais les charges ne peuvent pénétrer les épaisses parois d'acier. Enfin, au bout d'un moment, le premier Tigre se décide à avancer et se présente à découvert devant la place de l'Hôtel de Ville. Un obus de 75 mm le touche sur le côté mais il ne semble pas en tenir compte. Il s'arrête et recule pour se mettre à l'abri. Un obus de* PIAT *tiré de près le touche à l'arrière, mais ne pénètre pas* » explique Cotton. Le char « touché au vif » fait alors un trou dans le bâtiment de l'angle de la rue pour se ménager un espace de tir en direction des chars anglais. Andrew Rae, un *trooper* de la *4th Troop* comprend la manœuvre. Il voit le canon de 88 puis la tourelle, tire et touche le Tigre qui recule mais n'est pas mis hors de combat. Le Tigre fait feu vers le *Cromwell* et le manque. Au même moment, il est touché par un obus de 17 livres et 2 obus de 75. Finalement le char allemand termine sa course, chenille arrachée près du *Panzer IV*. Le *Panzer IV* était immobilisé devant la chapellerie Jouin et « le café du Centre » Fessard à peu près devant les numéros 37 et 39 de l'actuelle rue Pasteur. Le 2e Tigre a été détruit à côté. Il semble s'agir du Tigre n° 112. Côte à côte, les deux chars allemands, bouchent presque complètement la rue et empêcheront les autres chars de passer.

Le troisième Tigre serait le char de Wittmann. Le lieutenant Cotton a vu le char arrêté, avant et canon dirigé vers Caen. Il a tout d'abord pensé que le Tigre descendait la rue en reculant comme pour se protéger d'une attaque anglaise venant de l'arrière. Le voyant immobile depuis un certain temps et le canon abaissé, devant le magasin Huet-Godefroy, il y a mis le feu. Personne n'est sorti du char, l'équipage l'avait abandonné. On peut se demander à quel moment il a été abandonné. Soit le matin, arrêté par un obus de canon antichar qui l'aurait touché à une chenille à son retour vers la cote 213, après avoir affronté Lockwood, soit l'après-midi, au moment où, arrêté par un projectile, il aurait essayé de faire demi-tour pour essayer de regagner les lignes allemandes. De toute évidence, selon la version allemande, aucun char de la 2e compagnie du bataillon 101, celle de Wittmann, n'a participé à l'attaque sur Villers l'après-midi, puisque le groupe était constitué de 2 Tigres de la 1re compagnie et d'un *Panzer IV* de la *Panzer-Lehr-Division*. Le Tigre auquel Cotton a mis le feu ne peut donc être que le char que Wittmann a utilisé le matin.

Le récit de ces diverses actions est basé sur un rapport de Cotton qui les a dirigées et même qui en a pris des photos avant de regagner son escadron. Il y a eu d'autres narrations, différentes les unes des autres. Toutes les unités présentes dans le centre de Villers en cet après-midi du 13 juin, ont participé à leurs réussites, les tankistes de la *4th Troop*, leurs *Cromwell* et leur *Firefly*, les fantassins du *7e Queen's* avec leurs canons antichars, leurs *PIAT* et même leurs mortiers. Mme Thorel se souvient « des hourras poussés par les servants du mortier installé chez

---

city, and a troop of Tigers towards the south. The first troop went down the Rue Clemenceau, skirted the Place Richard-Lenoir, arriving in the Rue Pasteur. It was around this time that a Tiger is thought to have knocked down a corner house just by entering the building to take out an anti-tank gun which was threatening it. In the Town Hall sector, Lieutenant Cotton and his crew, on foot, and Horne, Grant and Bramall, in their tanks, waited until the German tanks, whose engines they could hear, passed in front of them where they had set their trap.

At the very moment the Panzer IV reached the spot, Corporal Horne did one of the bravest things that Cotton had ever seen. This is how the lieutenant tells the story : "He came out of a side street and entered the German column in the middle, knocked out one of their tanks, and then dashed up a side street before the German tanks behind him knew what had happened." The Tigers replied by firing armor-piercing shells at the British tank through the houses, then ceased fire, presumably thinking they had knocked out the Cromwell.

The first Tiger moved into the square, presenting its left flank to Bramball's Firefly which hit it with a 17-pdr shell on the level of the left track and the "Tiger went free-wheeling down the street out of control, plunged through a shop front [Lepiètre's clothes store, the current n° 64 Rue Pasteur] and was followed in by some of our infantry chucking hand grenades. None of the Germans escaped from that."

Too close to a possible target and hindered by the smoke, Bramall had to adopt an empirical technique to adjust his range. This involved lining up on a precise point through the daylight of the empty tube, immobilizing the gun in this position and then firing when the target reached this set point. In fact, the point must have been on the frontage of one of the buildings in the Rue Pasteur, something like a doorway or window.

According to quite a few authors, the Tiger that was knocked out belonged to Oscha (Adjutant) Ernst Krieg of 1st Company. However this name does not appear in the list of the company's tank crews, or in the battalion's for that matter. On the other hand, the name Oscha Heinrich Ernst, tank 113, does appear in the list of killed at Villers on June 13th 1944.

Two tanks had passed. There must be one more between the two side streets covered by the British tanks and anti-tank guns. Cotton went on foot to see what was going on in the Rue Pasteur, and, as the rain began to fall, cool, calm and collected, he stood under an umbrella he had picked up in the nearby ruins. What he in fact saw were two Tigers one behind the other, one with its gun turned towards Caen, and he wondered how he was going to coax them down towards their target. An intelligence officer of 7th Queen's in position in a building used a technique learnt at infantry school, namely to throw a blanket soaked in gasoline at the tank and set fire to it. The tank responded by charging at the window from which the attack had come, demolishing part of the wall of the building.

The two Tigers were «being persecuted from the rooftops by a British unit armed with a Piat gun, which bounced concussion shells of the top of their tanks. The tops of the tanks were too thick to be knocked in, but after a while one of the tanks could no longer stand it and made a dash for the intersection, where a 75mm shell cracked it. For a moment crossing the street it seemed to take the shell in its stride" explained Cotton. Stung, the tank then made a hole in the building on the street corner through which to open up on the British tanks. One member of no. 4 Troop, Andrew Rae, figured out what it was up to. He saw

M. Doublet et du pouce levé par l'un d'eux pour manifester sa joie auprès des civils réfugiés dans la cave ». (L'action de Cotton a été relatée dans un article signé Tom Treanor dans le *News Chronicle* du 19 juin 1944).

Pour atteindre leur but et contrôler ainsi l'accès aux routes qui aboutissent à la place Jeanne d'Arc, les routes de Caumont, de Vire, d'Aunay, Möbius et Wittmann ont envoyé deux sections de chars. En effet, la longueur des canons de 88 des Tigres les empêchant de manœuvrer facilement dans les rues étroites et les ruelles, il leur semblait plus rationnel de les faire évoluer et progresser là où l'habitation est plus clairsemée, c'est-à-dire, dans la partie sud de la ville, à travers les jardins depuis la voie ferrée jusqu'au boulevard Joffre et la route d'Aunay.

C'est dans ce secteur que le *7ᵉ Queen's,* grâce à ses canons antichars, a pu revendiquer trois Tigres mis hors de combat et deux autres touchés. Le matin, vers 10-11 heures, une section de *Queen's* avait installé une pièce de 57 mm antichar dans le petit chemin qui longe la cour de l'atelier de menuiserie de M. Bertou. La chenillette qui la tractait, était placée à l'abri de la maison d'habitation, hors de vue depuis le boulevard Joffre. Le canon était dirigé vers la rue Jeanne Bacon de façon à pouvoir la contrôler et la prendre en enfilade, tout véhicule apparaissant boulevard Joffre, rue Jeanne-Bacon, rue Emile-Samson et même rue Saint-Germain pouvait être atteint.

De cette position trois Tigres ont été touchés, deux ont été mis hors de combat, le 3ᵉ a pu être récupéré. L'un, après avoir traversé un herbage et un ou deux jardins, apparaît rue Jeanne-Bacon entre deux pavillons. Touché immédiatement et incontrôlable il traverse la rue puis le jardin de M. Bertou et s'arrête en flammes sur le trottoir boulevard Joffre. Ce devait être le char n° 122 dont le chef de char, le sergent Arno Salamon bien que blessé, a pu s'échapper. Rétabli, Salamon est devenu chef d'un char *Tiger II* et sera le seul à ramener de France en septembre 1944 son *Königstiger*. Il sera ensuite versé en

La Troop du lieutenant Bill Cotton cinq jours après les combats.
1. De gauche à droite, les *Troopers* Arthur Nelson et William Leonard, le *Sergeant* A. Gordon.
2. Le lieutenant Bill Cotton (à gauche) avec son équipage (de gauche à droite), le *L/Corporal* Hodgeson, le *Trooper* W.Jones, le *L/Corporal* Payne et le *Trooper* Humphreys. Un personnage de *cartoon* a été peint à l'avant du char. (Photos IWM.)

Lieutenant Bill Cotton's troop five days after the battle.
1. *From left to right, Troopers Arthur Nelson and William Leonard, Sergeant A. Gordon.*
2. *Lieutenant Bill Cotton (left) with his crew (left to right), L/Corporal Hodgeson, Trooper W. Jones, L/Corporal Payne and Trooper Humphreys. A cartoon character has been painted on the front of the tank. (Photos IWM.)*

décembre à la 1re compagnie du bataillon de chars lourds 501.

Un autre qui descendait la rue Emile-Samson venant de la place du marché est mis hors de combat au carrefour de la rue Jeanne-Bacon à 200 mètres environ de l'emplacement du canon. Personne ne semble en être sorti. André Marie, charcutier, a entendu les munitions exploser durant une bonne partie de l'après-midi.

Le troisième qui venait de la place du marché par la rue Saint-Germain, 100 mètres plus loin, seulement touché, a pu être remis en état.

Le résultat marquant de cette attaque de chars allemands pour la reprise de Villers est que leurs pertes ont été de 6 chars Tigre, 3 dans la ville, 2 rue Jeanne-Bacon, 1 près de la gare, 2 *Panzer IV* et quelques autres endommagés, en partie à cause des canons antichars de 57 mm à munitions « sabot » des fantassins du *7th Queen's*. Les Allemands ont cru que la puissance de leurs panzers aurait été suffisante pour investir complètement une ville sans s'appuyer sur une infanterie d'accompagnement, tout comme les Anglais le matin ont laissé derrière leurs chars leur compagnie d'infanterie motorisée dont les soldats avaient été instruits, tels des « voltigeurs », pour combattre avec eux et les protéger. En effet, la défense rapprochée des chars étant faible, l'infanterie doit être là pour nettoyer le terrain de l'infanterie ennemie et l'occuper.

Le lieutenant-colonel Gordon qui commandait le *7th Queen's* a relaté dans son rapport, écrit le lendemain, le travail effectué par ses compagnies. Il a rapporté qu'un *« travail particulièrement excellent avait été fait par le Major French et une partie de sa compagnie C et par le capitaine Beard et une partie des carriers qui traquaient les chars depuis les rues transversales et les maisons et les attaquaient avec des PIAT et des charges creuses. La compagnie A était aux abords de la gare »*.

the 88 mm gun, then the turret, fired and hit the Tiger which recoiled but was not disabled. The Tiger fired back at the Cromwell and missed. At the same time, it was hit by a 17-pdr shell and two 75 mm shells. The German tank ended up with its track torn off, next to the Panzer IV. The Panzer IV was brought to a standstill in front of Jouin's hat shop and Fessard's Café du Centre in front of what is now number 37 and 39 Rue Pasteur or thereabouts. The second Tiger was destroyed nearby. This would seem to have been Tiger n° 112. Side by side, the two German tanks were almost completely blocking the street, preventing the other tanks from passing.

The third Tiger would have been Wittmann's tank. Lieutenant Cotton saw the tank stationary, with its gun facing forward towards Caen. At first he thought that the Tiger was reversing down the street to protect itself from a British attack from the rear. But after seeing it was still motionless after a while with its gun lowered, in front of Huet-Godefroy's store, he set fire to it. Nobody baled out, the crew had already abandoned the tank. One may wonder exactly when it had been abandoned. Either that morning, stopped by an anti-tank gun shell that hit it in the track as it returned to Point 213, after confronting Lockwood, or during the afternoon, when it might have tried to turn round and get back to the German lines after being stopped by a projectile. Clearly, according to the German version, no tank belonging to Wittmann's 2nd Company, 101st Battalion, took part in the afternoon attack on Villers, since the group consisted of 2 of the 1st Company's Tigers and one of the Panzer-Lehr-Division's Panzer IVs. The Tiger Cotton set fire to can therefore only be the tank that Wittmann used in the morning.

The account of these various actions is based on a report by Cotton who directed them and even took a few photographs before rejoining his squadron. There were other different versions. All the units present in the center of Villers on that afternoon of June 13th had a hand in their successes, the no. 4 Troop tank crews, their Cromwells and Fireflies, the infantrymen of the 7th Queen's with their anti-tank guns, PIATs and even mortars. Madame Thorel remembers "cheers from the men serving the mortar at Monsieur Doublet's and one of them giving a happy thumbs up to the civilians sheltering in the basement". (Cotton's action was reported in an article signed Tom Treanor in the News Chronicle of June 19th 1944).

Après que la rue principale eut été « *embouteillée, il fut ordonné aux compagnies de reprendre leur position initiale pour couvrir les sorties du village. Il fut estimé que la tâche était très difficile car les hommes étaient très dispersés dans les maisons d'un bout à l'autre de la ville. Plus tard dans l'après-midi, la menace des chars allemands avait été brisée et de sérieux problèmes consistaient maintenant en quelques poches d'infanterie qui essayaient de trouver leur route à travers le village. La compagnie A avait été refoulée de sa position initiale et il se trouvait une large trouée entre le sud et la bordure nord-est qui était couverte par aucune de nos compagnies et aucune compagnie était disponible pour le faire* ».

L'avant-garde de la *7th Armoured Division* n'est pas seule à subir la contre-attaque allemande. Le reste de la division restée en réserve, route de Caumont jusqu'à Amayé-sur-Seulles, doit répondre aux assauts des Allemands au nord et au sud de la route. Le *11th Hussars* est là pour protéger la colonne. C'est ainsi qu'à Tracy-Bocage, il doit intervenir pour déloger un nid de mitrailleuse lourde.

L'attaque de panzers sur Villers ayant échoué, les Anglais sont toujours maîtres de la ville mais leur zone se rétrécit et les Allemands ne peuvent laisser derrière leurs lignes un tel îlot de résistance sans réagir. Les Britanniques savent que des renforts ennemis arrivent de tous côtés. Du nord, la *Panzer-Lehr-Division* a envoyé en plus des *Panzer IV* de la 6ᵉ compagnie du IIᵉ bataillon du régiment 130, deux canons de 88, 3 canons de campagne et quelques troupes d'arrière-garde rassemblées par le lieutenant-colonel Kurt Kauffmann. Du sud, vont arriver dans l'après-midi, venant de Briouze, deux bataillons de grenadiers de la *2. Panzer-Division*. L'objectif, de cette division autrichienne du général von Lüttwitz, est de renforcer l'aile gauche de la *Panzer-Lehr-Division* lorsque ses chars l'auront rejoint et de tenter une percée entre le secteur américain et le secteur anglais pour séparer les armées alliées.

En fin d'après-midi, la situation devient préoccupante pour les Britanniques. Au sud-ouest, la *2. Panzer-Division*, même dépourvue de ses chars mais avec le support d'artillerie lourde, attaque en force routes d'Aunay et de Vire avec l'intention évidente de couper la route à une éventuelle retraite alliée.

Le lieutenant K.S. Berelowitz, des forces sud-africaines défend ce secteur, avec sa section de l'escadron B. Il inflige de lourdes pertes à l'assaillant qui réussit à s'approcher à quelques mètres de son char. Sautant de son tank, il court, sous un feu nourri, vers une position d'infanterie proche et obtient l'assistance d'un détachement de mortiers qu'il ramène avec lui, place sur le terrain et qu'il dirige jusqu'à ce que, devant l'importance de ses pertes, l'ennemi soit contraint à la retraite. Par son action, le lieutenant a sans aucun doute sauvé l'axe central de la brigade et la *Military cross* lui sera attribuée plus tard par le commandant en chef du *21st Army Group*.

Le lieutenant-colonel Gordon continue ainsi : « *de petits éléments d'infanterie se sont glissés à proximité des chars du 4th CLY. La route principale depuis l'entrée du village jusqu'au reste de la Brigade est entièrement sous le feu de l'artillerie lourde qui fait sauter un détachement entier de mortiers et un carrier* ».

La rue principale est sous le feu d'un *Panzer IV* de la *Panzer-Lehr-Division* placé dans le virage du haut de la rue G. Clemenceau jusqu'à ce qu'il soit détruit. Dans le secteur de la gare, une section du *7th Queen's* est faite prisonnière. Dans la confusion qui s'installe, les fantassins qui se battent sous la pluie et qui subissent des tirs d'obus de gros calibre réclament des renforts mais ne les obtiennent pas et ne

To reach their target and control access to the roads ending up in the Place Jeanne d'Arc, the Caumont, Vire and Aunay roads, Möbius and Wittmann sent in two tank troops. With their long 88 guns, the Tigers were not easy to drive through the narrow streets and lanes, and so it made more sense to take them forward along a less built-up part of the town, i.e. to the south, through gardens from the railway line to the Boulevard Joffre and the Aunay road.

It was in this sector that the 7th Queen's claimed to have destroyed three Tigers and hit two more with their anti-tank guns. That morning, at around 10-11, a troop of Queen's had set up a 57 mm anti-tank gun in the lane skirting the yard of Monsieur Bertou's joinery shop. The Bren gun carrier towing it was put in a safe place behind the house, out of sight from the Boulevard Joffre. The gun was aimed towards the Rue Jeanne Bacon to cover the street and enfilade any vehicle entering the Boulevard Joffre, the Rue Jeanne Bacon or the Rue Emile Samson; even the Rue Saint-Germain came within range.

From this position three Tigers were hit, two were put out of action, and the third was recovered. One came out between two houses in the Rue Jeanne Bacon after passing through a meadow and a garden or two. It was hit immediately and spun out of control, crossing the street and then Monsieur Bertou's garden, coming to a stop in flames on the sidewalk of the Boulevard Joffre. This must have been tank n° 122 whose commander, Sergeant Arno Salamon, was able to bale out although wounded. On recovering, Salamon became commander of a Tiger II tank and in September 1944 became the only man to bring his Königstiger back from France. He was assigned in December to the 1st Company, 501st Heavy Tank Battalion.

Another, coming down the Rue Emile Samson from the marketplace, was knocked out at the corner of Rue Jeanne Bacon about 200 meters from where the gun was. Nobody seems to have baled out. Pork-

3. Bill Cotton exhibe un trophée et porte un blouson de la *Luftwaffe* sur lequel est encore accroché une croix de fer. (IWM.)

3. Bill Cotton exhibits a trophy and wears a Luftwaffe jacket on which an Iron Cross is still pinned. (IWM.)

Le « 634 » est alors détruit dans le virage. (BA.)

Tank "634" was then destroyed in the bend. (BA.)

peuvent contenir l'infiltration allemande. Les pertes anglaises sont inquiétantes d'autant que les poches extérieures à la ville s'amenuisent et que de nouvelles troupes allemandes arrivent. Villers est difficile à défendre et à conserver.

Un petit détachement s'est aventuré dans la direction de Vire. Au hameau de la poste, à Maisoncelles-Pelvey, il est stoppé brutalement et laisse sur le terrain un char, un véhicule léger et une moto. Il est obligé de revenir par Villers, le bourg de Tracy étant occupé par les Allemands qui observent la campagne du haut du clocher.

« La question se posait de savoir si nous devions tenir la localité cette nuit ou non. Sans davantage de compagnies d'infanterie, ce n'était pas possible et le commandant de la Brigade donna des ordres pour que les bataillons et les escadrons du 4th CLY se replient sur la hauteur à l'ouest du village d'où nous étions partis le matin et ensuite rejoignent sur une position de base ferme, le reste de la 22ᵉ Brigade Blindée. Ce retrait était organisé et couvert par le 4th CLY. Il s'ensuivit un bombardement assez intense le long de la route mais les pertes furent très faibles », rapporte le lieutenant-colonel Gordon.

Le *Major* Aird qui commande l'escadron B prend le commandement du *4th CLY* et le *Major* E.P. Mac Coll devient commandant en second tandis que le capitaine F.A. Jarvis prend la tête de l'escadron B.

L'état-major prend sa décision et le *Brigadier* Hinde donne le signal du repli vers l'ouest de Villers. Il est demandé aux derniers chars de l'escadron B et à l'escadron C resté en réserve, d'organiser le départ et d'en fixer l'heure, afin de les couvrir par des tirs d'obus de 25 livres anglais et des tirs de calibre 155 américains expédiés depuis les environs de Caumont. Le tir de barrage sera dirigé vers la zone des virages de la côte de la Vierge noire.

Pendant ce temps, les Allemands qui ont fait beaucoup de prisonniers les rassemblent route de Caen avant de les diriger vers l'arrière. Malheureusement les Américains ont allongé leur tir et cinq soldats de la *Rifle Brigade* qui se jettent dans les fossés pour se protéger sont tués par leurs gardiens qui ont mal interprété leur geste.

Place Jeanne d'Arc, Stan Lockwood a reçu l'ordre du *Major* Aird de prendre la tête de la colonne de chars et de fantassins. Il devra se mettre en route à une heure précise. Lockwood se met en place, démarre et se dirige vers la route de Caumont. Soudain, pour une raison inconnue, le char s'arrête et bloque la rue étroite à cet endroit, le conducteur ne peut remettre le char en route. C'est l'affolement, de courte durée heureusement. Bob Moore qui suit, réagit immédiatement et glisse son *Firefly* devant celui de Lockwood. Les deux chefs de char sautent à terre, décrochent les câbles d'acier fixés sur la coque de leurs engins et les fixent aux crochets de remorque prévus à cet effet au bas de la tourelle du char en panne. Bob Moore dit à son conducteur de démarrer et le convoi peut se mettre en route. En 15 secondes, la route est dégagée et l'opération de retraite peut continuer malgré les tirs des mitrailleuses allemandes qui ne sont pas restées inactives pendant la durée de cet incident. Bob qui était titulaire de la *Military Medal* devait être coutumier de telles actions, son copain Stan aimait le rappeler.

Desmond Gordon replie ses hommes jusqu'aux environs de la cote 134, sur la route de Caumont après le hameau du Mesnil de Tracy-Bocage. Sa mission est de fixer quelque temps, pour recueillir les hommes de ses compagnies, éparpillées dans Villers et les fantassins de la *Rifle Brigade* dispersés le matin. Il retrouve à cet endroit, une base défensive solide constituée du *5th RTR* commandé par Joe Lever et

butcher Andre Marie heard its ammunition exploding for most of the afternoon.

The third tank, which came from the marketplace down the Rue Saint-Germain, 100 meters further on, was just hit and was repaired.

The major outcome of this German tank attack for the recapture of Villers was that it had cost 6 Tiger tanks, 3 in the town, 2 in the Rue Jeanne Bacon, 1 near the station, 2 Panzer IVs and a few others damaged, partly thanks to the 7th Queen's infantry's 6 pdr. anti-tank guns firing "sabot" ammunition. The Germans believed that the power of their panzers would be sufficient to gain full control of the town without the need for supporting infantry, just as the British that morning had left their motorized infantry company behind their tanks which the soldiers had been trained like "acrobats" to fight alongside and guard. In fact, tanks being poor defenders at close quarters, the infantry needed to be there to clear the area of enemy infantry and to occupy it.

Lieutenant-Colonel Gordon, commander of the 7th Queen's, noted the work carried out by his companies in his report submitted the following day. He reported how "some particularly fine work was done by Major French and a party from 'C' Coy and Capt. Beard and a party from the Carriers in stalking these tanks through the side streets and houses and tackling them with PIATs and sticky bombs."

"Having made certain that the enemy tanks in the main street had been well and truly bottled up, Coys were ordered to get into the original dispositions to cover the approaches into the village. This was found to be a most difficult task, as personnel of Coys had got very scattered throughout the houses in the village. In the meantime 'A' Coy were sweeping the ground in the further outskirts and were directed on the railway station".

The vanguard of the 7th Armoured Division was not the only unit to face the German counter-attack. The remainder of the division held in reserve along the Caumont road as far as Amayé-sur-Seulles had to fend off German attacks to the north and south of the road. The 11th Hussars were there to guard the column, and were called to action at Tracy-Bocage to dislodge a heavy machine-gun nest.

The panzer attack on Villers having failed, the British continued to hold the town but their lodgement was dwindling and the Germans could not just leave this pocket of resistance behind their lines without doing something about it. The British knew that enemy reinforcements were arriving from all directions. From the north, the Panzer-Lehr-Division sent, in addition to the Panzer IVs of the 6th Company of the 2nd Battalion 130th Regiment, two 88 mm guns, 3 field guns and some rearguard troops brought together by Lieutenant-Colonel Kurt Kauffmann. Arriving that afternoon from Briouze further south were two grenadier battalions of the Panzer-Division. The objective of this Austrian division commanded by General von Lüttwitz was to reinforce the Panzer-Lehr-Division's left flank when his tanks joined up, and to attempt to break out between the American and British sectors in order to drive a wedge between the Allied armies.

Late that afternoon, the position became alarming for the British. To the south-west, the 2. Panzer-Division, although without its tanks but with the heavy artillery in support, attacked in force along the Aunay and Vire roads with the clear intention of cutting off a possible Allied withdrawal.

Lieutenant K.S. Berelowitz, of the South-African forces, was defending this sector, with his section of "B" Squadron. He inflicted heavy losses on the attacker who managed to come within a few yards of his

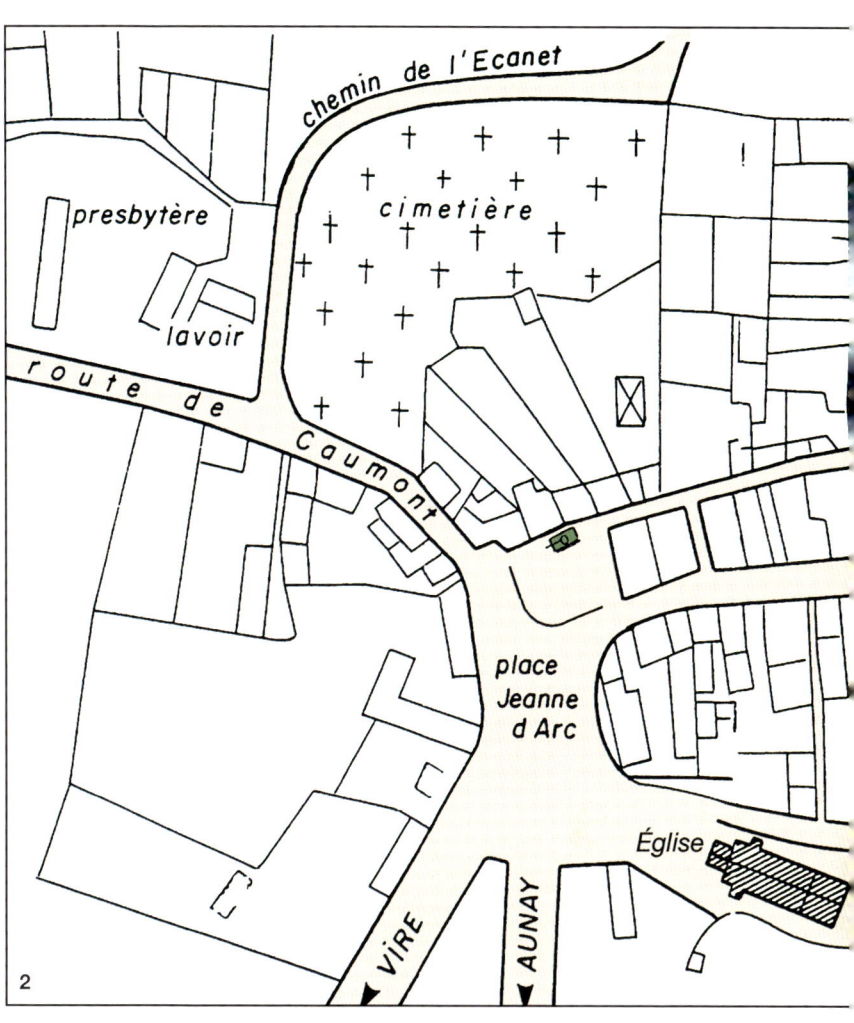

*In the hurry to withdraw, B Squadron 4th CLY abandoned one of its Cromwells in the Place Jeanne d'Arc. The tank had probably broken down (**photo 1** - After the Battle). Here we see a detail of the map of the bottom of Villers (**2** Coll. H. Marie) locating the tank's position, with a present-day photograph (**3**) of the building in front of which the tank came to a halt, as reconstructed after the bombardment. (EG/Heimdal.)*

Dans la précipitation de la retraite, le *B Squadron* du *4th CLY* a abandonné un de ses Cromwell place Jeanne d'Arc. On peut supposer que l'engin était en panne (photo 1 - *After the Battle*). Nous voyons un détail du plan du bas de Villers (**2** coll. H. Marie) situant l'emplacement de ce char et la vue actuelle (**3**) du bâtiment devant lequel le char était arrêté, reconstruit après le bombardement. (EG/Heimdal.)

de deux batteries du *5th RHA*, celle commandée par Tim Lanyon et celle commandée par Andrew Burn, installées le long du ruisseau du Coudray. Cette base aurait peut-être pu être maintenue si elle avait reçu du renfort d'unités d'infanterie. Elle restera en place jusqu'au lendemain.

A 20 heures, l'infanterie allemande attaque depuis le bourg de Tracy le flanc gauche de la colonne qui se replie. L'escadron C du *4th CLY*, commandé par le capitaine K.H. Hiscock, se porte en avant pour briser cette offensive de l'ennemi qui devra reculer en subissant de lourdes pertes. Ce n'est qu'à 22 h 30 que la situation sera rétablie et que l'escadron C pourra reprendre sa position de réserve.

---

tank. He jumped out of his tank, ran under heavy fire to a nearby infantry position, and obtained help from a mortar detachment which he brought back with him, put in position and coordinated until the enemy was forced to withdraw after sustaining heavy losses. Through his action, the lieutenant had unquestionably saved the brigade's center line, and he was later awarded the Military Cross by the Commander-in-Chief of the 21st Army Group.

Lieutenant-Colonel Gordon continues as follows: "small infantry elements had worked their way round to within close proximity of the tank of 4 CLY. Bn.HQ was almost pinned to the ground and the road leading into the village from the remainder of the Bde. was under quite heavy Arty fire, which knocked out a complete detachment of the Mortars and 'brewed up' a carrier".

The main street came under fire from one of the Panzer-Lehr-Division's Panzer IVs placed in the bend at the top of the Rue Clemenceau, until it was destroyed. In the station sector, a troop of 7th Queen's was taken prisoner. In the ensuing confusion, infantry fighting in the rain under large caliber shellfire called out for reinforcements, but as none were forthcoming they failed to contain the German infiltration. The British losses were all the more worrying as the pockets outside the town were shrinking and fresh German troops were arriving on the scene. Villers was hard to defend and hold onto.

A small detachment ventured in the direction of Vire. At the hamlet La Poste, at Maisoncelles-Pelvey, it was stopped in its tracks and had to leave behind a tank, a light vehicle and a motorcycle. It made its way back via Villers, as Tracy was occupied by the Germans, who could survey the area from the top of the bell-tower.

"A decision had to be made as to whether we could hold the village that night or not; without further infantry Coys it did not appear possible and the Bde.Comd. issued orders for the Bn. and Sqn 4 CLY to withdraw on to the high ground west of the village from where we had started that morning and then join up in a firm base position with the remainder of 22 Armd.Bde. This withdrawal was successfully carried out covered by the 4 CLY. It attracted a certain amount of shelling along the road but casualties were very few", reported Lieutenant-Colonel Gordon.

Major Aird, commander of "B" Squadron, assumed command of the 4th CLY and Major E.P. McColl became second-in-command, while Captain F.A. Jarvis took over command of "B" Squadron.

Headquarters made its decision and Brigadier Hinde gave the signal to withdraw to west of Villers. The last tanks of "B" Squadron, and of "C" Squadron held in reserve, were requested to organize the retreat and set a time for it, in order to cover them with British 25-pdr and American 155 mm shellfire from the Caumont area. The barrage was to be aimed at the bends up the Vierge Noire hill.

Meanwhile on the Caen road, the Germans were gathering up the numerous prisoners they had taken, before sending them off to Venoix. Unfortunately, the Americans lengthened their range and five men of the Rifle Brigade were killed by their guards who misunderstood their reason for throwing themselves into the ditch: to take cover.

In the Place Jeanne d'Arc, Stan Lockwood was ordered by Major Aird to lead the column of tanks and infantry. It was to start out at a precise time. Lockwood got into position and set off towards the Caumont road. Suddenly, for no apparent reason, the tank stalled and blocked the narrow street at this point, and the driver was unable to get the tank going.

Pages 1
103, 104

Pages 106, 107, 108, 109, 110, 111.

Firefly sergent Lockwood

chars cadron B 4 CLY

Ecoles

It was panic stations, but fortunately not for long. Bob Moore just behind reacted quickly and slipped his Firefly in front of Lockwood's. The two tank commanders jumped to the ground, unhooked the steel tow ropes fastened onto the hull of their machines and hitched them to the purpose-designed tow hook at the bottom of the turret of the broken-down tank. Bob Moore called to his driver to start and so got the convoy moving. 15 seconds later, the road was clear and the withdrawal operation could proceed despite continuous German machine-gunfire throughout this incident. For Bob, who had already won a Military Medal, such deeds were routine, as his buddy Stan liked to point out.

Desmond Gordon withdrew his men to the area around Point 134, on the Caumont road after the hamlet of Le Mesnil at Tracy-Bocage. His assignment was to hold out for a while until he could gather up the men of his companies who were scattered in Villers, and the Rifle Brigade infantry who had spread out during the morning. Here he found a firm defensive base made up of the 5th RTR led by Joe Lever and two batteries of the 5th RHA, those commanded by Tim Lanyon and Andrew Burn, in position along the Coudray stream. This base might perhaps have been held had it received some infantry units in reinforcement. As it is, it stayed in position until the next day.

At 20.00, from Tracy itself the German infantry attacked the left side of the column as it pulled out. "C" Squadron 4th CLY commanded by Captain K.H. Hiscock went on ahead to break this offensive by the enemy, who were forced back and sustained heavy casualties. The situation was not restored until 22.30, when "C" Squadron was able to return to its position in reserve.

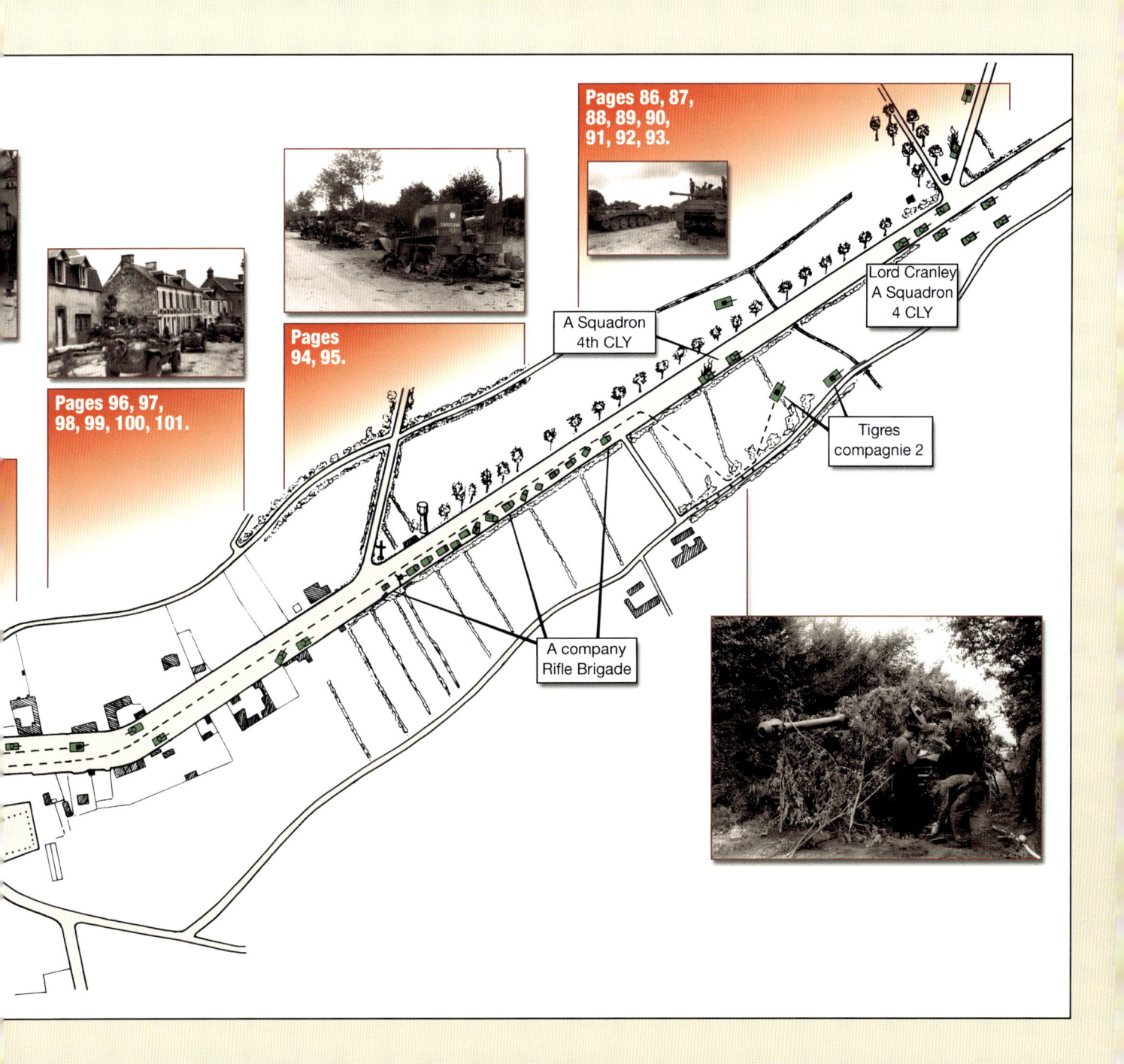

La cote 213

1. Sur cette carte, on voit la position de la 2ᵉ compagnie de Wittmann à proximité de la colonne britannique se dirigeant vers Caen et, de l'autre côté de la route nationale, la 1ʳᵉ compagnie de Möbius. On voit aussi la direction de l'attaque de Wittmann. (Heimdal.)

Point 213

1. On this map, we see the position of Wittmann's 2nd Company near the British column moving towards Caen, and Möbius's 1st Company on the other side of the highway. We also see the direction of Wittmann's attack. (Heimdal.)

2, 3 and 5. On these three photographs, we see Cromwell tanks of A Squadron 4th County of London Yeomanry (4th CLY, also known as the Sharp-Shooters). Its commander was Lieutenant-Colonel Cranley and near Point 213 it came under German artillery fire and the tanks were damaged and abandoned. Their crews were captured following the arrival of the 1st Company commanded by SS-Hauptsturmführer Möbius. (Bundesarchiv photographs.)

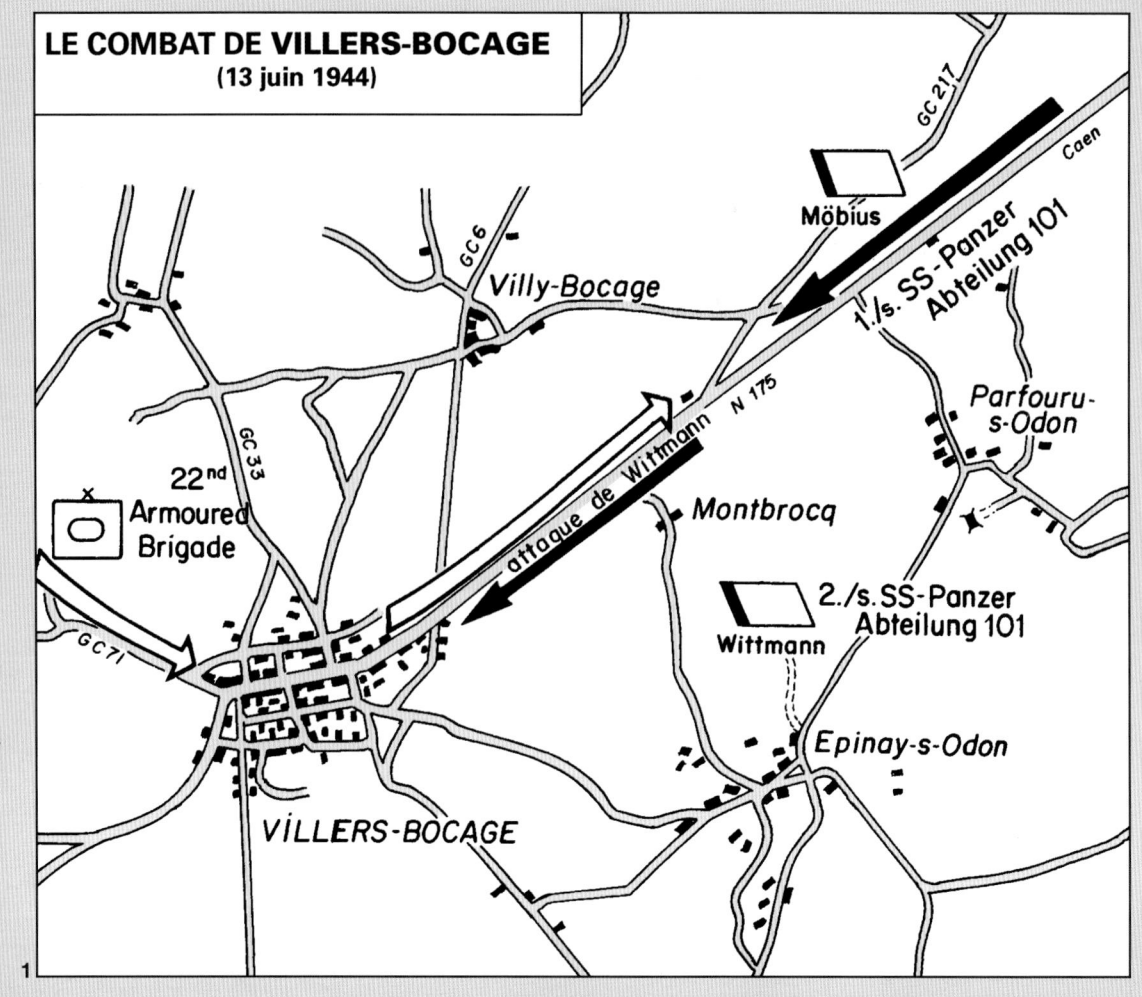

LE COMBAT DE VILLERS-BOCAGE
(13 juin 1944)

2, 3 et 5. Sur ces trois photos, nous voyons des chars Cromwell du *A Squadron* du *4th County of London Yeomanry* (*4th CLY*, on les appelle aussi « *Sharp-Shooters* »). Ils étaient commandés par le lieutenant-colonel Cranley et ont subi près de la cote 213 le feu de l'artillerie allemande et ont été endommagés et abandonnés. Leurs équipages ont été capturés après l'arrivée de la 1ʳᵉ compagnie commandée par le *SS-Hauptsturmführer* Möbius. (Photos Bundesarchiv.)

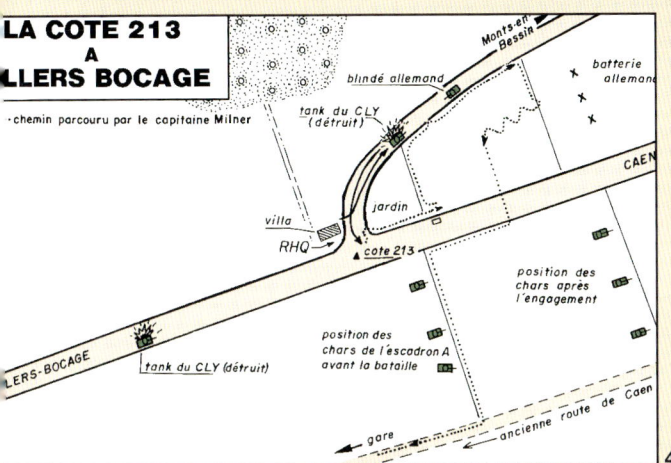

4. Ce plan reprend les indications données sur le croquis du capitaine Milner (voir page suivante) avec l'indication des combats sur la cote 213. (Heimdal.)

4. This map carries the indications given on Captain Milner's sketch (see following page) and shows the engagements on Point 213. (Heimdal.)

6. Le magazine « Signal » a ainsi présenté l'action de Wittmann avec son Tiger dévalant vers Villers-Bocage en détruisant sur son chemin des chars britanniques. (Coll. Heimdal.)

6. This is how the magazine "Signal" presented Wittmann's action with his Tiger sweeping down on Villers-Bocage destroying British tanks as it went. (Heimdal Coll.)

4th CLY, A Squadron, N°4 Troop

Cote 213

1. Nous voyons ici le char Sherman Firefly qui était commandé par le sergent Norman Jones, tourelle tournée vers l'arrière, « à six heures ». On aperçoit en arrière la maison située près de l'embranchement de la petite route menant vers Monts-en-Bessin. (BA.)

2. Le même endroit actuellement, la maison a été modifiée. (E. Groult/Heimdal.)

3. Croquis réalisé par le capitaine Milner, commandant en second de la *A Company* de la *Rifle Brigade* expliquant « sa » bataille de Villers-Bocage. Il montre le secteur de la cote 213. En haut (à l'est) se trouvent des chars du *A Squadron* du *4th CLY* puis, dans le cercle, les chars détruits du *RHQ Party* (groupe de commandement) du *4th CLY* — ceux qu'on voit sur les photos de ces deux pages — puis des chars détruits du *CLY* et, enfin en bas (à l'ouest en bas de la cote), des engins de la *A Company* de la *Rifle Brigade*, moins des éléments sur la cote 213.

4. Nous voyons un autre aspect du char Firefly détruit du sergent Jones. Son avant est tourné vers Villers-Bocage et le canon est dirigé vers Caen. Il s'agit peut-être du char qui a fait un tête à queue pour répondre à Wittmann puis qui a fait pivoter sa tourelle pour diriger son canon vers Caen. Nous voyons

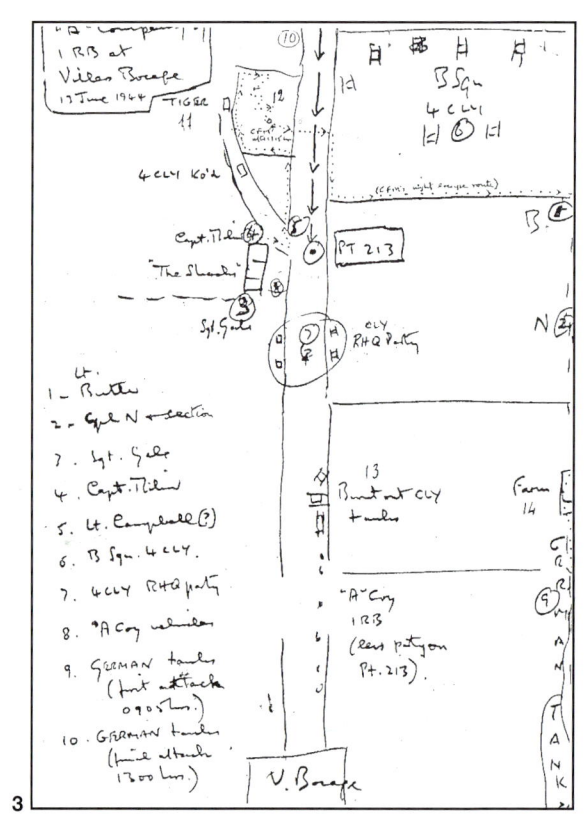

Point 213

1. Here we see the Sherman Firefly tank commanded by Sergeant Norman Jones, turret turned rearwards to "six o'clock". In the background is the house near the junction of the minor road to Monts-en-Bessin. (BA.)

2. The same place today, the house has been converted. (E Groult/Heimdal.)

3. Sketch drawn by Captain Milner, second-in-command of "A" Company The Rifle Brigade explaining "his" battle of Villers-Bocage. It shows the Point 213 sector. At the top (to the east) are the tanks of A Squadron 4th CLY, then in the circle, the destroyed tanks of 4th CLY's RHQ Party — the ones in the photographs on these two pages — then the CLY's destroyed tanks, and finally, at the bottom (to the west at the foot of the hill), vehicles of "A" Company The Rifle Brigade, less elements on Point 213.

4. Another view of Sergeant Jones' destroyed Firefly tank. Its front is turned towards Villers-Bocage and the gun is directed towards Caen. This may be the tank that wheeled round to answer Wittmann, and then swung round its turret to aim its gun towards Caen. We also see the inscription "Alla Keefek" (written in an Anglo-Arab pidgin peculiar to the men of the 7th Armoured Division) painted on the rear of the turret. (BA.)

5. Another photo taken from Point 213 with the Firefly tank on the right near the house and outbuildings. The destroyed Cromwells can be seen on the left. (BA.)

6. Side view of the house near Point 213 and the crossroads and outbuildings seen in the preceding photograph. (EG./Heimdal.)

aussi l'inscription « *Alla Keefek* » (écrite dans un sabir anglo-arabe propre aux hommes de la *7th Armoured Division*) peinte à l'arrière de la tourelle. (BA.)

5. Autre photo prise de la cote 213 avec le char Firefly sur la droite près de la maison et de ses dépendances. On aperçoit à gauche les chars Cromwell détruits. (BA.)

6. Vue latérale de la maison située près de la cote 213 et du carrefour et les dépendances aperçues sur la photo précédente. (EG./Heimdal.)

7. Depuis la cote 213 actuellement, la route dévale vers l'ouest, vers Villers-Bocage. (EG/Heimdal.)

7. From Point 213 as it is now, the road runs down westwards to Villers-Bocage. (EG/Heimdal.)

Près de la cote 213.
Sur ces photos, nous voyons d'autres blindés détruits près de la cote 213. Celui que nous voyons ci-dessous appartenait au *5th RHA*, c'était un char d'observation d'artillerie avancée comme ses marquages l'indiquent (*Captain* Roy Dunlop). Un tankiste du bataillon lourd (probablement de la 1ʳᵉ compagnie) examine les épaves de ces chars : le char d'observation d'artillerie déjà aperçu et le « Shufti Cush ». (BA.)

Near Point 213.
On these photographs, we see more tanks destroyed near Point 213. The one shown under belonged to the 5th RHA, its markings indicate that it was a forward artillery observation tank (Captain Roy Dunlop). A heavy tank battalion crew member (probably the 1st Company) examines the wreckage of these tanks: the artillery observation tank already seen and the "Shufti Cush". (BA.)

Ci-dessous : le corps du *Major* Scott, qui commandait l'escadron « A » du *4th CLY* gît dans le fossé ; il fut tué par des éclats d'obus. (BA.)

Below: the body of Major Scott, commander of "A" Squadron 4th CLY, lies in the ditch; he was killed by shrapnel. (BA.)

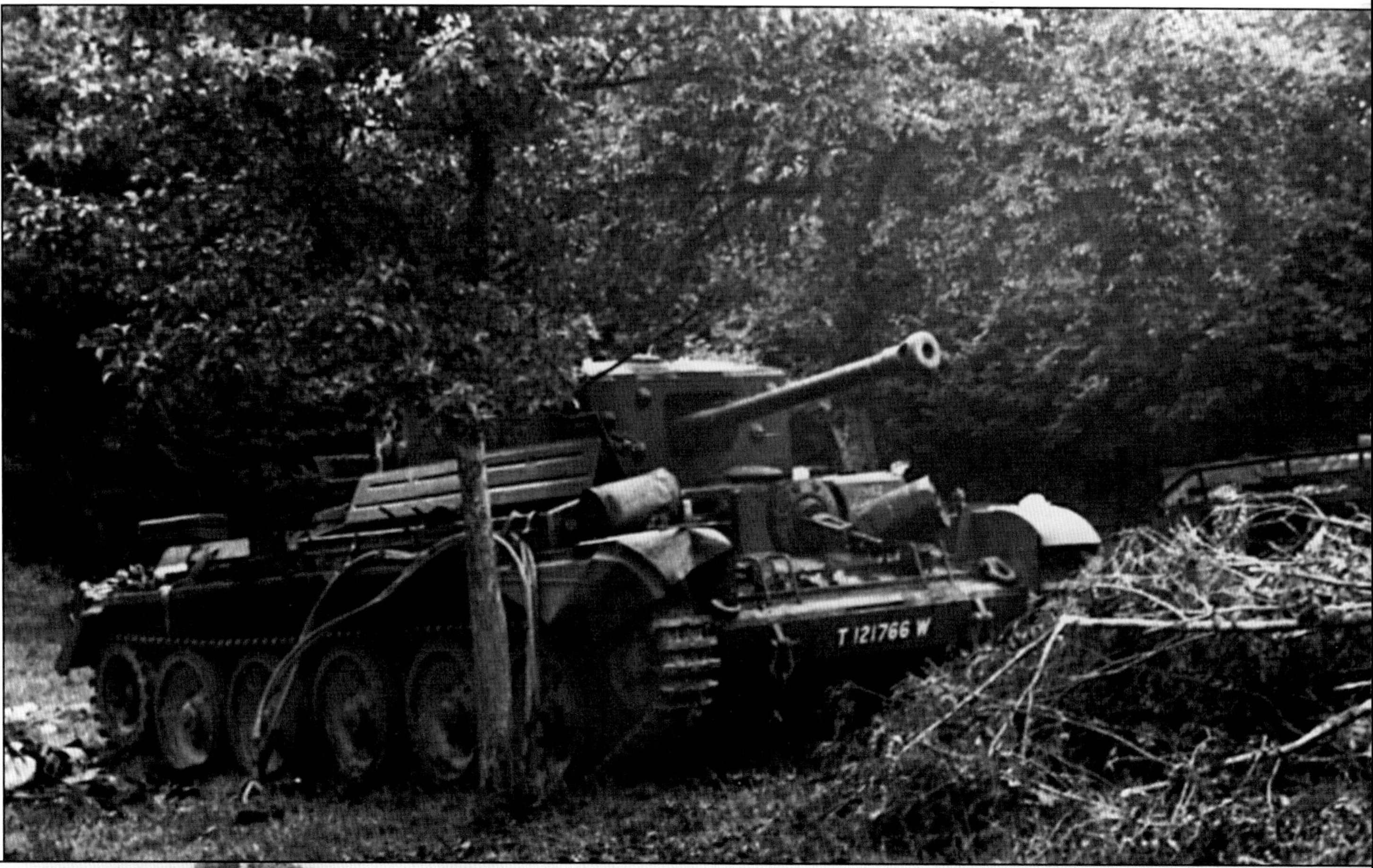
T 121766 W

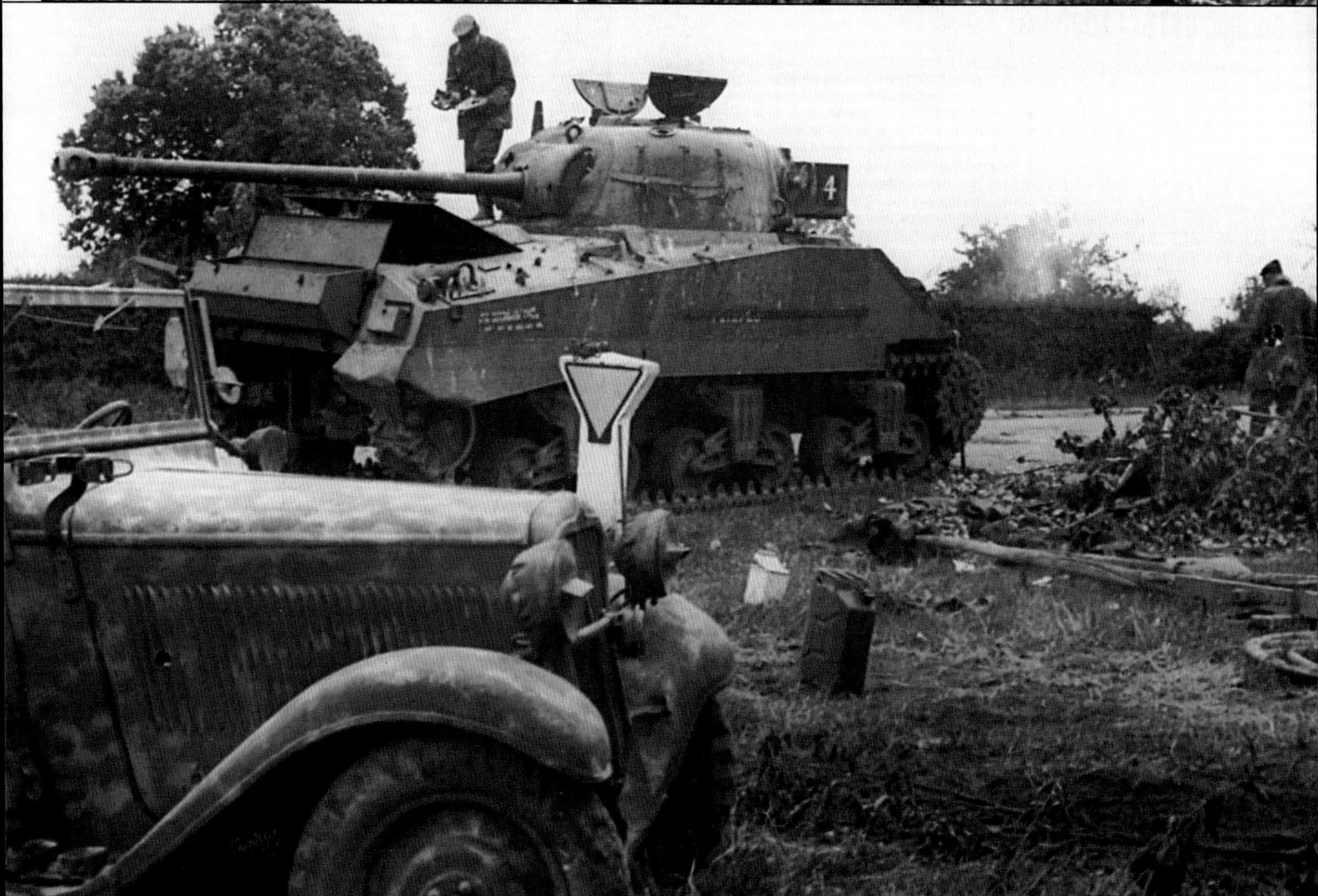
4

Toujours sur la cote 213

Nous voyons ici d'autres photos prises de la cote 213. Tout d'abord des chars Cromwell abandonnés dans le pré situé à l'est de la cote puis d'autres aspects du char Firefly et des chars Cromwell à proximité. Sur la photo du bas de la page ci-contre, on aperçoit un véhicule allemand d'état-major détruit par le lieutenant Bill Garnett lorsque le « A » *Squadron* a atteint la cote 213. (BA.)

Again at Point 213

Here we see more photographs taken from Point 213. First Cromwell tanks abandoned in the field to the east of the hill, followed by more views of the nearby Firefly tank and Cromwells. We see also (opposite page) the German staff car shot by Lieutenant Bill Garnett as « A » Squadron reached the point 213. (BA.)

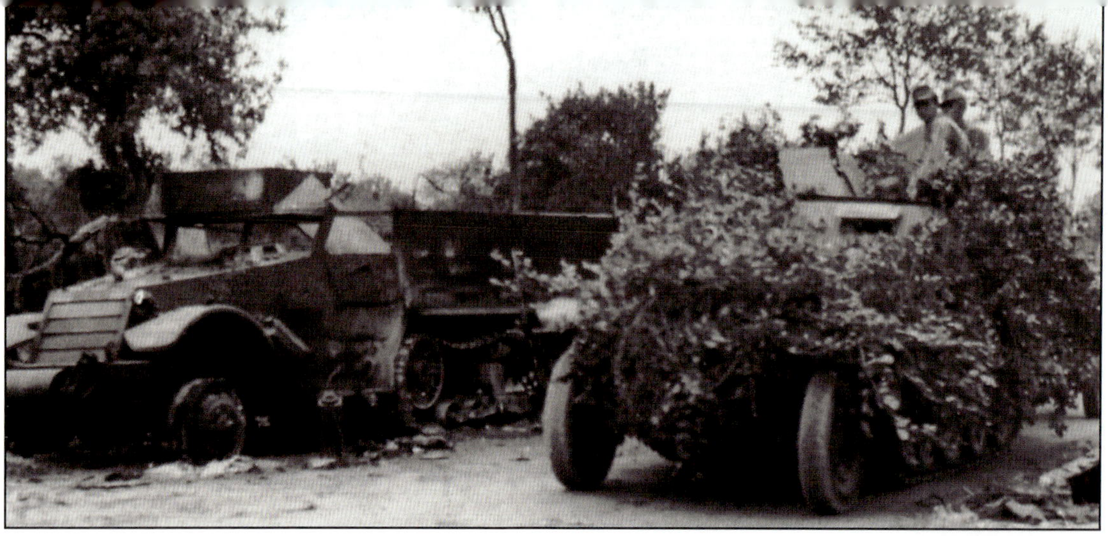

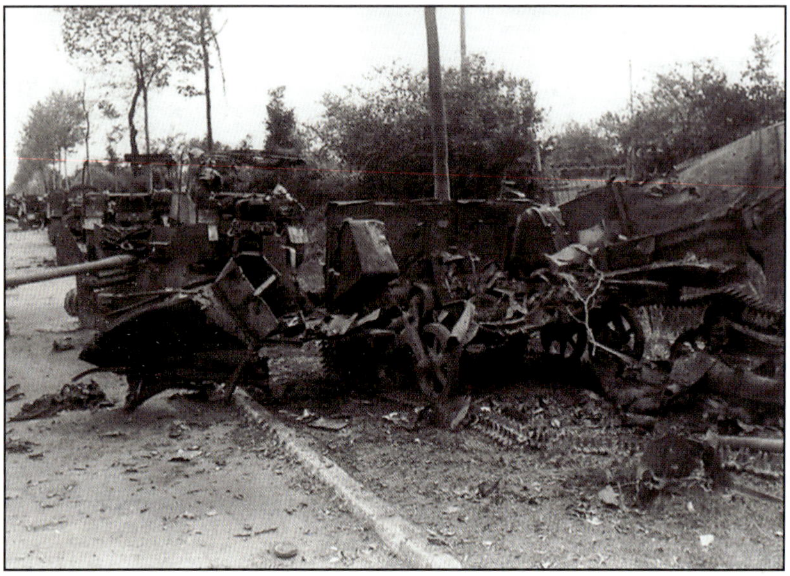

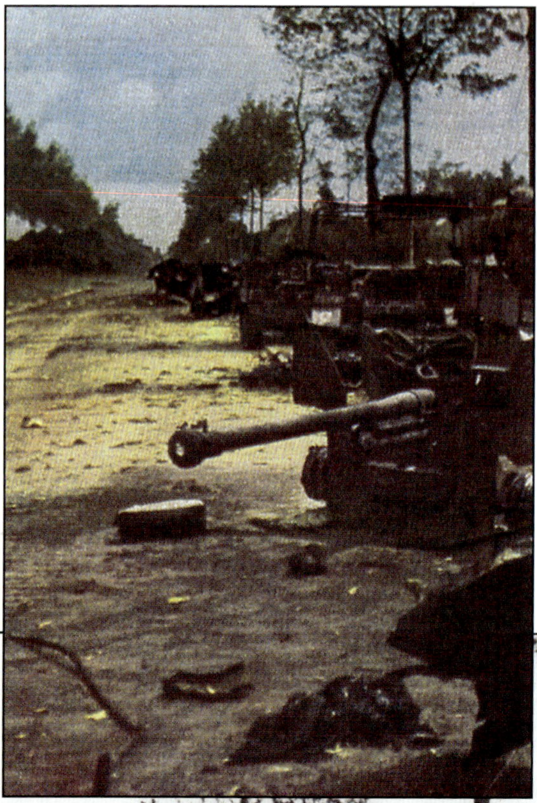

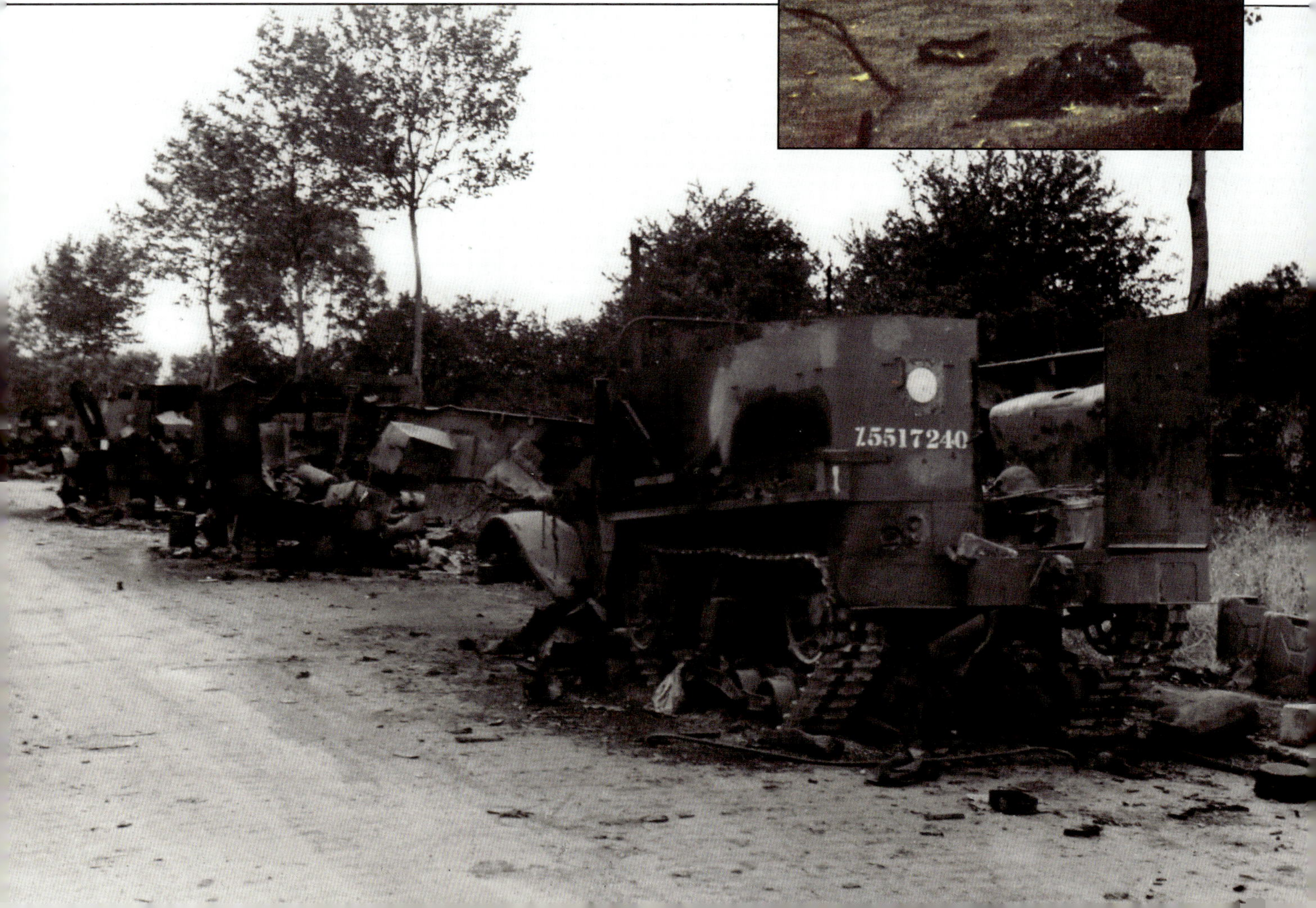
Z5517240

Pertes de la *Rifle Brigade*.

Suivons le correspondant de guerre allemand. En ayant quitté la cote 213 et en descendant vers l'ouest, vers le bourg de Villers-Bocage, nous tombons là sur la colonne de véhicules détruits de la *Rifle Brigade*. Ces photos nous montrent des half-tracks, des Loyd-Carriers et des Bren-Carriers, les premiers en tête de colonne et les *Carriers* derrière. Un canon antichar de 57 a été mis en place par les fantassins anglais ; il est dirigé vers la route de Bayeux, vers le calvaire. Nous voyons aussi l'entrée du bourg de Villers avec le fameux panneau « Caen 24 km ». Deux des photos sont en couleurs et proviennent de l'hebdomadaire « Signal ». (Photos BA et coll. Heimdal. pour les photos en couleurs.)

Rifle Brigade losses.

Let's follow the German war correspondent. Leaving Point 213 and going down westward towards Villers-Bocage, we come to the column of destroyed Rifle Brigade vehicles. These photographs show half-tracks and Loyd and Bren-carriers, with the half-tracks at the head of column and the carriers in the rear. A 6 pounder anti-tank gun has been set up by British infantry; it is aimed at the Bayeux road, towards the roadside cross. We also see the road into Villers with the famous signpost "Caen 24 km". Two of the photographs are in color and are taken from the weekly magazine "Signal". (BA photographs and Heimdal Coll. for the color photographs.)

95

96

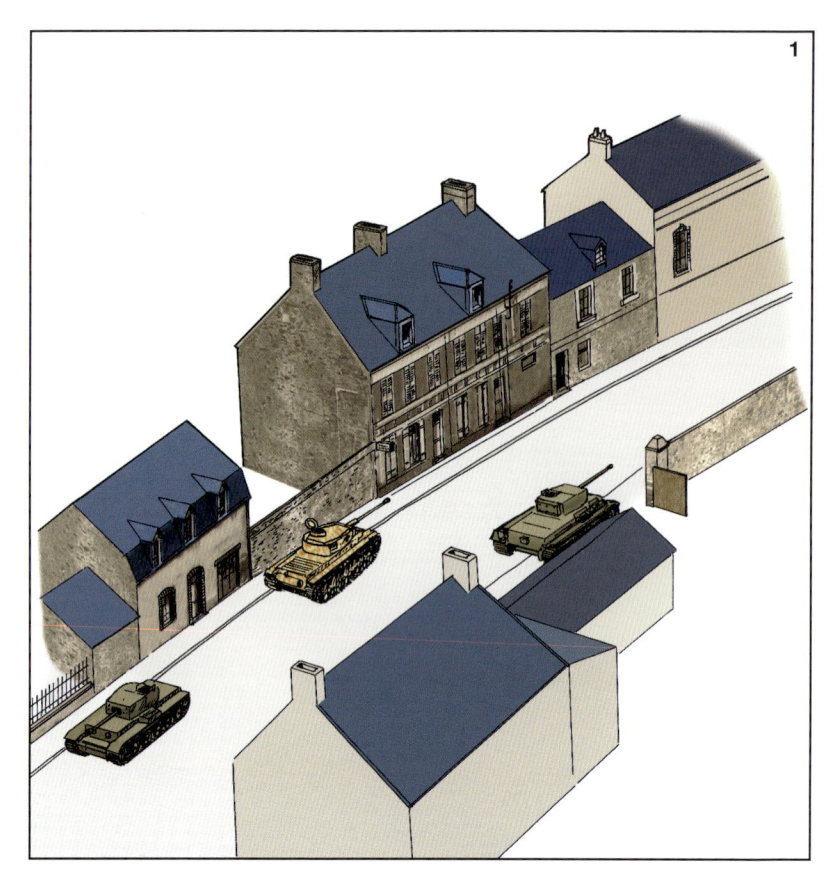

remarquons l'écusson de la famille von Schönburg-Waldenburg. Le prince von Schönburg-Waldenburg commandait le II° bataillon (équipé de Panzer IV, la *II./Panzer-Lehr-Regiment 130*) jusqu'à sa mort le 9 juin 1944. Il a ainsi fait peindre les armes de sa famille sur les tourelles des chars de son bataillon. (Photos BA, documents Heimdal.)

The road into Villers

Three tanks were destroyed in the bend in the Rue Clemenceau, going down towards the center of Villers-Bocage. They were, first, a Cromwell tank (marked "4" on the map in the following two-page spread) then a Panzer IV ("634") of II./Panzer-Lehr-Regiment 130 (photograph 2) destroyed there after it came to support the heavy battalion. Lastly, facing towards the town center, is Captain Dyas's wrecked Cromwell ("3" on the plan). On these photographs (3 and 4), we see several Schwimmwagen of I SS-Panzerkorps coming down the Rue Clemenceau and passing in front of the wrecks of the three tanks. Notice on the wreck of Panzer IV "634", the von Schönburg-Waldenburg family shield. Prince von Schönburg-Waldenburg was in command of the 2nd Battalion (equipped with Panzer IVs, II./Panzer-Lehr-Regiment 130) until his death on June 9th, 1944. He had his family's coat of arms painted on the turrets of his battalion's tanks. (BA photographs, Heimdal documents.)

Entrée du bourg

Dans la courbe de la rue Clemenceau, en descendant vers le centre du bourg de Villers-Bocage, trois chars ont été détruits. Il s'agit tout d'abord d'un char Cromwell (« 4 » sur le plan de la double page suivante) puis d'un Panzer IV (le « 634 ») de la *II./Panzer-Lehr-Regiment 130* (photo 2) venu appuyer le bataillon lourd et détruit là. Enfin, dirigé vers le centre du bourg, se trouve l'épave du Cromwell (« 3 » sur le plan) du capitaine Dyas. Sur ces photos (3 et 4), nous voyons plusieurs Schwimmwagen du *I. SS-Panzerkorps* descendre la rue Clemenceau en passant devant les épaves des trois chars. Sur l'épave du Panzer IV « 634 », nous

Char du capitaine Pat Dyas

Nous sommes maintenant devant la « Ferme Guéroult » où a été détruit le char Cromwell du capitaine Pat Dyas (« 3 » sur le plan) alors qu'il se dirigeait vers le centre de Villers-Bocage. (BA.) Les deux photos actuelles nous montrent la « Ferme Guéroult » très transformée actuellement : la maison sur la rue a disparu, le mur a été abaissé et l'entrée élargie. (EG Heimdal.). La maquette de la ville avant 1944 situe l'emplacement de ces lieux.

2. Cromwell, Lt. Cloudsley-Thompson.
3. Cromwell, Pat Dyas.
4. Cromwell, RSM Holloway.
5. Pz IV « 634 ».

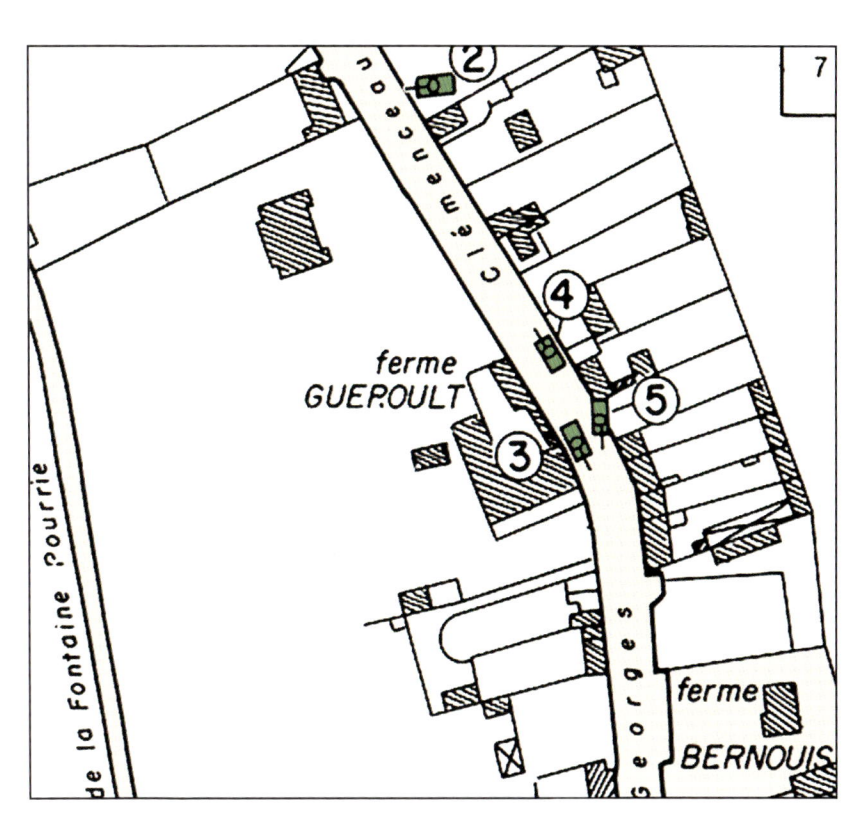

Captain Pat Dyas's tank
We are now in front of "Guéroult Farm" where Captain Pat Dyas's Cromwell tank was destroyed ("3" on the map) as it was heading towards the center of Villers-Bocage. (BA.) The two presentday photographs show "Guéroult Farm" now completely transformed: the house on the street has gone, the wall has been lowered and the entrance widened. (EG/Heimdal.). These places figure in the model of the town prior to 1944.

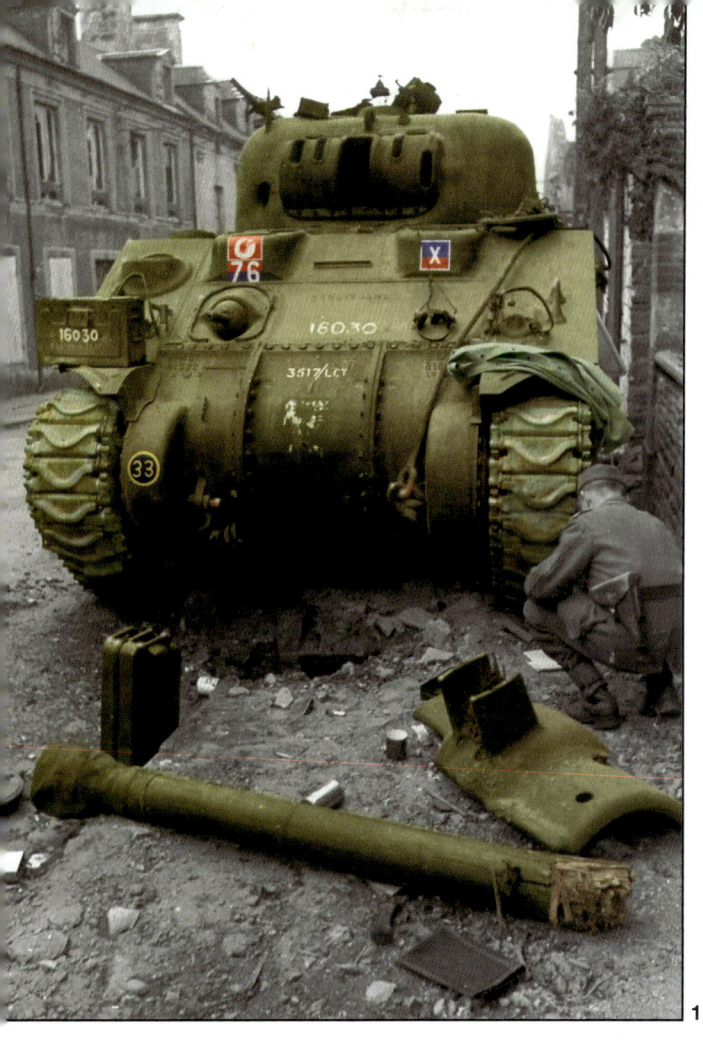

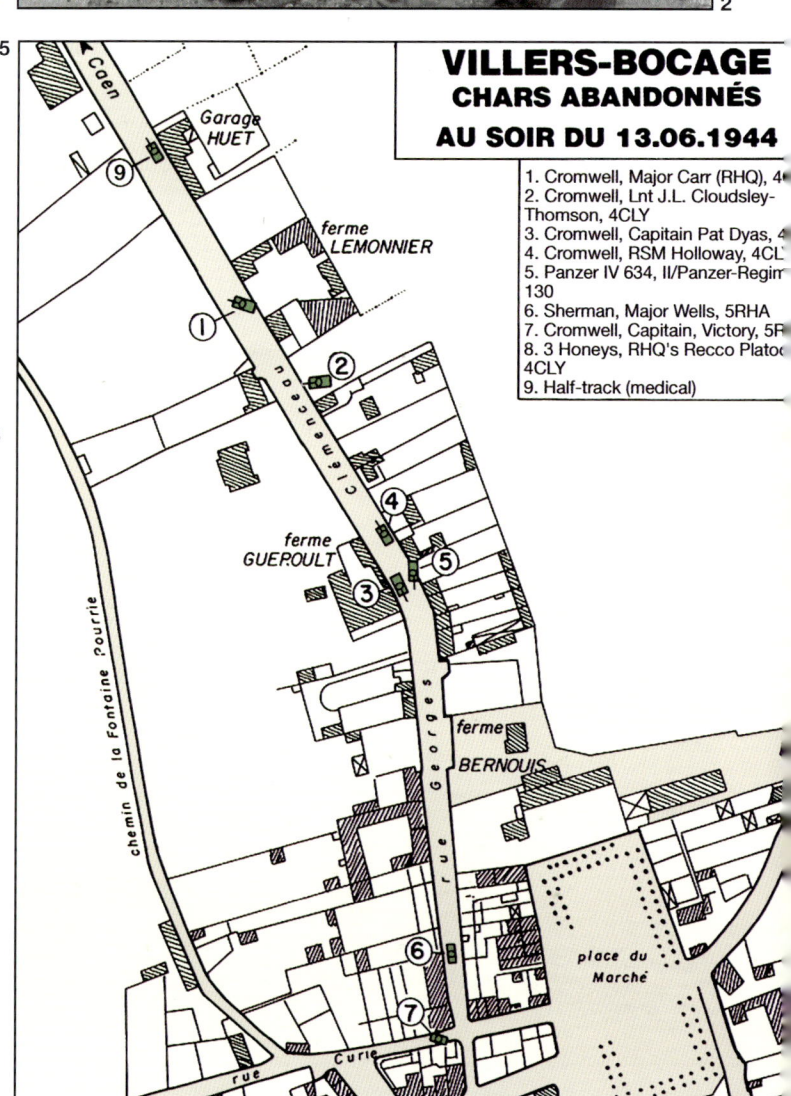

VILLERS-BOCAGE
CHARS ABANDONNÉS
AU SOIR DU 13.06.1944

1. Cromwell, Major Carr (RHQ), 4
2. Cromwell, Lnt J.L. Cloudsley-Thomson, 4CLY
3. Cromwell, Capitain Pat Dyas, 4
4. Cromwell, RSM Holloway, 4CLY
5. Panzer IV 634, II/Panzer-Regim 130
6. Sherman, Major Wells, 5RHA
7. Cromwell, Capitain, Victory, 5R
8. 3 Honeys, RHQ's Recco Plato 4CLY
9. Half-track (medical)

Rue Clemenceau et rue Curie

Un des chars Sherman d'observation (**photo 1**) du *5th RHA* détruit (*Major* Wells). Il était muni d'un canon fatice en bois visible par terre sur la photo. Ce canon en bois a été l'objet de moqueries de la part des soldats allemands et a rendu perplexe les Français qui le voyaient pour la première fois. Grâce à ce subterfuge, ce char ne se distinguait pas des autres ce qui lui évitait de devenir une cible privilégiée pour l'ennemi qui connaissait l'importance d'un poste d'observation en première ligne. Le point d'impact de l'obus que lui a décoché Wittmann est visible à gauche de la tourelle.

La photo suivante (**photo 2**) nous montre l'arrière de ce char (« 6 » sur le plan) puis l'Hôtel du Bras d'Or en allant vers le centre, un cadavre au premier plan. La photo suivante (**3**) nous montre le même endroit avant guerre avec l'Hôtel du Bras d'Or puis toujours le même endroit après le bombardement, avec le char. Le plan (**5**) montre les sept épaves dans le haut de Villers-Bocage. (BA **1 et 2**, Heimdal : **3 et 4**.)

Dans la route de Saint-Louet (rue Curie actuelle), le char Cromwell du *Captain Victory (5th RHA)* a brûlé et endommagé une maison qu'il a incendiée (**photo 6**, « 7 » sur le plan). Le bombardement du 30 juin a ravagé tout ce secteur et détruit les maisons restées intactes comme le montre ce cliché d'un correspondant de guerre britannique (**photo 7**) puis le 5 août au même endroit. (BA : **6**, IWM : **7**.)

Rue Clemenceau and Rue Curie

*One of the 5th RHA's destroyed observation Shermans (**photo 1**). It was provided with a dummy gun made of wood which can be seen in the photograph on the ground. The German soldiers laughed at this wooden gun and it puzzled the French, who had never seen one before. Through this subterfuge, the tank looked no different from any other, thus avoiding it becoming a prime target for the enemy, who knew the importance of an observation post in the front line. The point of impact of the shell which Wittmann let fly at it is visible to the left of the turret.*

*The following photograph (**photo 2**) shows the rear of this tank ("6" on the map) then heading towards the center, the Hôtel du Bras d'Or, and a corpse in the foreground. The next photograph (**3**) shows us the same spot as it was before the war, with the Hôtel du Bras d'Or, then the same spot once more, with the tank after the bombardment. The map (5) show the seven wrecks at the top end of Villers-Bocage. (BA **1** and **2**, Heimdal: 3 and **4**.)*

*In the Route de Saint-Louet (the present Rue Curie), the Cromwell tank belonging to Captain Victory (5th RHA) has brewed up and damaged a house it has set fire to (**photo 6**, "7" on the map). The bombardment of June 30th devastated the entire sector and destroyed any houses still standing as this picture by a British war correspondent shows (photo 7), and again the same spot on August 5th. (BA: **6**, IWM: **7**.)*

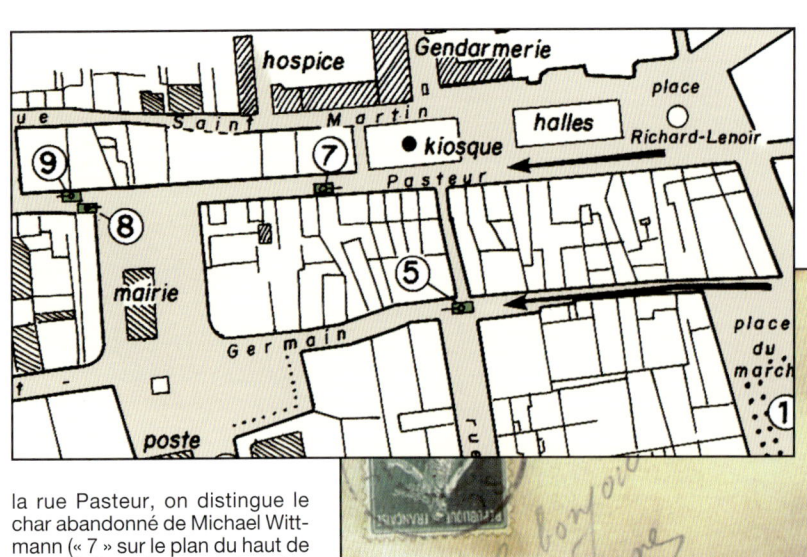

Vers la rue Pasteur

Au débouché de la rue Clemenceau, nous arrivons à la rue Pasteur, avec le kiosque qui fait face aux halles comme nous le voyons sur cette photo **(1)** prise avant la guerre. Un Schwimmwagen de la *SS-Panzer-Abteilung 101* passe entre les halles, situées derrière le photographe, et le kiosque qu'on distingue derrière les arbres **(photo 2)**. Plus loin, dans la rue Pasteur, on distingue le char abandonné de Michael Wittmann (« 7 » sur le plan du haut de la page). Il a été stoppé devant le magasin Huet-Godefroy. Puis le photographe s'avance dans la rue **(photos 3 et 4)**. On voit que Wittmann a abandonné son Tiger alors qu'il repartait dans la direction de Caen. Des soldats allemands déambulent dans la rue. Sur la dernière photo on aperçoit dans le fond les autres chars allemands détruits (« 8 » et « 9 » sur le plan) vers la hauteur de mairie. L'immeuble aux volets fermés était occupé par la gendarmerie militaire allemande. On remarquera aussi la pompe à essence. (Collection H. Marie : **1**, BA : **2, 3, 4**.)

Towards the Rue Pasteur

*Emerging from the Rue Clemenceau, we come into the Rue Pasteur, with the bandstand facing the market as we see it on this photograph **(1)** taken before the war. One Schwimmwagen of SS-Panzer-Abteilung 101 passes between the market, located behind the photographer, and the bandstand which can be made out behind the trees **(photo 2)**. Further down the Rue Pasteur, we see the tank abandoned by Michael Wittmann ("7" on the map at the top of the page). It was brought to a halt in front of Huet-Godefroy's store. Then the photographer moves on along the street **(photos 3 and 4)**. It can be seen how Wittmann left his Tiger when setting off again in the Caen direction. There are German soldiers walking up and down the street. On the last photograph the other destroyed German tanks ("8" and "9" on the map) can be seen in the background on a level with the town hall. The building with the closed shutters was occupied by the German military police. Notice also the gas pump. (Coll. H. Marie: **1**, BA: **2, 3, 4**.)*

1. Nous sommes toujours au même endroit et deux soldats du *I.SS-Panzer-Korps* patrouillent dans la rue Pasteur. Ils posent ici pour le correspondant de guerre à proximité de l'épave du char de Wittmann et de la pompe à essence.

2. Un cliché en couleurs est pris presqu'en même temps, il servira à illustrer une couverture du magazine « Signal » n° 16 de fin 1944.

3. Autre vue des soldats allemands patrouillant dans ce secteur de la grande rue après les combats.

4. Le même endroit en regardant dans l'autre direction, vers 1930. (BA : **1 et 3**, Coll. Heimdal : **2**, Coll. H. Marie : **4**.)

1. Still at the same spot, two soldiers of I.SS-Panzer-Korps patrol in the Rue Pasteur. They pose here for the war correspondent near the wreck of Wittmann's tank and the gas pump.

2. A color picture was taken almost at the same time, and was used to illustrate the cover of issue n° 16 of the magazine "Signal" late in 1944.

3. Another view of the German soldiers patrolling in this main street sector after the fighting.

*4. The same spot looking in the other direction, c. 1930. (BA: **1** and **3**, coll. Heimdal: **2**, coll. H. Marie: **4**.)*

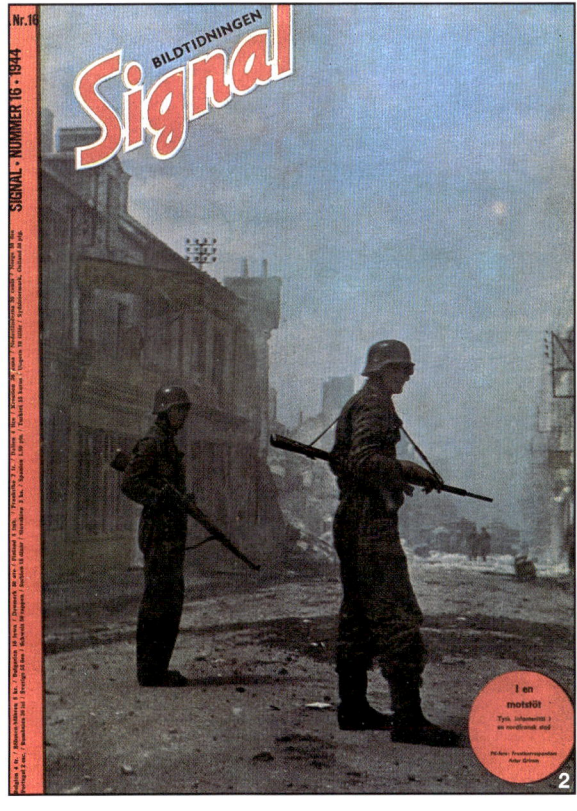

3

2

4

105

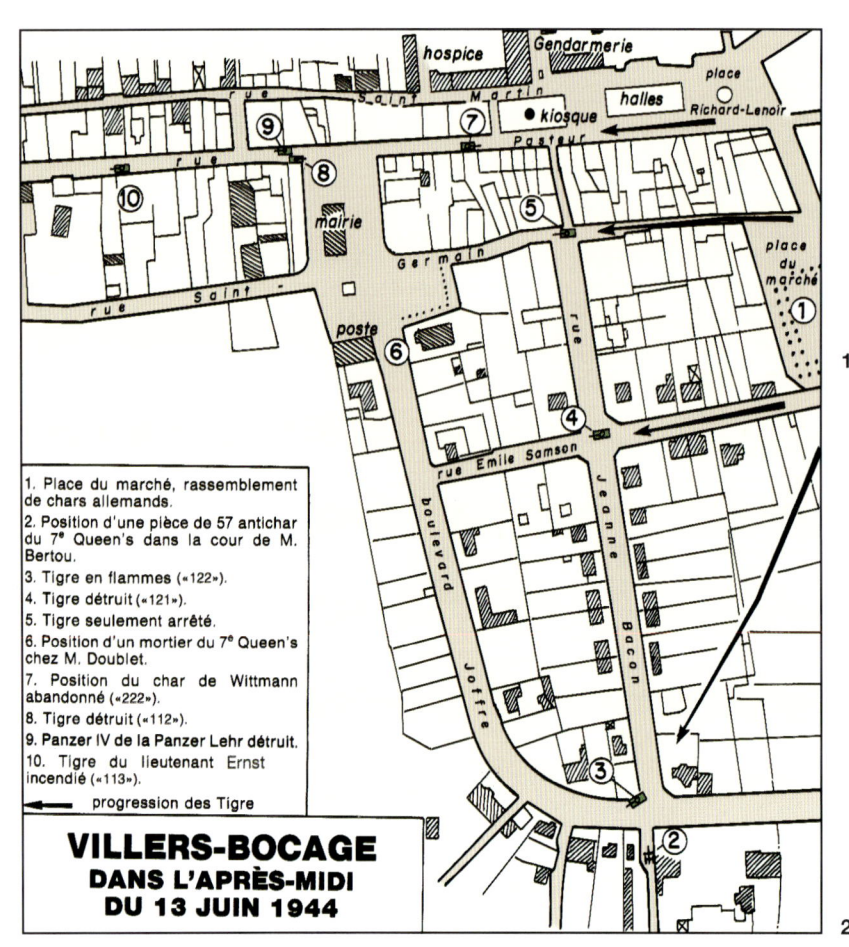

1. Place du marché, rassemblement de chars allemands.
2. Position d'une pièce de 57 antichar du 7ᵉ Queen's dans la cour de M. Bertou.
3. Tigre en flammes («122»).
4. Tigre détruit («121»).
5. Tigre seulement arrêté.
6. Position d'un mortier du 7ᵉ Queen's chez M. Doublet.
7. Position du char de Wittmann abandonné («222»).
8. Tigre détruit («112»).
9. Panzer IV de la Panzer Lehr détruit.
10. Tigre du lieutenant Ernst incendié («113»).
← progression des Tigre

VILLERS-BOCAGE DANS L'APRÈS-MIDI DU 13 JUIN 1944

résisté aux bombes. On voit le même endroit (on reconnaît les immeubles sur la gauche) vers 1930 sur la **photo 5**. Le cliché **6** nous montre le Panzer IV en plan rapproché. (BA : **2, 3** et **6**, IWM : **4**, B. Cotton : **1**, H. Marie : **5**.)

*Two German tanks **(photo 1)**, a Panzer IV of II./Panzer-Lehr-Regiment 130 ("9" on the map) and a Tiger ("8" on the map) were destroyed in the Rue Pasteur, after the Town Hall square. The Panzer IV was knocked out in the afternoon by a Cromwell tank of B Squadron. It was leading the way, followed by the two Tigers ("8" and "10" on the map). On the front of the destroyed Tiger beside it, we recognize the tactical marking of Möbius's 1st Company. This photo was taken by Captain Bill Cotton immediately after the battle; the Tiger is still smoking. The second photograph **(2)** was taken later by a German war correspondent, there are some damaged buildings, as in the next photograph **(3)** showing the rear of these two tanks. This second photograph **(4)** was taken by a British war correspondent from the same angle on August 8th, 1944, after the bombardment of June 30th. The buildings were crushed as was the Panzer IV, whereas the robust Tiger withstood the bombs well. We see the same spot (the buildings on the left are recognizable) in c. 1930 on **photo 5**. Picture 6 is a close-up of Panzer IV. (BA: **2, 3** and **6**, IWM: **4**, B. Cotton: **1**, H. Marie: **5**.)*

Deux chars allemands **(photo 1)**, un Panzer IV de la *II./Panzer-Lehr-Regiment 130* (« 9 » sur le plan) et un Tiger (« 8 » sur le plan) ont été détruits dans la rue Pasteur, après la place de la mairie. Le Panzer IV a été détruit dans l'après-midi par un char Cromwell du *B Squadron*. Il avançait en tête, suivi par les deux Tiger (« 8 » et « 10 » du plan). Sur l'avant du Tiger détruit à côté de lui, on reconnaît le signe tactique de la 1re compagnie, celle de Möbius. Cette photo a été prise par le *Captain* Bill Cotton juste après le combat ; le Tiger fume encore. A l'extrême droite on distingue, sous le canon du Tiger, le char de Wittmann. La seconde photo **(2)** a été prise plus tard par un correspondant de guerre allemand, des immeubles sont endommagés, comme sur la photo suivante **(3)** où l'on voit l'arrière de ces deux chars. Cette seconde photo **(4)** sous le même angle a été prise le 8 août 1944, après le bombardement du 30 juin, par un correspondant de guerre britannique. Les immeubles ont été écrasés ainsi que la Panzer IV alors que les robustes Tiger ont bien

Ces cinq clichés montrent le Tiger (« 10 » sur le plan) qui s'est le plus avancé dans la rue Pasteur, dépassant le Panzer IV détruit. Il s'agit d'un type E très tardif qui appartenait à la 1re compagnie (**photos 1, 2 et 3**). La tourelle a été calcinée par l'incendie qui a ravagé l'engin. On voit, sur la photo **5**, comment la chenille gauche du Tiger « 113 » a « mordu » dans le bâtiment. En arrière : le « 112 ». Ces photos ont été prises quelques jours après le 13 juin. Comme le montre une autre photo prise peu après les combats par le *Captain* Bill Cotton, des immeubles ont été entre-temps endommagés le 14 juin par un premier bombardement. Sur la **photo 4**, on aperçoit dans le fond les autres chars (« 8 » et « 9 ») détruits. (Photos BA.)

*These five pictures show the Tiger ("10" on the map) which has advanced furthest down the Rue Pasteur, passing the destroyed Panzer IV. It is a very late E type which belonged to the 1st Company (**photos 1, 2 and 3**). The turret has been charred by the fire that devastated the machine. On **photo 5**, we see how the Tiger's "113" left track has "bitten" into the building. Behind : the "112". These photographs were taken a few days after June 13th. As shown in another photo taken by Captain Bill Cotton shortly after the battle, some buildings had been damaged since then, on June 14th, in an initial bombardment. In **photo 4** the other destroyed tanks ("8" and "9") can be seen in the background. (Photos BA.)*

108

4

5

Rue Jeanne-Bacon

Les chars Tiger de la 1re compagnie ont contre-attaqué en éventail à travers Villers-Bocage dans l'après-midi du 13 juin. Celui-ci **(photo 1)** a été détruit au carrefour rue Jeanne-Bacon et rue Emile-Samson (« **4** » sur le plan). Le tube du canon est rentré à l'intérieur de la tourelle et la chenille gauche a été touchée au niveau du barbotin, la roue dentelée solidaire du moteur qui entraîne la chenille. Ce barbotin **(photo 2)** a été conservé à proximité de l'endroit où cet engin a été détruit (ainsi qu'un patin de chenille) **(photo 3)**. Ce char appartenait à la 1re compagnie. On voit sur la photo actuelle **(4)** son itinéraire et l'endroit où il a été détruit. Il a brûlé et sa tourelle est noircie. En fait ce char Tiger a été détruit par un canon antichar de 57 mm installé (« **2** » sur le plan) par les *Queens* dans la cour de M. Bertou qu'on voit actuellement **(photo 5)** avec une flèche indiquant la direction du tir. Cette pièce antichar prenait la rue Jeanne-Bacon en enfilade comme on le voit sur une photo actuelle **(6)** avec le sens du tir et le point d'impact sur le Tiger (« **4** »). Sur une photo réalisée avant la guerre **(7)**, prise au même endroit en regardant vers le nord, on voit que la maison de gauche existait déjà, celle de droite a été transformée et agrandie. Un second Tiger (« **3** » sur le plan) a été aussi incendié. (BA : **1**, EG/Heimdal : **2, 3, 4, 5, 6, plan**, Coll. H. Marie : **7**.)

Rue Jeanne-Bacon

*The 1st Company's Tigers fanned out for a counter-attack through Villers-Bocage during the afternoon of June 13th. This one **(photo 1)** was destroyed at the junction between the Rue Jeanne-Bacon and the Rue Emile-Samson ("4" on the map). The gun barrel has been pushed back inside the turret and the left track has been hit at the sprocket-wheel, the notched wheel that meshes with the engine and engages the track. This sprocket-wheel **(photo 2)** was preserved near the place where the vehicle was destroyed*

along with a piece of track *(photo 3)*. This tank belonged to the 1st Company. This present-day photograph *(4)* shows its route and exactly where it was destroyed. It is burnt up and the turret is blackened. In fact this Tiger tank was destroyed by a 6 pounder anti-tank gun set up ("2" on the map) by the Queens in the yard of Monsieur Bertou seen here *(photo 5)*, with an arrow indicating the direction of fire. This anti-tank gun enfiladed the Rue Jeanne-Bacon as can be seen in a recent photograph *(6)* with the direction of the shooting and the point of impact on the Tiger ("4"). On a photograph taken before the war *(7)*, at the same place but looking north, we see that the house on the left already existed, the one on the right-hand side was transformed, 1944 and was replaced by the one in the recent picture. A second Tiger ("3" on the map) also brewed up. (BA: 1, EG/Heimdal: 2, 3, 4, 5, 6, map, coll. H. Marie: 7.)

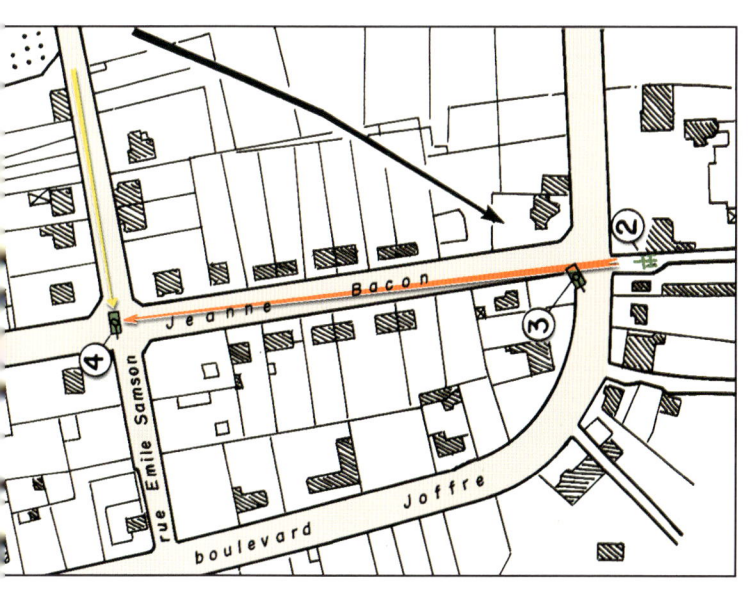

Les prisonniers britanniques

Après la bataille, la police militaire SS achemine les nombreux prisonniers de guerre de la *7th Armoured Division* qui ont été rassemblés à Villers-Bocage. On les voit ici **(photo 1)** traverser le village de Vieux, en venant d'Esquay, passant sur la place d'Armes qu'on voit vers 1900 **(photo 2)** et actuellement **(photo 4)**. Le photographe s'avance pour prendre sous un autre angle le bâtiment qu'ils longent **(photo 5)**, et actuellement **(photo 3)**, les arbres ont disparu. Les prisonniers continuent d'avancer dans Vieux **(photo 7)**. Actuellement **(photo 6)**, la maison du fond et le mur de gauche sont conservés mais la route a été élargie et le mur de droite a disparu. Ils passent maintenant devant l'église de Vieux **(photo 9)**, les lieux sont restés inchangés **(photo 8)**. (BA : **1, 5, 7, 9**. EG./Heimdal : **2, 3, 4, 6, 8**.)

British prisoners

After the battle, the SS military police conveys the numerous prisoners-of-war of the 7th Armoured Division rounded up at Villers-Bocage. Here they are seen (photo 1) passing through the village of Vieux, coming from Esquay, crossing the parade ground, seen in c. 1900 (photo 2) and today (photo 4). The photographer has moved forward to take the building they are skirting from another angle (photo 5), and today (photo 3), the trees have gone. The prisoners continue to advance through Vieux (photo 7). Today (photo 6), the house at the bottom and the wall on the left are stilll standing but the road has been widened and the wall on the right has gone. They now pass in front of the church at Vieux (photo 9), the place has not changed at all (photo 8). (BA: 1, 5, 7, 9. EG./Heimdal: 2, 3, 4, 6, 8.)

113

Les prisonniers britanniques

La carte ci-contre montre qu'ils ont quitté Villers-Bocage pour rejoindre Caen en évitant la route nationale, trop dangereuse, pour emprunter des petites routes, par Esquay, Vieux et Amayé-sur-Orne, ainsi que ces photos le montrent. Escortés par la police militaire SS **(photo 3)**, ils progressent dans la campagne d'un village à l'autre, croisant parfois des civils **(photo 2)**. Ils arrivent maintenant à Amayé-sur-Orne et marquent une pause devant l'église **(photo 4)**, la route a été longue. Actuellement, les lieux **(photo 5)** sont inchangés, même les ifs sont toujours soigneusement taillés. (Photos BA : **2, 3, 4**. Coll. Heimdal : **1 et 5**.)

British prisoners

*The map opposite shows that they left Villers-Bocage for Caen avoiding the main highway, as being too dangerous, and taking the back lanes via Esquay, Vieux and Amayé-sur-Orne, as these photographs show. Escorted by SS military police **(photo 3)**, they advance through the countryside from one village to the next, passing the odd civilian **(photo 2)**. They are now arriving at Amayé-sur-Orne and take a break in front of the church **(photo 4)**, they have come a long way. Today, the spot **(photo 5)** is unchanged, even the yews are still neatly trimmed. (Photographs BA: **2, 3, 4**. Coll. Heimdal: **1** and **5**.)*

Avant de traverser l'Orne, les prisonniers britanniques marquent une pause à proximité de l'église d'Amayé-sur-Orne (**photos 1** et **2, 3** et **4, 5** et **6**). Les lieux sont demeurés intacts, sauf la gouttière du bâtiment agricole qui a été déplacée. Près de la gare de Villers-Bocage, une section du *1/7 Queens* a été faite prisonnière. Mais on reconnaît aussi sur cette photo des hommes de la *Rifle Brigade* (**photo 8**) ou de la *Royal Artillery*, probablement du *5th RHA* (**photo 9**). Ces hommes vont ensuite traverser l'Orne pour rejoindre Rennes où les prisonniers de guerre alliés étaient triés avant

de continuer sur l'Allemagne jusqu'à un camp de prisonniers de guerre. A proximité, un véhicule (un Kübelwagen) patrouille non loin du château de Fontaine-Etoupefour **(photos 11 et 10)** où auront lieu de très durs combats dans les semaines à venir. (BA : **1, 3, 5, 7, 8, 9, 11**. EG/Heimdal : **2, 4, 6, 10**.)

*Before crossing the Orne, the British prisoners take a break near the church at Amayé-sur-Orne **(photos 1 and 2, 3 and 4, 5 and 6)**. The place has remained intact, except for the gutter on the farm building which has been moved. A section of the 1/7 Queens was captured near Villers-Bocage station. But on this photograph we also recognize men of the Rifle Brigade **(photo 8)** or the Royal Artillery, probably 5th RHA **(photo 9)**. These men are about to cross the Orne on their way to Rennes where Allied prisoners-of-war were sorted before being moved on to a POW camp in Germany. Near by, a vehicle (Kübelwagen) patrols in the vicinity of the château de Fontaine-Etoupefour (photos 11 and 10) where a very hard battle was to be fought in the weeks that followed. (BA: **1, 3, 5, 7, 8, 9, 11**. EG/Heimdal: **2, 4, 6, 10**.)*

4 Tragique dénouement
Tragic ending

Les Britanniques se regroupent

Beaucoup de soldats anglais ne vont pas rejoindre leurs lignes aussitôt. Quelques-uns profitent de la nuit pour regagner leur unité, d'autres vont mettre plusieurs jours, ravitaillés et dirigés par des civils qui n'étaient pas sans ignorer le sort qui était réservé à ceux qui étaient surpris à aider des Britanniques. D'autres n'auront pas cette chance, ils seront repris et faits prisonniers, surtout ceux qui ont essayé de rejoindre en allant vers le nord en suivant les vallées de la Seuline ou de la Seulles.

Le capitaine Christopher Milner commandant en second de la 1re compagnie de la *Rifle Brigade* est de ceux qui ont rejoint le matin du 14 juin. De son retour, il a écrit un récit détaillé dont voici l'essentiel.

Encerclé avec les membres du poste de commandement avancé de la cote 213, il a pu rester posté derrière la maison Ruelle pour surveiller vers le nord sans être inquiété et ce n'est que lorsqu'il a vu les Tigres de la compagnie de Möbius arriver qu'il a réalisé la situation. Il s'est alors jeté dans le jardin de M. Ruelle situé à l'angle de la route de Monts-en-Bessin et de là dans un champ de foin voisin où il est resté allongé tout l'après-midi. Il a eu le temps de réfléchir et de trouver un moyen de sortir de sa position inconfortable, sans manger, ni boire, et au risque d'être atteint par les éclats des obus que les Anglais envoyaient généreusement. Il pensa tout d'abord traverser la route de Monts, gagner le petit bois de Villy et prendre la direction de Villers, mais la route étant un espace découvert, il y renonça et décida d'attendre la nuit avant de chercher à s'enfuir.

« *Enfin, il commença à faire nuit et je partis en rampant doucement vers la grand route*, raconte-t-il. *Mon plan était d'attendre que tout soit net, de m'élancer à travers la grand route, vers un champ où les traces des Tigres étaient parallèles à la route et, me souvenant de la carte, tourner vers l'ouest en direction de Villers-Bocage, essayer de prendre un chemin qui me mènerait au sud ou à l'intérieur de la ville, là où étaient nos troupes.*

Enfin, vers minuit, il y eut un moment de calme. Je sprintai sur la route, plongeai dans la haie du côté opposé et rampai le long d'une autre haie jusqu'à ce que je trouve les traces des Tigres qui me conduisirent à un chemin qui menait vers l'ouest. Je l'empruntai et arrivai aux voies de garage de la gare. Durant ce trajet, j'avais contourné une petite unité d'Allemands qui dormaient par terre en ronflant et j'étais rentré dans une cour de ferme où un chien aboya et où je m'étais désaltéré à une auge remplie d'eau potable ».

Au lever du jour, il gagna le Mesnil de Tracy après avoir traversé les routes d'Aunay et de Vire, longé dans un herbage la haie qui borde la route de Caumont, traversé la Seuline sur un pont en bois, avant de pouvoir avaler enfin « *une grande gamelle de bacon et un pot de thé* ».

Le lieutenant Cloudsley-Thomson, dont le char avait été détruit lors de l'attaque du QG le matin par Wittmann, s'est réfugié avec ses quatre hommes d'équi-

The British regroup

Many British soldiers did not join their lines at once. Some rejoined their unit under cover of darkness, others took several days, after receiving supplies and directions from civilians who were well aware of the fate reserved for anyone caught helping the British. Others were not so lucky, as they were recaptured and taken prisoner, especially those who tried to link up by moving northwards following the Seuline or Seulles valleys.

Captain Christopher Milner, second-in-command of the 1st Company the Rifle Brigade, was among those who linked up on the morning of June 14th. On his return, he wrote up a detailed account of what happened.

Encircled with the members of the advanced headquarters on Point 213, he contrived to stay posted behind the Ruelle house to keep a clear view to the north, and it was only when he saw the Tigers of Möbius's Company arriving that he realized what was going on. He then threw himself into Monsieur Ruelle's garden on the corner of the Monts-en-Bessin road, and from there into a nearby hay field where he lay all afternoon. He had plenty of time to think and find a way out of his predicament, with nothing to eat or drink, and with the risk of being hit by shrapnel from British shells which were falling thick and fast. His first idea was to cross the Monts road, reach the small Villy wood and head towards Villers, but he decided against taking the open road and to wait until nightfall before trying to escape.

"At last it began to get dark and I started to crawl slowly towards the main road again [...] My plan was now to wait until all was clear, dash across the main road, over the intervening field to the parallel track (that is to say the Panther tanks' track) and, once again remembering the map, turn back in a westerly direction towards Villers Bocage and try to take a course which would bring me round the south side of the town instead of the north – or into the town itself if that was where our own troops were. At last, perhaps around midnight [...] I was able to sprint across the road, dive into the hedge on the far side and creep along the inside of the hedge until I hit the track. [...] so I stood up and walked gingerly along the grass track, back towards the town. Suddenly I heard [...] a section of German soldiers who had dug a broad shallow trench [...] were all sound asleep [...] I don't think that even a dog barked [...] I found a trough full of drinkable water and I remember taking off my beret [...] and it held water remarkably well, I had an extremely good drink."

At daybreak he reached Le Mesnil de Tracy after crossing the Aunay and Vire roads, skirting a meadow under the hedge along the Caumont road, crossing the Seuline over a wooden bridge, before at last being able to tuck into "a large billy-can full of bacon, plus a mug of tea".

Lieutenant Cloudsley-Thomson, whose tank had been destroyed at the time of Wittmann's morning attack on the HQ, took cover in a cellar with his four crew members. There, the following day, he received sup-

page dans une cave. Là, le lendemain, il est ravitaillé par un commis boucher de Raymond Anne, Emile Buisson qui leur conseille d'attendre la nuit avant de partir. Les cinq hommes passent la journée du 14 dans leur refuge et partent dans la nuit du 14 au 15. Ils se dirigent vers le nord, contournent Villy, traversent la Seulles vers Saint-Vaast-sur-Seulles en se déplaçant uniquement la nuit. Affamés, assoiffés mais sains et saufs, ils arrivent dans les lignes du 50ᵉ Régiment d'Infanterie anglaise du côté de Hottot-les-Bagues dans la nuit du 16 au 17 juin.

Dans le bourg, un soldat est resté deux jours dans une cave où André Marie, charcutier, lui a apporté à boire et à manger. Le troisième jour, il était parti. A la gare, Monsieur Julienne a caché deux soldats sous des wagons, les a ravitaillés et les a accompagnés jusqu'à la « Vierge Noire », route de Caumont. Harry Hopkins, de la *Rifle Brigade* et trois hommes de sa section sont restés cachés dans un poulailler route de Caen, derrière le calvaire jusqu'à la nuit du 13 au 14 juin. Ces quelques retours au sein des unités britanniques ne sont pas les seuls.

Après le départ des unités anglaises, les Allemands ont continué à rechercher des combattants ou des blessés, dans les ruines ou à l'intérieur des habitations où ils se faisaient ouvrir parfois les portes revolver au poing. Il est heureux qu'aucun Villérois n'ait été victime de représailles. Pourtant, une fois, une famille aurait pu payer cher l'aide qu'elle portait à un soldat anglais. En effet, route de Saint-Louet, un soldat s'était réfugié dans le foin d'une grange où des poules avaient l'habitude de pondre leurs œufs. Un enfant allait lui porter des aliments et revenait avec l'assiette vide. A un moment, il est sorti de la grange avec l'assiette à l'instant même où une patrouille allemande passait. Heureusement pour lui et sa famille, l'assiette était pleine... d'œufs qu'il avait ramassés avant de revenir.

D'autres soldats ont eu moins de chance, c'est le cas du sergent Ken Weitman, chef d'un *Firefly* de l'escadron A dont la capture a été racontée dans le livre « *Carpiquet Bound* » de W. Allen. Son *Firefly* bloqué sur la route nationale et détruit, il parvient à s'extraire de son char en feu et sous les tirs de mitrailleuses allemandes, réussit à se mettre à l'abri sous un couvert où il retrouve un fantassin de la *Rifle Brigade*. Pendant deux jours et deux nuits, les deux hommes évitent l'ennemi et tentent de trouver leur route vers les abords de Villers sans aucune carte et avec des Allemands partout. Le deuxième jour, ils deviennent affamés et, en quête de nourriture, se dirigent vers une cour de ferme déserte. Tandis qu'ils cherchent des comestibles, le fermier et son épouse reviennent soudain dans les bâtiments apparemment fouillés par les Allemands. Ken essaie de converser dans un français scolaire avec le fermier mais la femme, effrayée et au bord de la crise d'hystérie, est terrifiée et essaie de les repousser. Avant de les quitter, le fermier les aide à trouver des œufs et de vieux vêtements. Les deux soldats estiment alors que le meilleur endroit pour passer la nuit est dehors dans les champs et, après avoir mangé un plat de farine et d'œufs à moitié cuits, ils s'installent dans un fossé. Comme ils se tapissent dans leur espace d'infortune, Ken se souvient avoir vu distinctement juste avant la tombée de la nuit, un garçon courir dans la cour de la ferme. Le matin, de très bonne heure, au réveil, ils voient des fantassins allemands debout devant eux, fusil en joue. Ken pense que le garçon a prévenu les Allemands de l'endroit de leur cachette. Les deux Anglais sont alors conduits dans un château où ils sont interrogés par un officier allemand d'allure efféminée et fortement parfumé. Assis à une grande table, il fait tourner son revolver dans ses mains très soignées, disant des phrases

plies from butcher Raymond Anne's apprentice Emile Buisson, who advised them to wait until nightfall before leaving. The five men spent the day of the 14th in their shelter and left during the night of the 14-15th. They moved northwards, by-passing Villy, crossing the Seulles near Saint-Vaast-sur-Seulles, and moving only at night. Famished and parched but safe and well, they reached the British 50th Infantry Regiment's lines in the Hottot-les-Bagues area during the night of June 16-17th.

In the village itself, a soldier stayed for two days in a cellar where pork butcher André Marie brought him food and drink. On the third day, he was gone. At the station, Monsieur Julienne hid two soldiers under rail cars, fed them and accompanied them to the «Vierge Noire» on the Caumont road. Harry Hopkins of the Rifle Brigade and three of his platoon stayed in hiding in a hen house behind the roadside cross on the Caen road until the night of June 13-14th. These were just some of the men who made it back to their British units.

After the British units left, the Germans continued to look around for fighting men or casualties, in the ruins or inside houses, where they sometimes had doors opened with their revolvers at the ready. Fortunately none of the townspeople at Villers were the victim of reprisals. However, one family almost paid dearly for helping a British soldier. This soldier had taken cover on the Saint-Louet road in the hay in a barn where hens were in the habit of laying their eggs. A child would take food out to him and bring back the empty plate. One time, he left the barn with the plate just as a German patrol was passing. Fortunately for him and his family, the plate was full of the eggs he had collected before coming back.

Other soldiers were less fortunate; one such was Sergeant Ken Weightman, an A" Squadron Firefly commander whose capture is recounted in W. Allen's book Carpiquet Bound. His Firefly having been blocked and destroyed on the main highway, he contrived to bale out from his burning tank under German machine-gunfire, and managed to take cover in a shelter where he found an infantryman of the Rifle Brigade. For two days and two nights, the two men evaded the enemy and tried to make their way towards the neighborhood of Villers without a map and with Germans everywhere. By the second day, they were really hungry and so headed towards a deserted farmyard in search of food. While they were looking for something to eat, the farmer and his wife suddenly returned to the buildings which had apparently been searched by the Germans. Ken tried to converse in school French with the farmer but the woman was terrified out of her wits and bordering on hysterics, and tried to get them to leave. But before they left, the farmer helped them to find some eggs and old clothing. The two soldiers decided the best place for them to spend the night was outside in the fields and, after eating a dish of flour and half-cooked eggs, they settled down in a ditch. As they cowered in their not very safe hiding-place, Ken remembered just before nightfall distinctly seeing a boy running in the farmyard. Next morning, very early, they woke up to see German infantrymen standing over them, taking aim with their rifles. For Ken, the boy must have told the Germans of their hiding-place. The two Britishers were then "marched off up the road to a château where they were questioned by an effeminate German officer who smelt strongly of perfume. He was sitting at an enormous table twirling his revolver on the highly-polished top, saying such things as 'Time has long passed sergeant, when prisoners only give their name and number', and 'Loony Hinde and your brigade have been captured' and 'You realize the British Government is evacuating London', a reference

dans le genre de, « *le temps semble long, sergent, quand les prisonniers donnent seulement leur nom et leur adresse, Lonney Hinde et votre brigade ont été capturés, réalisez-vous que le gouvernement anglais évacue Londres* », allusion possible à l'envoi des premières bombes volantes dirigées le 13 juin vers la capitale anglaise, « *cédez et racontez tout* » étaient ses dernières paroles. Ken Weitman fut conduit dans un camp de prisonniers de guerre où parmi les premiers captifs qu'il rencontra fut un camarade *sharpshooter* fait prisonnier en Egypte.

Le soir du 13 juin, la *7th Armoured Division* quitte Villers. Les Villérois voient de nouveau les Allemands occuper leur ville. Ils ont espéré pendant une journée que la guerre allait se terminer rapidement pour eux. Ils ne seront délivrés que le 4 août mais, avant cette date, ils vont connaître de nouveaux drames, subir des bombardements, compter des morts et des blessés parmi la population, et partir en exode pour fuir les combats avant de revenir chez eux.

Le 14 juin, bataille d'Amayé-sur-Seulles

Durant la nuit du 13 au 14 juin, les artilleurs anglais et allemands continuent de pilonner Villers, bien que la ville ait été abandonnée par les combattants. Par contre les civils qui n'ont pas gagné la campagne vont passer la nuit dans leurs tranchées ou leurs caves et quelques-uns vont être les victimes des tirs. Peu avant minuit, des obus tombent sur une des tranchées creusées à l'Hospice, près de la gendarmerie. Sont tués le gendarme Paul Blin, son épouse Aline et Emile Leroy âgé de 24 ans, fils de l'adjudant. Ils s'ajoutent aux autres victimes de Villers tuées le 13, M. Raymond Quettier et M. Louis Letellier, tués rue G. Clemenceau, Mme Cuiller, place du marché. Les autres tués seront, le 14 Mme Marie Piel, route de Vire, Mlle Ernestine Auriac, M. Ferdinand Sendon, Michel Dallet, 11 mois, tué par une balle « perdue » dans la cour de M. Bernouis, le 15, M. François Pochat, tué dans la rue Clemenceau, près de son domicile par une rafale de mitrailleuse, le 16, trois victimes dont deux n'ont pas été retrouvées, Mme Augustine Poisson et M. Emile Nodot, la troisième étant Mme Aline Faucon.

Une partie seulement de la division a quitté la ligne de front le matin du 13 dans les environs de Livry. Certaines unités qui ont combattu à Villers et qui ont reçu l'ordre de se replier sont encore loin de leur base de départ. Le matin du 14, il y avait encore des Anglais au Mesnil de Tracy. Ces troupes vont devoir rejoindre après s'être regroupées sur les hauteurs autour d'Amayé-sur-Seulles, sur la « Butte du Moulin à vent » dont le point culminant est de 193 mètres, entre la route qui mène à Caumont et le hameau des Bruyères. Il s'agit pour les Britanniques d'installer une position défensive, une *brigade box* et de rassembler leurs unités dans un quadrilatère de 2 000 mètres environ. Cette position va leur permettre de contrôler Villers bien sûr, mais aussi de protéger les troupes qui rejoignent la division et vont emprunter la route qui mène au hameau de Briquessard, situé à trois kilomètres environ. Cette route, « véritable cordon ombilical », aurait pu être coupée la veille par une unité de la *2. Panzer-Division*, constituée en particulier d'éléments du 38e bataillon de chasseurs de chars équipée de 75 à canon long qui avait atteint la cote 174 sur la route qui va de Cahagnes au hameau de la Tringale.

Dans la « *box* » les positions sont les suivantes : à l'ouest, le *5th RTR*, au sud le *1st RTR*, le *5th Queen's* et le *4th CLY*, à l'est le *7th Queen's*, au nord, le long de la route de Caumont, le *1st bataillon* de la *Rifle*

possibly to the launching of the first flying bomb, on 13th June, aimed at the capital. *'Give up and tell all'* was his final remark. Ken Weightman landed up in a POW camp where one of the first men he met was a fellow Sharpshooter taken prisoner in Egypt".

The 7th Armoured Division left Villers on the evening of June 13th. The people of Villers had the Germans occupying their town once again. They just hoped that some day soon the war would be over for them. They were not liberated until August 4th and meanwhile fresh drama awaited them, further bombing raids were to cause deaths and casualties among the population, and they would all have to flee the battlefield before eventually returning home.

June 14th, the battle of Amayé-sur-Seulles

During the night of June 13-14th, the British and German artillery continued to pound Villers, although the town had been abandoned by the fighting troops. There were however civilians who had not moved out into the countryside bu spent the night in their trenches or cellars, and some of them were caught in the crossfire. Shortly before midnight, shells fell on one of the trenches dug at the hospice, next to the gendarmerie. A gendarme, Paul Blin, his wife Aline and the adjudant's son Emile Leroy, aged 24, were all killed. They came after the other fatal casualties at Villers of the 13th, Messrs Raymond Quettier and Louis Letellier, killed in the Rue Clemenceau, and Madame Cuiller in the market place. The other killed were, on the 14th: Madame Marie Piel, on the Vire road, Mademoiselle Ernestine Auriac, Monsieur Ferdinand Sendon, Michel Dallet, aged 11 months, killed by a stray bullet in Monsieur Bernouis' yard; on the 15th, Monsieur François Pochat, killed by a burst of machine-gun fire in the Rue Clemenceau close to his home; on the 16th, three dead, two of whom, Madame Augustine Poisson and Monsieur Emile Nodot, were never found, the third being Madame Aline Faucon.

Only a part of the division left the front line in the Livry area on the morning of the 13th. Certain units which had fought in Villers and received orders to withdraw were still a long way from their starting line. On the morning of the 14th, there was still some British troops at Mesnil de Tracy. They had to link up after mustering on the high ground around Amayé-sur-Seulles, on the Butte du Moulin à Vent (Windmill Hillock) culminating at 193 meters, between the road to Caumont and the hamlet of Les Bruyères. It was a matter of the British setting up a defensive position or brigade box, and gathering their units within a quadrilateral some 2,000 meters in size. This position would of course give them control of Villers, but also protect the troops joining the division and taking the road to the hamlet of Briquessard, about three kilometers away. This road, a veritable lifeline, could have been cut off the previous day by a unit of 2. Panzer-Division, comprising mostly elements of the 38rd Tank Destroyer Battalion equipped with long-barreled 75mm guns and which had reached Point 174 on the road from Cahagnes to the hamlet of La Tringale.

In the "box" the positions were as follows: to the west, the 5th RTR; to the south the 1st RTR, 5th Queen's and 4th CLY; to the east the 7th Queen's; to the north, along the Caumont road, the 1st Battalion the Rifle Brigade, in the center the 8th and 11th Hussars, the 54th RHA, and the HQ tactical brigade.

All gardens, orchards, sunken lanes and hedgerow undergrowth were filled with soldiers, tanks, guns, trucks carrying ammunition and food, all sitting targets for the enemy artillery and air force. The Ger-

Brigade, au centre le *8th* et le *11th Hussars*, le *5th RHA*, la section tactique du QG.

Tous les jardins, les vergers, les chemins creux, les pieds de haie sont couverts de soldats, de chars, de canons, de camions de munitions et de nourriture, ce qui constitue des cibles parfaites pour l'artillerie et l'aviation adverses. Les Allemands tiennent trois côtés de la « Box » et utilisent haies et chemins pour s'approcher au plus près des Britanniques, à 100 mètres, même par endroits à 50 mètres.

Mais la situation est maintenant inversée. De défenseurs, ils sont devenus attaquants. Eux qui ont si bien su depuis le 6 juin retarder l'avance des alliés en leur infligeant de lourdes pertes en hommes et en chars en utilisant pour leur défense le moindre chemin, le moindre talus, le moindre mur pour y abriter une mitrailleuse, un canon antichar, un *panzerfaust*, en se servant des haies épaisses, des taillis, des hauts arbres pour placer leurs tireurs d'élite, ils vont devoir attaquer des adversaires qui ont su tirer des leçons de leurs déboires et s'adapter rapidement à un terrain inhabituel pour eux.

La *Panzer-Lehr-Division* qui a envoyé des troupes au nord de Villers participe à la poursuite des unités anglaises en retraite. Elle vient de Sermentot, d'Anctoville. Au sud, la *2. Panzer-Division* a pris les positions qui lui étaient fixées et son QG étant à Brémoy, elle peut attaquer sur un secteur assez important venant de Coulvain, de Cahagnes.

Le soir du 13, ce qu'il reste de la première compagnie du bataillon de chars lourds, la *SS-Pz.-Abt. 101*, s'est fixé au sud de Villers. Le 14, les Tigres au nombre de 3 vont soutenir les grenadiers de la *2. Panzer-Division* qui assiègent la « *Brigade Box* » sur les hauteurs d'Amayé-sur-Seulles. A l'est de Cahagnes, l'attaque qu'elle va mener vers la route d'Amayé-Briquessard sera contenue par l'artillerie alliée.

La deuxième compagnie de Wittmann quant à elle, est à l'est de Villers, cote 213. Ces troupes allemandes ont profité de la nuit pour remettre en état leurs chars endommagés la veille et des mauvaises conditions atmosphériques qui ont empêché les avions anglais de sortir.

Il serait important que la *7th Armoured Division* puisse s'installer solidement dans la « *Box* » et créer une base avancée derrière le front, mais il va lui falloir résister aux attaques allemandes qu'elle va devoir subir de trois côtés au moins.

Dans le secteur de Tilly, la *50th Infantry Division* essaie de fixer la *Panzer-Lehr-Division*. Elle a demandé l'intervention de l'aviation anglaise, en l'occurrence le 11e escadron de chasseurs-bombardiers pour lui venir en aide, mais elle ne marque aucun progrès. A Amayé, les Britanniques ne peuvent se dégager et doivent se contenter de défendre. Il leur faudrait au moins une brigade d'infanterie supplémentaire, mais le général Bucknall ne la demande pas à la 2e Armée.

Au lever du jour, le *Major* Bill Apsey qui commande la compagnie I du 1er bataillon de la *Rifle Brigade* en position le long de la route de Caumont au nord de la « *Box* », reçoit l'ordre d'aller inspecter le nord de la route, d'examiner et de jauger la force de l'ennemi. Il constate que les Allemands sont infiltrés dans un petit bois et qu'ils constituent une menace sérieuse. Le QG de la *Brigade* demande alors à l'artillerie américaine, installée près de Caumont, et à chaque batterie anglaise du secteur, de déverser leurs obus sur la zone suspecte.

Un peu avant 10 heures, rapporte Desmond Gordon, « *il devient clair que l'infanterie ennemie avance à travers l'épais couvert vers nos deux compagnies de tête du 7th Queen's. Pour s'y opposer le bataillon*

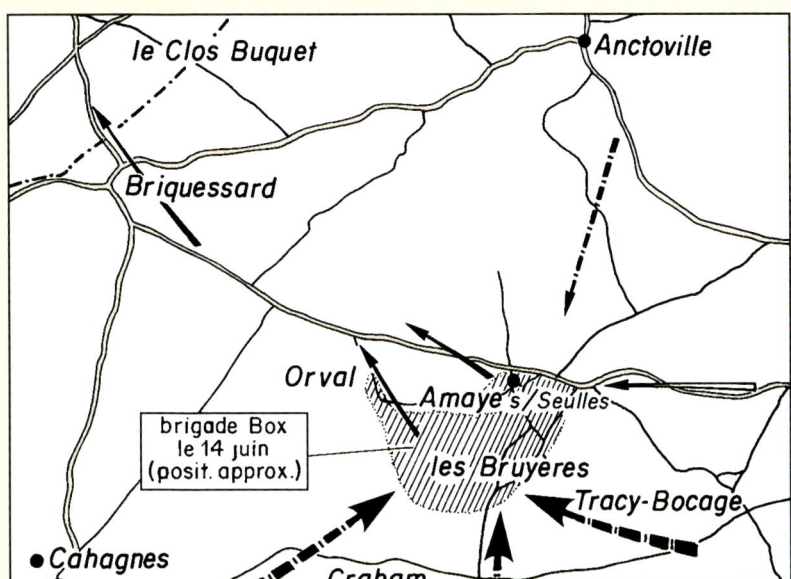

mans held three sides of the "box" and used the hedges and narrow lanes to get in close to the British, at 100 meters range, as little as 50 meters in places.

But the boot was now on the other foot. From being defenders, they were now on the attack. After being so successful since D-Day in delaying the Allied advance and inflicting heavy losses in men and tanks by gaining defensive advantage from the slightest lane, embankment or wall to conceal a machine-gun, anti-tank gun or Panzerfaust, while making use of the thick hedges, coppices and tall trees to place their marksmen, they now had to take on an enemy who had learnt from his mistakes and quickly adjusted to what was for him unfamiliar terrain.

The Panzer-Lehr-Division, which had sent troops north of Villers, took part in the pursuit of the withdrawing British units. It came from Sermentot and Anctoville. In the south, 2. Panzer-Division took up its assigned positions, and with its HQ at Brémoy, it was able to attack over quite a broad sector from Coulvain and Cahagnes.

On the evening of the 13th, what was left of the first company of the heavy tank battalion, SS-Pz.-Abt. 101, stood south of Villers. On the 14th, the 3 Tigers gave support to the grenadiers of 2. Panzer-Division besieging the Brigade Box on the high ground at

1. Là se trouvait l'artillerie britannique *(5th RHA)*, installés le long du ruisseau du Coudray.

2. Repli sur la *Brigade Box* au sud d'Amayé.

1. Two batteries of the 5th RHA were in position along the Coudray stream.

2. Withdrawal into the Brigade Box south of Amayé.

121

met en action ses mortiers et les obus de 25 livres du 5th RHA. A 11 heures, une attaque précise se matérialise sur la gauche devant la compagnie C et une bataille intense d'armes légères fait rage pendant deux heures avec l'assistance importante de mitrailleuses et de chars ».

L'attaque allemande est une attaque de fantassins uniquement, des *panzergrenadiers* qui sont fauchés au moment où ils approchent des chars et ces scènes provoquent des sentiments d'horreur chez certains tankistes britanniques. « *A ce moment*, continue Gordon, *les choses deviennent sérieuses quand une section paraît être en danger, le flanc gauche est sur le point d'être contourné. Mais la compagnie C contre-attaque en utilisant des fusils-mitrailleurs et des grenades et restaure la situation. Le Quartier Général de la Brigade, situé à environ 20 yards à l'intérieur de la ligne de front, risquait d'être contourné si la compagnie C ne l'avait pas secourue sur son flanc gauche. La raison pour laquelle le bataillon du QG déployait un tel courage en étant si près du front était que c'était le seul endroit où il ne pouvait être vu de l'ennemi. La compagnie C en particulier avait eu une action très précise ce jour-là, en dépit de la fatigue de ses hommes. Ils avaient livré un combat loyal et avaient définitivement repoussé le boche, lui infligeant de lourdes pertes* ».

Le *Brigadier* Hinde pense que la position de la *« Brigade Box »* peut être tenue et qu'elle pourrait être utilisée comme point de départ pour reprendre l'initiative mais ce n'est pas l'avis du commandement anglais. De leur côté, les Allemands ne veulent pas laisser une telle menace derrière leurs lignes. A 16 heures, Hinde est en conférence avec Erskine au QG de la *7th Armoured Division* près de Livry. Là, il apprend que la *50th Infantry Division* ne fait pas de progrès dans le secteur de Tilly et qu'elle ne pourra ni le secourir, ni relever ses hommes. Ainsi, il serait préférable de redresser la ligne de front et de replier la 22ᵉ Brigade Blindée sur Livry. La retraite est inéluctable. Le départ de la première unité est fixé à 23 h 45, à la tombée de la nuit afin de profiter de l'obscurité pour rallier le reste de la division. La retraite sera couverte par des tirs d'artillerie et par l'intervention de bombardiers autour de la « *Box* » pour empêcher les Allemands d'intervenir et pour couvrir le ronflement des moteurs et le bruit des chenilles des blindés. L'opération s'appellera « *Aniseed* », « graine d'anis » en français. « *Peu de temps après que les ordres ont été transmis pour annoncer cette retraite de nuit, l'ennemi lançait une attaque lourde sur l'arrière-garde de la* Brigade Box *avec deux bataillons d'infanterie de la* 2. Panzer-Division *et environ 30 chars. Le QG du Bataillon Tactique, le 5th RHA et le 5th Queen's sont en grande partie encerclés tandis que la 7th Queen's subit pour la première fois de la journée un lourd bombardement de son secteur. En plus, la compagnie A, sur la droite, engage des patrouilles de fantassins ennemis* ». Une attaque venant de la *Panzer-Lehr-Division* au nord touche la *Rifle Brigade*. Ainsi, vers 21 heures, le « tonnerre » éclate de tous côtés.

Le QG de la Brigade obtient le soutien d'une *AGRA (Army Group Royal Artillery)* et de trois groupes d'artillerie américaine basés dans la région de Caumont qui vont déclencher des tirs défensifs d'obus de 155 mm. Dans la « *Box* », faute de place, le *5th RHA* n'a pu déployer que quelques batteries de ses canons de 105 automoteurs. La batterie G tire vers le sud à vue avec des obus fusants sur les *panzergrenadiers* qui arrivent presque à découvert à moins de 400 mètres de canons. La batterie K tire vers l'est. Le lieutenant-colonel Goulburn, commandant du *8th Hussars* écrit dans son journal privé « *Notre infanterie fait le feu avec des armes portatives légères qui*

Amayé-sur-Seulles. East of Cahagnes, the attack it launched towards the Amayé-Briquessard road was contained by the Allied artillery.

Meanwhile, Wittmann's second company was east of Villers at Point 213. These German troops had taken advantage of the night and poor weather conditions which prevented British planes from taking off to repair their tanks damaged the day before.

It was crucial for the 7th Armored Division to be able to gain a firm foothold in the Box and create a forward base behind the front, but to achieve this it had to see off German attacks coming from at least three sides.

In the Tilly sector, the 50th Infantry Division tried to pin down the Panzer-Lehr-Division. It asked for the RAF to intervene, in fact the 11th Fighter-Bomber Squadron to come to its assistance, but it failed to make any progress. In Amayé, the British failed to pull clear and it was all they could do to defend. They would have needed at least another infantry brigade, but General Bucknall did not ask for one from the 2nd Army.

At daybreak, Major Bill Apsey, commander of "I" Company of the 1st Battalion the Rifle Brigade in position along the Caumont road in the north of the Box, was ordered to go and inspect north of the road, to examine and assess the enemy's strength. He reported that the Germans had infiltrated a small wood and posed a serious threat. The Brigade HQ then requested the American artillery near Caumont and every British battery in the sector to rain their shells down on the suspect zone.

Shortly before 10.00 hours, reports Desmond Gordon, "it became clear that enemy infantry were feeling their way forward through the thick cover towards our two leading Coys. Both mortars and Arty were used to good effect. By 1100 hrs a definite attack had materialized on the left forward Coy ('C') and a small arms battle of great intensity raged for the next two hours, with the MGs of the tanks lending valuable assistance."

The German attack was an infantry attack only, Panzergrenadiers who were mown down on approaching the tanks in scenes that caused feelings of horror among some British tank crews. «At one stage», continues Gordon, "matters looked serious when one platoon was overrun and there appeared to be a danger of the left flank being turned; but 'C' Coy counterattacked using sten guns and grenades and restored the situation. Bn.HQ was within 20 yds of the front line and would have been in great danger of being run over if 'C' Coy had not secured the left flank. The reason why Bn.HQ displayed such courage in being so close to the front line was that it was the only place where they were not overlooked from the flanks. 'C' Coy in particular fought a very fine action that day and despite their weariness their tails were very much up. They had fought a square fight and very definitely driven the Hun off with heavy loss".

Brigadier Hinde thought that the Brigade Box position could not be held and that it could be used as a starting point to regain the initiative, but this was not the British command's opinion. The Germans for their part were unwilling to leave such a threat behind their lines. At 16.00 hours, Hinde conferred with Erskine at the 7th Armoured Division's HQ near Livry. There he was told that the 50th Infantry Division was making no progress in the Tilly sector and would be unable either to help him or to relieve his men. Accordingly, it became advisable to straighten up the front line and to pull back the 22nd Armored Brigade to Livry. Retreat was inevitable. The first unit was due to set off at 23.45, at nightfall in order to move under cover of darkness and link up with the rest of the division.

ont été distribuées aux soldats et les chars avec leurs mitrailleuses... L'ennemi fait une trouée dans nos positions d'artillerie qui demande le support de chars, j'envoie ce qu'il reste de la section de reconnaissance Presque aussitôt après, l'ennemi fait irruption à travers le front dans le secteur des Queen's à l'est, j'envoie mon chef de troupe du RHQ sur la route pour arrêter tous les chars qui s'enfoncent directement vers le milieu de la Box... Il me reste trois chars au RHQ... Le QG de l'escadron B est salement touché, un de ses officiers est tué, et Punch, commandant de l'escadron B est blessé... La bataille continue dans un incessant fracas infernal. Enfin le feu commence à s'apaiser puis bientôt meurt complètement. L'attaque a été repoussée ». Elle a duré environ une heure.

La tentative de l'armée allemande au cours de la journée s'est soldée par un véritable échec et ses pertes sont importantes en matériel et en hommes qui se sont battus parfois, d'une façon fanatique, sans beaucoup de couverture et de préparation d'artillerie. Les Allemands ont trouvé en face d'eux des Britanniques qui, bien à l'abri, ont utilisé superbement leurs armes individuelles légères, leurs mitrailleuses, leurs mortiers, les chars d'une Brigade Blindée et, ne pouvant déployer que peu d'artillerie, ont pu bénéficier de l'aide de batteries de l'armée américaine et de bombardiers de la *Royal Air Force*. Le résultat est que les Allemands ont perdu une vingtaine de *Panzer IV* et de 700 à 800 hommes tués ou blessés.

Les Britanniques ont eu de leur côté, des pertes assez légères en comparaison de celles de leurs adversaires, une quarantaine de tués, 8 au *5th Queen's*, 6 au *7th Queen's*, 7 à la *Rifle Brigade*, 5 au *5th RHA*, 3 au *8th Hussars*, 3 au *5th RTR* et 9 au *1st RTR*. Le *5th RTR* a perdu 3 *Cromwell*.

La Brigade va pouvoir se retirer mais, par précaution, l'heure de la retraite est retardée d'une heure, il faut réorganiser le départ, rassembler les unités, laisser des troupes en arrière-garde pour défendre le village et couvrir la retraite.

Vers une heure et demie du matin, la première unité du *7th Queen's* prend le départ pour Livry en compagnie du *5th RTR*. Les deux régiments laissent derrière eux les compagnies C et D de l'infanterie avec un escadron de blindés pour constituer l'arrière-garde qui restera en position jusqu'à environ 2 heures. Les fantassins gagneront leurs nouvelles positions derrière les lignes anglaises, montés sur le capot des chars. A Livry, ils retrouveront le *6th Queen's* du lieutenant-colonel Mickael Forrester resté en réserve à l'arrière pour défendre le front.

Après sa rencontre avec le lieutenant-colonel Woods, commandant du *5th Queen's* pour préparer le départ, le lieutenant-colonel Goulburn retourne à son QG. Il est 22 h 30. Un quart d'heure plus tard, un flot important de véhicules commence à quitter le village. Son tour de partir viendra vers 3 heures, à la suite de la *Rifle Brigade* et du *4th CLY*. L'escadron B toutefois suivra plus tard, il restera en arrière-garde et emportera sur les coques de ses chars vers un refuge plus sûr, les fantassins d'une compagnie du *5th Queen's* fatigués et sommeillants.

Un régiment de chars était resté dans la « Box », le *1st RTR*. Il restera en alerte jusqu'à l'aube et, avant de partir, il s'assurera que la Brigade Blindée est bien rentrée à l'abri. Le matin du 15 juin, il ne reste du bourg d'Amayé-sur-Seulles, qu'une maison encore debout ainsi que le clocher de l'église qui servait de repère pour les réglages de l'artillerie. Les Britanniques partis, les Allemands le feront sauter. Au sud, au hameau des Bruyères, tout le terrain est sans dessus dessous, ce ne sont que des arbres déchiquetés et des immeubles réduits à des pans de murs

The withdrawal was covered by artillery fire and the intervention of bombers around the Box to keep back the Germans and to drown the drone of engines and the rumbling of tank tracks. The operation was codenamed Aniseed. "Shortly after orders had been issued for this night withdrawal the enemy put in a heavy attack on the rear of the Bde.Box with two Bns. of infantry supported by about 30 tanks. Tac Bde.HQ, 5 RHA and 1/5 Queens were mainly involved, though the Bn. had, for the first time that day, some quite heavy shelling of the area. In addition, 'A' Coy on the right were engaged with infantry patrols". An attack from the Panzer-Lehr-Division in the north hit the Rifle Brigade Thus, at around 21.00 hours, the «thunder» burst on all sides.

Brigade HQ obtained support from one AGRA (Army Group Royal Artillery) and three American artillery groups based in the Caumont area and these launched a defensive barrage of 155 mm shells. Lacking room in the Box, the 5th RHA could only deploy some of its motorized 105 mm gun batteries. "G" Battery sight fired air shells at Panzergrenadiers arriving from the south almost in the open within less than 400 meters of the guns. "K" Battery first eastwards. The 8th Hussars commander Lieutenant-Colonel Goulburn, wrote in his diary, "our infantry are firing every small arms weapon they have got and the tanks their Besas. At this moment [...] the enemy are breaking into the gun positions and they want tank support. I send [...] what was left of the Recce Troop. Almost immediately afterwards a very frightened man arrives from the Queens on the East sector to say that enemy had broken through on their front. I send my RHQ troop leader [...] down the road to stop any tanks breaking right into the middle of the "box". I am now left with three RHQ tanks. [...] B Sqn HQ has been badly stonked. Philip de May has had his legs blown off and is dead. Punch is wounded. The battle continues and there is no abatement in the noise. Fire begins to slacken and soon dies away altogether. The attack has been beaten off. It lasted about an hour."

The German Army's attempt during the day was a decided failure, with significant losses in equipment and in men who sometimes fought fanatically, without much cover or preliminary artillery fire. The Germans found themselves facing British soldiers who were well covered and made superb use of their small arms, their machine-guns and mortars, the tanks of an Armored Brigade and, being unable to deploy much artillery, received help from U.S. Army batteries and Royal Air Force bombers. The outcome was that the Germans lost about twenty Panzer IVs and 700 to 800 men killed or wounded.

The British on their side, sustained only light casualties compared to the enemy, with forty killed: 7th Queen's 8, 7th Queen's 6, the Rifle Brigade 7, 5th RHA 5, 8th Hussars 3, 5th RTR 3 and 1st RTR 9. The 5th RTR lost 3 Cromwells.

The Brigade was able to withdraw, but as a precaution, the retreat was put back an hour to make time to reorganize the departure, gather the units together, leave rearguard troops to defend the village and cover the retreat.

At around one thirty a.m., the first 7th Queen's unit set off for Livry with the 5th RTR. The two regiments left behind infantry companies "C" and "D" with a tank squadron to form the rearguard which was to remain in position until around 02.00 hours. The infantrymen rode on the top of the tanks up to their new positions behind the British lines. In Livry, they found Lieutenant-Colonel Michael Forrester's 6th Queen's standing in reserve in the rear to defend the front.

encore debout, au milieu de trous de bombes et d'obus. Cinquante ans après, tout a été reconstruit et les maisons de style moderne, qui ont remplacé les anciennes fermes, font deviner l'étendue du champ de bataille d'alors.

Au premier abord, on peut constater que l'opération *Perch* a apporté aux alliés un léger avantage territorial. L'avancée américaine sur Caumont est maintenant protégée vers l'est par suite de la présence de l'armée anglaise à Livry. Toutefois, l'important nœud de communication de Briquessard, hameau de Livry, est resté allemand. Les Anglais conseillent d'ailleurs aux habitants de quitter la zone et d'aller se réfugier quelques kilomètres vers le nord. Beaucoup suivront le conseil et reviendront le jour soigner leurs animaux restés dans les fermes. Approximativement, la ligne de front est la suivante : depuis Caumont libéré par les Américains le 13, elle passe au nord de Briquessard, ensuite par le hameau du Buquet, les hameaux du Bosq Renard et de la Couarde puis, à l'ouest de Torteval. Cette zone va rester sensible un certain temps et restera sous la menace de l'armée allemande jusqu'au déclenchement de l'opération « *Blue coat* » fin juillet prochain.

Les raisons d'un échec

Pour beaucoup de commentateurs, la bataille de Villers-Bocage aura été un « *épisode lamentable* » pour les Britanniques et l'opération *Perch* un réel échec. Toutefois nous avons vu que si la tentative de contournement de la *Panzer-Lehr-Division* avait échoué elle avait permis de faire subir à l'armée allemande des pertes importantes en chars, principalement en Tigres le premier jour, et en hommes dans la bataille d'Amayé-sur-Seulles que l'on ne peut dissocier de la bataille de Villers.

Un ancien combattant de la *11th Armoured Division* en 1944 et de la *7th Armoured Division* en 1945, Patrick Delaforce, devenu écrivain, a résumé ainsi en quelques mots cette bataille dans le titre du 5e chapitre d'un de ses livres « *Churchill's desert rats : Villers-Bocage, Defeat, a Draw and Victory* » (*Defeat*, défaite le matin du 13 juin, *Draw*, match nul l'après-midi du 13 juin, *Victory*, victoire le 14 juin).

On peut aussi dire que l'armée allemande a subi un échec puisque la *2. Panzer-Division* n'a pas atteint le but qui lui était fixé : séparer les armées américaine et anglaise. Les Anglais avaient une supériorité numérique en hommes et en matériel et, malgré cela, l'opération *Perch* a été un échec. La réussite de cette action aurait porté un coup sévère à l'armée allemande, surtout à la *Panzer-Lehr-Division* et peut-être précipité la fin des combats.

Tout d'abord, la *7th Arm. Div.* était la seule division à être engagée dans cette entreprise et certains de ses régiments étaient incomplets. De plus les unités qui ont mené l'attaque à Villers-Bocage ne constituaient qu'une avant-garde qui, ne rencontrant aucune opposition au début, s'était aventurée un peu trop rapidement en avant. Le gros de la division suivait, avançant plus lentement le long d'une route assez étroite, bien protégée sur ses flancs par deux régiments de reconnaissance le *8th* et le *11th Hussars* alors que les unités de tête s'étaient avancées pratiquement sans protection.

De bonne heure, le matin du 13 juin, l'objectif de la division était atteint, l'occupation d'un des points culminants du Bocage, la cote 213 et la côte des Landes, sur la route de Caen. Son but était d'établir et de constituer une espèce d'enclave derrière le front allemand. Un escadron de blindés du *4th CLY* était en place, suivi d'une compagnie de fusiliers de la *Rifle Brigade* qui devait l'aider à prendre position.

After his meeting with 5th Queen's commander, Lieutenant-Colonel Woods, to prepare for the departure, Lieutenant-Colonel Goulburn went back to his HQ. This was at about 22.30. Fifteen minutes later, a great stream of vehicles began to leave the village. His turn to leave came at around 03.00 hours, behind the Rifle Brigade and the 4th CLY. "B" Squadron was to follow later however and stay in the rearguard, to carry on the hulls of its tanks the infantrymen of a company of the 5th Queen's half-asleep with exhaustion towards somewhere safer.

A tank regiment, the 1st RTR, had stayed behind in the Box. It remained on the alert until dawn and before leaving it made sure that the Armoured Brigade had got back safely. On the morning of June 15th, all that was left of Amayé-sur-Seulles was just one house still standing, and the church spire, used by the artillery to find its range. Once the British had left, the Germans blew it up. To the south, at the hamlet of Les Bruyères, the whole area was devastated, with nothing but shattered trees and buildings reduced to sections of still upright wall amid the bomb and shell craters. Fifty years on, everything has been rebuilt and the modern houses which replaced the old farms give some idea of the size of the battlefield.

At first sight, we note how Operation Perch gave the Allies a slight territorial advantage. With the presence of the British Army at Livry, the American advance on Caumont was now protected on its eastern flank. However, the major crossroads at Briquessard, in the hamlet of Livry, remained in German hands. The British in fact advised the local inhabitants to leave the zone and seek shelter a few kilometers further north. Many followed this advice and came back during the day to tend their livestock left on the farms. The front line was roughly as follows: from Caumont liberated by the Americans on the 13th, it passed north of Briquessard, then by the hamlet of Le Buquet, the hamlets of Bosq Renard and La Couarde, then west of Torteval. This was to remain a sensitive area for some time to come and under threat from the German Army until the launching of Operation Bluecoat late in July.

The causes of failure

For many commentators, the Battle of Villers-Bocage proved a "lamentable episode" for the British and Operation Perch a real failure. However we have seen how, while the attempt at skirting the Panzer-Lehr-Division failed, it did nevertheless make it possible to inflict heavy tank losses on the German Army, mainly Tigers on the first day, and men in the Battle of Amayé-sur-Seulles, which cannot be disassociated from the Battle of Villers.

A veteran of the 11th Armored Division in 1944 and the 7th Armoured Division in 1945, Patrick Delaforce, who later became a writer, summarized this battle in the following few words in the title of Chapter 5 of one of his books: "Churchill's Desert Rats: Villers-Bocage, Defeat, a Draw and Victory" (defeat on the morning of June 13th; a draw on the afternoon of June 13th; victory on June 14th).

The German Army can also be said to have suffered a setback inasmuch as 2. Panzer-Division failed to achieve its assigned objective of driving a wedge between the American and British armies. The British had numerical superiority in men and equipment, and for all that, Operation Perch was a failure. The success of that operation would have struck a severe blow to the German Army, especially the Panzer-Lehr-Division, and maybe even brought the battle to an end.

Un deuxième escadron devait les rejoindre ainsi qu'un régiment d'infanterie, le *7th Queen's*, qui devait aussitôt occuper le terrain, préparer des abris, placer ses hommes et ses bouches à feu, canons antichars, PIAT, mortiers et remplacer les fusiliers.

Cependant, côte des Landes, se trouvait le lieutenant M. Wittmann et une partie de sa compagnie de Tigres. Personne ne l'avait prévu. Il était là un peu par hasard puisqu'il devait ce jour-là prendre un peu de repos pour réparer les dommages causés à ses chars par les attaques de l'aviation alliée depuis son départ de Beauvais et attendre que ce qu'il restait de sa compagnie le rejoigne.

Depuis leur arrivée en Normandie, les Britanniques devaient savoir qu'une attaque de chars dans le Bocage ne devait pas se faire sans accompagnement d'infanterie de support. En ce matin de juin, ils ont pu constater qu'une telle erreur coûte cher. Lorsque des tanks sont seuls en avant, coupés de leurs fusiliers qui les suivent, ils sont exposés aux tirs d'obus perforants des chars ennemis, aux tirs des canons antichar ou aux tirs des charges creuses des *panzerfaust* de l'infanterie. De son côté, l'infanterie qui avance seule sans le soutien de chars ou d'artillerie est menacée par les mitrailleuses et les canons à munitions explosives des tanks adverses et des tirs de barrage de l'artillerie. Les Allemands ont eu à souffrir eux aussi de ces erreurs, l'après-midi du 13 lorsque les Tigres ont affronté sans leurs grenadiers les obus antichars de 57 mm et les charges creuses des *PIAT* utilisés avec efficacité par les fantassins du *7th Queen's* et le lendemain lorsqu'ils ont lancé sans suffisamment de soutien des vagues de fantassins contre la « *Brigade Box* ».

Les Anglais, mal informés, mal renseignés, ont oublié qu'ils avaient devant eux une armée bien entraînée, bien organisée, invisible mais présente, utilisant chaque parcelle de terrain pour se camoufler et s'implanter afin de constituer de petites unités difficiles à déloger. D'ailleurs, lorsque l'escadron B du *4th CLY* et le *7th Queen's* veulent sortir de Villers pour aller secourir l'escadron A, ils ne peuvent franchir l'espace découvert situé entre la lisière du bourg et la côte des Landes sous le feu des mitrailleuses des panzers, des armes individuelles, des « *panzerfaust* » et des obus de mortier des panzergrenadiers.

Le capitaine Christopher Milner, de la *Rifle Brigade* a écrit dans un livre intitulé « *Les Rats du désert et la Guerre* » : *si quelqu'un avait eu le temps de s'arrêter, les petits groupes d'habitants qui étaient là, dans la rue, et les pompiers bien en évidence sur une place de la ville, auraient pu prévenir qu'il y aurait des problèmes.* » Plus loin, il regrette que le matin du 13, la section de reconnaissance du lieutenant Allen Matter ait été absente. « *Cet officier a prouvé par la suite, avant d'être tué, qu'il était un commandant de carriers très capable et plein de ressources et que le résultat de la bataille aurait pu être bien différent si le commandant des forces avait bénéficié des renseignements que ce peloton de reconnaissance lui aurait apportés. S'il avait été disponible, pour protéger le sud de la colonne, les chars allemands auraient pu être repérés beaucoup plus vite puisque la piste parallèle à notre route que les ennemis suivaient, aurait pu permettre une meilleure couverture aux véhicules surbaissés. On peut spéculer sur un résultat fantastiquement différent de la bataille de Normandie si la percée avait été assez réussie pour que la brigade des* Queen's *puisse suivre à travers la brèche et s'enfoncer derrière les lignes allemandes.*

Le résultat aurait été différent si les chefs de section avaient été avec leurs hommes et les chars allemands auraient été vulnérables à nos soldats derrière ces

First of all, the 7th Arm. Div. was the only division committed in this action and some of its regiments were under-strength. Moreover the units that led the attack at Villers-Bocage were no more than the vanguard which had forged ahead rather too quickly after at first encountering no opposition. The bulk of the division followed, advancing more slowly along a pretty narrow road, well protected on its flanks by two reconnaissance regiments, the 8th and 11th Hussars, whereas the leading units had advanced with practically no protection.

Early on the morning of June 13th, the division's objective was achieved, the occupation of one of the highest hills in the Bocage, Point 213 and the Côte des Landes, on the Caen road. Its aim was to establish and constitute a kind of enclave behind the German front. The 4th CLY had a tank squadron in place, followed by a company of the Rifle Brigade to help it get into position. A second squadron was to join them, along with an infantry regiment, the 7th Queen's, which was to occupy the area immediately, prepare shelters, position its men and ordnance, anti-tank guns, PIATs, mortars, and relieve the riflemen.

However, on the Côte des Landes were Lieutenant M. Wittmann and part of his Tiger company, something nobody had bargained for. He was there more by chance than by design, as that day he was supposed to be getting some rest and repair the damage caused to his tanks by Allied air attacks since leaving Beauvais, and also to wait for what remained of his company to catch up.

Since arriving in Normandy, the British should have learned that a tank attack in the Bocage was not to be attempted without supporting infantry. On this June morning, they found out just how costly such an error could be. When tanks went on ahead alone, cut off from their riflemen following behind, they were exposed to shellfire piercing the enemy armor, anti-tank gunfire and shaped charges fired by the infantry's Panzerfausts. Likewise, infantry advancing alone without the support of tanks or artillery came under threat from machine-guns and enemy tank guns firing explosive ammunition and artillery barrages. The Germans also suffered the consequences of these same mistakes on the afternoon of the 13th when without their grenadiers the Tigers faced 6 pounder anti-tank shells and PIAT shaped charges used to great effect by the infantrymen of the 7th Queen's, and again the following day when they launched waves of infantrymen against the Brigade Box without adequate support.

The British, with poor information and intelligence, forgot that they were facing a well-trained, well-organized army, unseen but very much there, using every field to camouflage themselves and dig in as small units hard to dislodge. Moreover, when "B" Squadron 4th CLY and the 7th Queen's tried to leave Villers to rescue "A" Squadron, they were unable to get across the open area between the edge of the town and the Côte des Landes when they came under fire from the panzer machine-guns, and Panzergrenadier small arms, Panzerfausts and mortar shells.

Captain Christopher Milner of the Rifle Brigade wrote in a book entitled «The Desert Rats and the War»: "if one had had time to top, the small groups of people standing about and "les pompiers" (the fire brigade) being much in evidence in one area, they might have warned of trouble". Further on, he regrets that on the morning of the 13th, Lieutenant Allen Mather's reconnaissance platoon was not there. "As this officer proved subsequently before he was killed, he was a particularly able and resourceful carrier commander and the outcome of the battle might have been very dif-

haies épaisses et ces talus du Bocage ou bien en arrière dans la ville où nos chars étaient piégés dans les rues étroites sans ces fusiliers « voltigeurs » qui avaient été formés pour combattre justement avec eux dans de telles conditions. »

Au Quartier Général du Haut Commandement, certains pensaient qu'une seule brigade aurait dû réussir. C'était méconnaître la puissance des forces blindées allemandes en particulier la force du Tigre et du canon de 88 mm. Les nouveaux chars *Cromwell* que certains pointeurs avaient à peine essayés au feu étaient beaucoup moins performants que le Tigre et les tankistes anglais regrettaient les *Shermans* qu'ils connaissaient mieux pour les avoir utilisés en Afrique du Nord et ils auraient souhaité disposer davantage de *Fireflies*.

Le général Dempsey disait que « *cette attaque aurait dû réussir, il rejetait la responsabilité de l'échec sur les généraux Bucknall et Erskine* » et avait espéré davantage d'imagination du *Brigadier* Hinde pour triompher.

Pour certains vétérans de la *7th Armoured Division*, il était intolérable de penser que d'une façon ou d'une autre, ils étaient responsables de la défaite et qu'ils avaient gaspillé une belle occasion, ce qui les conduisait à méditer utilement sur la tactique d'une formation blindée anglaise. Enfin, d'autres pensaient que les généraux Montgomery et Dempsey auraient dû s'intéresser personnellement à une bataille aussi importante.

Bilan

Les tankistes de la *7th Arm. Div.* connaissaient le Tigre et le craignaient pour l'avoir rencontré en Afrique. Ils savaient que son blindage avant était si épais qu'aucun de leurs obus ne pouvait le pénétrer et que son canon de 88 mm était d'une efficacité redoutable. Ils savaient aussi qu'il était moins bien protégé sur les flancs et au niveau des chenilles. Le mythe de l'invulnérabilité du Tigre après la destruction de 6 chars lors des combats dans Villers avait été remis en question. Les Britanniques savaient déjà que des chars avaient été mis hors de combat par des bombes et des roquettes lors d'attaques aériennes et maintenant ils avaient constaté que les obus des *Fireflies*, les obus de 57 mm et les charges creuses des *PIAT*, tirés d'assez près et bien ajustés pouvaient immobiliser le *Panzer VI* et le détruire.

Un autre événement a permis de transformer l'échec initial en satisfaction, la venue de la *2. Panzer-Division*. Dès son arrivée, elle a permis d'aider les grenadiers de la *SS-Panzer-Abteilung 101* et le détachement de la *Panzer-Lehr-Division* venu à la rescousse et de neutraliser la percée de la *7th Arm. Div.* à Villers. Mais, le lendemain, elle a été littéralement stoppée dans son action lorsqu'elle s'est heurtée au véritable « hérisson » blindé que constituait la « *Brigade Box* » à Amayé-sur-Seulles. La *2. Pz.-Div.* qui devait renforcer le *Panzer-Lehr-Division* sur sa gauche et séparer les armées américaine et anglaise à la limite des deux secteurs d'influence aux environs de Port-en-Bessin en attaquant vers Balleroy et la forêt de Cerisy, n'a pu réussir sa contre attaque. On peut dire que la « bataille défensive » livrée par la brigade s'est transformée en une « victoire défensive » et rappeler que la *7th Arm. Div.* ne s'était pas trouvée dans une telle situation depuis El Alamein en 1942. Cette résistance anglaise et cet échec allemand arrivaient à un moment important de la guerre en Normandie, à un moment où le front se stabilisait.

ferent if the force commander had had the benefit of information from this very experienced reconnaissance platoon to draw upon, [...]. Had they been available to protect the southern flank of the regimental group as it swung from south to east, not only might the German tanks have been rumbled much earlier (since the parallel track which the enemy followed would have provided even better cover for the low-slung carriers). [...] One can even speculate upon a fantastically different outcome of the Normandy battle if only this break-through been successful enough to enable the Queen's Brigade to follow through the gap and dig in behind the German lines. Things would also have been different had the platoon commanders been with their men [...] (the German tanks would have been terribly vulnerable to intelligent soldiers behind those thickly banked bocage hedges [...]) or back into the town, where their presence could have made all the difference to the CLY, who were trapped in the narrow streets without the riflemen stirrup-hangers who had been trained to work with them in just such conditions."

At the High Command headquarters, some felt that just one brigade ought to have succeeded. This was to ignore the strength of the German armored forces and particularly the mighty Tiger and the 88mm gun. Some of the gun-layers had hardly had a chance to test the gun on the new Cromwells, and these tanks were much less powerful than the Tiger; the British tank crews missed the Shermans, with which they were more familiar, having used them in North Africa, and they would also have liked to have more Fireflies.

According to General Dempsey, the attack "should have succeeded"; he held Generals Bucknall and Erskine responsible for its failure, also saying he had hoped for more imagination from Brigadier Hinde to win the day.

Some 7th Armoured Division veterans found it intolerable to think that one way or another they were responsible for the defeat and had squandered a golden opportunity, providing food for thought on the tactics of an British armored formation. Others felt that Generals Montgomery and Dempsey should have taken a personal hand in such an important battle.

Assessment

The 7th Arm. Div.'s tank crews knew and feared the Tiger for having faced it in Africa. They knew about its front armor being so thick that none of their shells could penetrate it, and the awesome effectiveness of its 88 mm gun. They also knew that it was less better shielded on the sides and on the level of the tracks. The myth of the Tiger's invulnerability was called into question following the destruction of 6 tanks during the engagements in Villers. The British already knew that some tanks had been knocked out by bombs and rockets during air raids and they now noted how, when aimed accurately and fired from fairly close range, the Firefly's shells, the 6 pdr. shells and the PIAT shaped charges could immobilize or destroy a Panzer VI.

Another event helped to turn initial failure into a cause for satisfaction: the arrival of the 2. Panzer-Division. Immediately on reaching the scene, it was able to help the grenadiers of SS-Panzer-Abteilung 101 and the detachment of the Panzer-Lehr-Division which had come to the rescue, and prevent the 7th Arm. Div. from breaking out at Villers. But, the next day, it was literally stopped in its tracks when it ran into this regular armored «hedgehog» which was the Brigade Box at Amayé-sur-Seulles. The 2. Pz.-Div., which was to reinforce the Panzer-Lehr-Division on its left and

Commentateurs et auteurs ont employé des expressions peut-être excessives pour décrire l'action de Michael Wittmann à Villers-Bocage telles que : « *grâce à l'intervention d'un seul homme...Wittmann a réalisé l'action isolée la plus dévastatrice de la guerre... Il a détruit en 5 minutes 25 véhicules britanniques... Un Tigre seul contre toute une brigade* ». Dans le rapport du général de corps d'armée SS Dietrich on peut lire : « *Au travers de cet acte déterminant contre un ennemi enfoncé dans ses propres lignes, agissant seul et de sa propre initiative avec un grand courage avec son char, il a détruit la plus grande partie de la 22ᵉ brigade blindée et sauvé le front du Iᵉʳ corps blindé SS de la menace d'un danger imminent* ». Son attaque a été une action peut-être irréfléchie mais courageuse d'un soldat ou bien celle d'un officier de panzer qui a évalué rapidement l'avantage qu'il pouvait tirer de sa position sans songer au danger potentiel qu'il affrontait. Détruire seul un escadron de chars, une compagnie d'infanterie, le Quartier Général d'un régiment blindé en peu de temps est un exploit impensable ou tout au moins rarement réalisé. Toutefois, il ne faut pas oublier en réalité, que la manœuvre osée du général Montgomery a été mise en échec par deux compagnies de chars lourds Tigre, de panzergrenadiers, d'un détachement de *Panzer IV*, de fantassins et d'artillerie de la *Panzer-Lehr-Division* et de grenadiers de la *2. Pz.-Div.* et non le fait d'un seul homme.

On pourra mettre à l'actif de Wittmann lors de son attaque surprise de la matinée, quelques chars de l'escadron A du *4th CLY*, Cromwell, Fireflies, les véhicules de la *Rifle Brigade*, half-tracks, carriers, canons antichars, les quatre *Cromwell* du Quartier Général du *4th CLY* et 3 chars *Honey* de la section de reconnaissance, le *Cromwell* d'observation du *5th RHA* rue Curie et le *Sherman* d'observation à canon en bois, rue G. Clemenceau, quelques véhicules blindés, le *half-track* médical et le scout-car de l'officier de renseignement.

Il a dû également participer avec les autres chars de sa compagnie au nettoyage de la côte des Landes, mais l'après-midi, c'est sur son conseil que le capitaine Möbius a lancé sa compagnie de Tigres sur Villers sans infanterie de soutien. On peut ainsi considérer qu'il est en partie responsable de la perte de 5 Tigres et qu'il a commis une erreur tactique qui a coûté cher à l'armée allemande. En outre, lui, l'as des panzers qui avait mis jusqu'alors 114 chars hors de combat, a été le chef de char du premier Tigre détruit au combat en Normandie.

Bien qu'il soit difficile de connaître exactement l'importance des pertes anglaises, en matériel détruit ou abandonné à Villers et la région, on peut admettre les chiffres donnés par le *War Diary* du 13 juin 1944, le journal de guerre.

Pour le *4th CLY*, 27 chars : 20 *Cromwell*, 4 *Fireflies*, 3 *Honey* ; 3 scout-cars, 1 *half-track*. Pour la *Rifle Brigade*, 9 *half-tracks*, 2 *Bren-carriers*, 4 *Loyd-carriers*, quelques véhicules légers et 2 antichars de 6 pounds. Pour le *5th RHA*, 1 *Sherman*, 2 *Cromwell*. Pour le *11th Hussars*, 9 véhicules blindés *Daimler* dont 3 à Villers même. Pour le *8th Hussars*, quelques chars légers.

De leur côté les Allemands ont perdu 6 Tigres dont 5 dans la ville même et au minimum 2 *Panzer IV*. Les 6 Tigres détruits sont les premiers Tigres mis hors de combat lors de la bataille. Par contre, le lendemain à Amayé-sur-Seulles, ils auraient perdu une vingtaine de chars, des *Panzers IV* de la *Panzer-Lehr-Division* touchés par les bombardements de la *RAF*, les tirs de l'artillerie anglaise et américaine et peut-être aussi de l'artillerie de marine croit-on, au cours

to drive a wedge between the American and British armies on the border between the two sectors in the Port-en-Bessin area by attacking towards Balleroy and Cerisy Forest, failed in this counter-attack. The «defensive battle» fought by the brigade may be said to have been turned into a «defensive victory», remembering that the 7th Arm. Div. had not been in such a situation since El Alamein in 1942. This British resistance and this German failure arrived at a crucial juncture of the war in Normandy, at a time when the front was being stabilized.

Commentators and writers have perhaps let themselves become a little carried away in describing Michael Wittmann's action at Villers-Bocage: «through the intervention of just one man...Wittmann carried out the most devastating isolated action of the whole war... In 5 minutes he destroyed 25 British vehicles... A single Tiger against an entire brigade». In SS Lt.-Gen. Dietrich's report we read: «Through this decisive action against an enemy who had penetrated his own lines, acting alone on his own initiative with great courage with his tank, he destroyed a large section of the 22nd Armored Brigade and saved I SS Armored Corps' front from the threat of imminent danger». His attack was perhaps the somewhat foolhardy action of a soldier or panzer officer who quickly assessed the advantage to be gained from his position without thinking of the potential danger that he faced. The feat of destroying single-handed a tank squadron, an infantry company, the headquarters of an armored regiment in next to no time is unthinkable, almost unheard of. However, it must be remembered that General Montgomery's daring move was actually thwarted not by just one man but by two companies of heavy Tiger tanks, Panzergrenadiers, a detachment of Panzer IVs, the infantry and artillery of the Panzer-Lehr-Division and the grenadiers of 2. Pz.-Div..

During his surprise attack that morning, Wittmann may be credited with the following kills: some tanks of Squadron A 4th CLY, Cromwells, Fireflies, Rifle Brigade vehicles, half-tracks, carriers, anti-tank guns, four 4th CLY Cromwells HQ and 3 Honey tanks belonging to the reconnaissance troop, the 5th RHA observation Cromwell in the Rue Curie and the observation Sherman with a wooden gun in the Rue Clemenceau, some armored vehicles, the medical half-track and the intelligence officer's scout car.

He also must have taken part with the other tanks of his company in mopping up the Côte des Landes, but that afternoon, it was on his advice that Captain Möbius launched his company of Tigers against Villers with no infantry in support. He may therefore be held partly responsible for the loss of 5 Tigers and for a tactical error that was to prove costly for the German Army. Moreover it was he, the panzer ace who had previously knocked out 114 tanks, who was commander of the first Tiger destroyed in the Battle of Normandy.

Although it is hard to know exactly the scale of British losses, in matériel destroyed or abandoned in and around Villers, we may accept the figures supplied by War Diary for June 13th, 1944.

4th CLY, 27 tanks: 20 Cromwells, 4 Fireflies, 3 Honeys; 3 scout cars, 1 half-track. Rifle Brigade, 9 half-tracks, 2 Bren-carriers, 4 Loyd-carriers, a few light vehicles and 2 6-pounder anti-tank guns. 5th RHA, 1 Sherman, 2 Cromwells. 11th Hussars, 9 Daimler armored vehicles including 3 in Villers itself. 8th Hussars, a few light tanks.

On the German side, losses came to 6 Tigers, including 5 in the town itself, and at least 2 Panzer IVs. The 6 Tigers destroyed were the first Tigers to be knocked out in the entire battle. On the other hand,

des actions des unités engagées dans la « *Brigade Box* ».

Quant aux pertes en hommes on peut les évaluer ainsi côté allié les 13 et 14 juin :

4th CLY : 2 officiers et 10 hommes tués, 10 officiers et 66 hommes blessés ou prisonniers. Le colonel Lord Cranley a été fait prisonnier à la côte de Landes.

Rifle Brigade : 4 officiers et 12 hommes tués, 48 hommes blessés, 85 prisonniers.

Queen's : le 13 juin, 21 hommes tués, 8 blessés et 47 prisonniers, durant les deux jours, il manquait 8 officiers et 120 hommes au *7th Queen's* dont 2 officiers et 8 hommes tués, au *5th Queen's*, 42 combattants étaient manquants.

Le *1st RTR* avait eu 2 officiers et 8 hommes tués, le *5th RTR* 1 officier et 3 hommes, le *8th Hussars* 2 officiers et 5 hommes, le *11th Hussars* 1 homme, le *5th RHA*, 4 hommes et le *Norfolk Yeomanry* 2 hommes.

Les pertes en hommes de la *SS-Pz.-Abt. 101* auraient été les suivantes :

Le 13 juin, 1ʳᵉ compagnie, 9 tués, 2 chefs de chars les adjudants H. Ernst et H. Swoboda, le sergent Zellner, le caporal Hruschka, le *Rttf* Hermani et 4 autres soldats. 10 blessés, le chef de char s/lieutenant Lukasius, le sergent Salamon, le sergent Langer et 7 hommes. 2ᵉ compagnie, 1 tué, le conducteur Schmidt et 3 blessés.

Le 14 juin, 1ʳᵉ compagnie, 3 blessés, 1 sous-officier et 2 hommes. 2ᵉ compagnie, 2 hommes blessés.

La *2. Pz. Div.* et la *Panzer-Lehr-Division* ont aussi perdu des combattants le 13 juin mais c'est surtout le 14 que ces deux divisions ont eu beaucoup de pertes. Ne parle-t-on pas en effet de 700 à 800 tués et blessés.

L'opération Perch s'était donc terminée par une retraite sur une base légèrement plus avancée que la base de départ. L'opération était hardie et aurait mérité d'être engagée avec plus d'unités pour pouvoir réussir complètement. La division des « rats du désert » aura souffert et le *4th CLY* été particulièrement « étrillé ». Il participera un mois plus tard, réduit à deux escadrons à l'opération *Goodwood* au sud de Caen, puis il sera dissous le 29 juillet, à la veille de l'opération *Bluecoat*, pour être réuni à un régiment « frère » le *3th CLY* et former le *3/4 CLY*.

En conclusion, on pourra dire manifestement que le responsable de cet échec de la tentative du général Montgomery est sans conteste le lieutenant Michael Wittmann, mais on ne pourra pas dire que ce dernier est le vainqueur de la bataille de Villers-Bocage

Qui était Michael Wittmann ?

Michael Wittmann est né le 22 avril 1914 à Vogelthal dans le Haut Palatinat. Son père, Johann, est cultivateur et Michael le seconde quelque temps à la ferme. En février 1934, il s'enrôle dans le service de travail volontaire, le précurseur du service du travail obligatoire du Reich. Après six mois de service du travail, il retourne à la ferme jusqu'à ce qu'il s'engage dans l'armée en octobre.

Après deux ans de service militaire au 19ᵉ régiment d'infanterie, il est démobilisé en septembre 1936 avec le grade de caporal. Le 1ᵉʳ avril 1937, il est volontaire pour la SS et accepté dans la *Leibstandarte SS Adolf Hitler, la LSSAH.*, une unité militaire de la Waffen-SS, un régiment renforcé à l'origine avant de devenir une division, dont l'un des bataillons est affecté à lar garde la Chancellerie.

A la déclaration de guerre de 1939, Wittmann est *Unterscharführer*, sergent SS, et combat en Pologne

the next day at Amayé-sur-Seulles, the Germans would seem to have lost some twenty tanks, Panzer IVs of the Panzer-Lehr-Division hit during RAF bombing raids or by British and American artillery fire and maybe also, some think, by naval artillery during action by the units committed in the Brigade Box.

As for losses in men, on the Allied side they may be assessed as follows for June 13th and 14th:

4th CLY : 2 officers and 10 men killed, 10 officers and 66 men wounded or captured. Colonel Lord Cranley was taken prisoner on the Côte des Landes.

Rifle Brigade : 4 officers and 12 men killed, 48 men wounded, 85 prisoners.

Queen's : on June 13th, 21 men killed, 8 wounded and 47 prisoners, over the two days, 8 officers and 120 men of 7th Queen's went missing including 2 officers and 8 men killed, and 42 fighting troops of 5th Queen's went missing.

1st RTR lost 2 officers and 8 men killed, 5th RTR 1 officer and 3 men, 8th Hussars 2 officers and 5 men, 11th Hussars 1 man, 5th RHA 4 men and the Norfolk Yeomanry 2 men.

SS-Pz.-Abt. 101's losses in men are thought to have been as follows:

June 13th, 1st Company, 9 killed, 2 tank commanders Adjudants H. Ernst and H. Swoboda, Sergeant Zellner, Corporal Hruschka, Rttf Hermani and 4 other ranks. 10 wounded, the tank commander 2nd Lieutenant Lukasius, Sergeant Salamon, Sergeant Langer and 7 men. 2nd company, 1 killed, driver Schmidt and 3 wounded.

June 14th, 1st Company, 3 wounded, 1 NCO and 2 men. 2nd Company, 2 men wounded.

The Pz. Div. and Panzer-Lehr-Division also lost fighting men on June 13th but these two divisions sustained most of their many casualties on the 14th. The figure of 700 to 800 killed or wounded has been mentioned.

Operation Perch thus ended in withdrawal to a base line slightly more advanced than the starting line. It was a bold move, one that deserved to have been launched with more units in order to achieve complete success. The «Desert Rats» division had a hard time of it and the 4th CLY in particular came in for a real "hiding". A month later, down to two squadrons, it took part in Operation Goodwood south of Caen, and was disbanded on July 29th, the day before Operation Bluecoat, to be merged with a «sister» regiment, the 3rd CLY, to form 3/4 CLY.

In conclusion, clearly it may be said that Lieutenant Michael Wittmann was unquestionably the person responsible for the failure of this attempt by General Montgomery, but Wittmann cannot be said to have won the Battle of Villers-Bocage.

Who was Michael Wittmann?

Michael Wittmann was born on April 22nd, 1914 at Vogelthal in the Upper Palatinate. His father, Johann, was a farmer and Michael helped him for a while the farm. In February 1934, he joined up for voluntary labor service, which later became the Reich compulsory labor service. After six months service, he went back to the farm until October, when he joined the army.

After two years of military service with the 19th Infantry Regiment, he was demobilized in September 1936 with the rank of corporal. On April 1st, 1937, he volunteered for the SS and was drafted into the Leibstandarte SS Adolf Hitler, LSSAH., a military unit of the Waffen-SS, a regiment initially reinforced before beco-

et dans la campagne de Grèce dans une unité blindée. Quand Hitler envahit la Russie en juin 1941, Wittmann sert dans une section de canons d'assaut de la *LSSAH*, la *1. SS-Panzer-Division*, *XIVᵉ Korps* blindé, au cœur de l'action dans le secteur sud du front. Dans cette section de canons d'assaut, les *Sturmgeschütz*, il montre une grande activité et il acquiert rapidement la réputation de rester ferme et déterminé. Le 12 juillet 1941, il est décoré de la Croix de fer de seconde classe et, quelques semaines plus tard, blessé pour la première fois, il est autorisé à porter le badge noir des blessés.

Le 8 septembre 1941, il reçoit la Croix de fer de première classe et, le 21 novembre, l'insigne des canons d'assaut. Lors d'un engagement au cours duquel il a devant lui 16 chars ennemis, il détruit 6 d'entre eux. Ce fait d'armes le fait remarquer et sa candidature est retenue pour entrer à l'école des jeunes officiers SS de Bad-Tölz, en Bavière en juillet 1942. Il est nommé *Untersturmführer*, sous-lieutenant, en décembre.

Au début de 1943, il entre à la 13ᵉ compagnie de chars lourds du 1ᵉʳ régiment blindé SS *(SS-Panzer-Regiment 1)*. Cette compagnie est composée de trois sections de chars lourds *Panzer VI*, Tigre, et d'une section de chars légers *Panzer III*. Wittmann est versé à cette dernière section. Après la bataille de Kharkov (ou Charkov), la compagnie est équipée com-

ming a division, and one of whose battalions was assigned to guard the Chancellery.

When war broke out in 1939, Wittmann was an SS sergeant or Unterscharführer, and fought with a tank unit in Poland and in the campaign in Greece. When Hitler invaded Russia in June 1941, Wittmann served with the LSSAH in an assault gun troop, the 1. SS-Panzer-Division, XIV Korps (armored), in the thick of the action in the southern sector of the front. With this formation, the Sturmgeschütz, he was extremely active and soon earned a reputation for remaining firm and determined. On July 12th, 1941, he was awarded the Iron Cross Second Class and, a few weeks later, after being wounded for the first time, was allowed to wear the black badge of the wounded.

On September 8th, 1941, he was awarded the Iron Cross First Class, and on November 21st, the assault gun insignia. During an engagement in which he faced 16 enemy tanks, he destroyed 6 of them. This feat of arms brought him into the limelight and he won a place at the school of young SS officers at Bad-Tölz in Bavaria in July 1942. He was promoted to Untersturmführer (second lieutenant) in December.

Early in 1943, he joined the 13th Heavy Tank Company of the 1st SS Armored Regiment (SS-Panzer-Regiment 1). This company comprised three troops

Document établi par le *SS-Oberführer* Wisch afin de justifier la demande d'attribution de la croix de chevalier de la croix de fer de Michael Wittmann. Il est daté du 10 juin 1944. (Coll. part.)

Document drawn up by SS-Oberführer Wisch in order to justify the nomination of Michael Wittmann for the Knight's Cross of the Iron Cross. It is dated June 10th 1944. (Coll. part.)

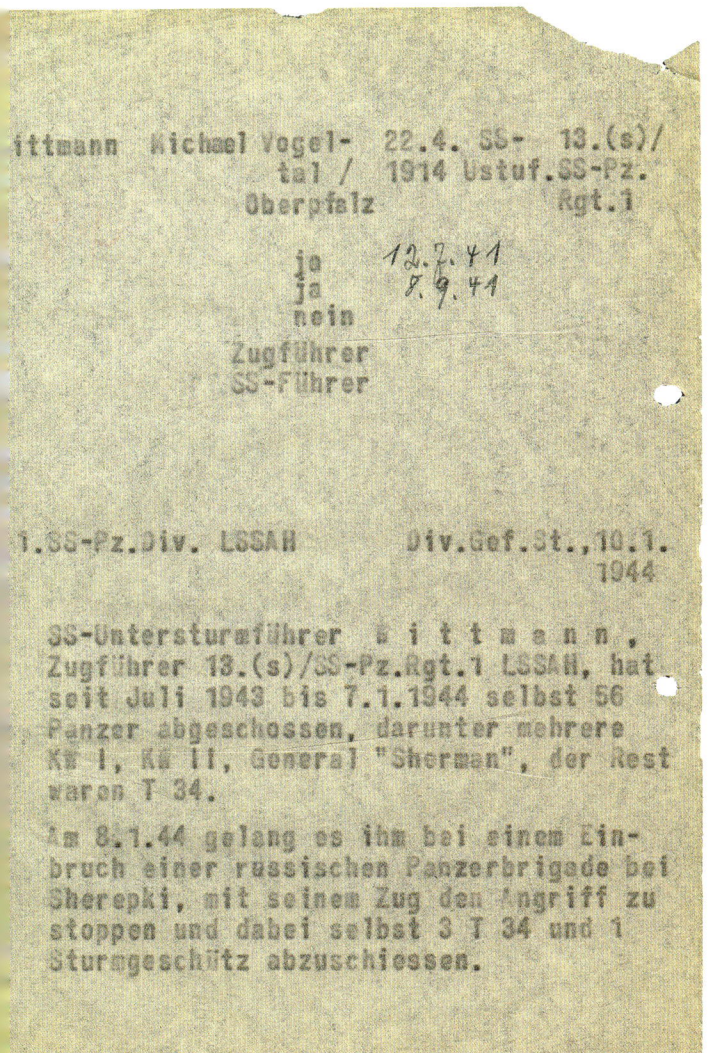

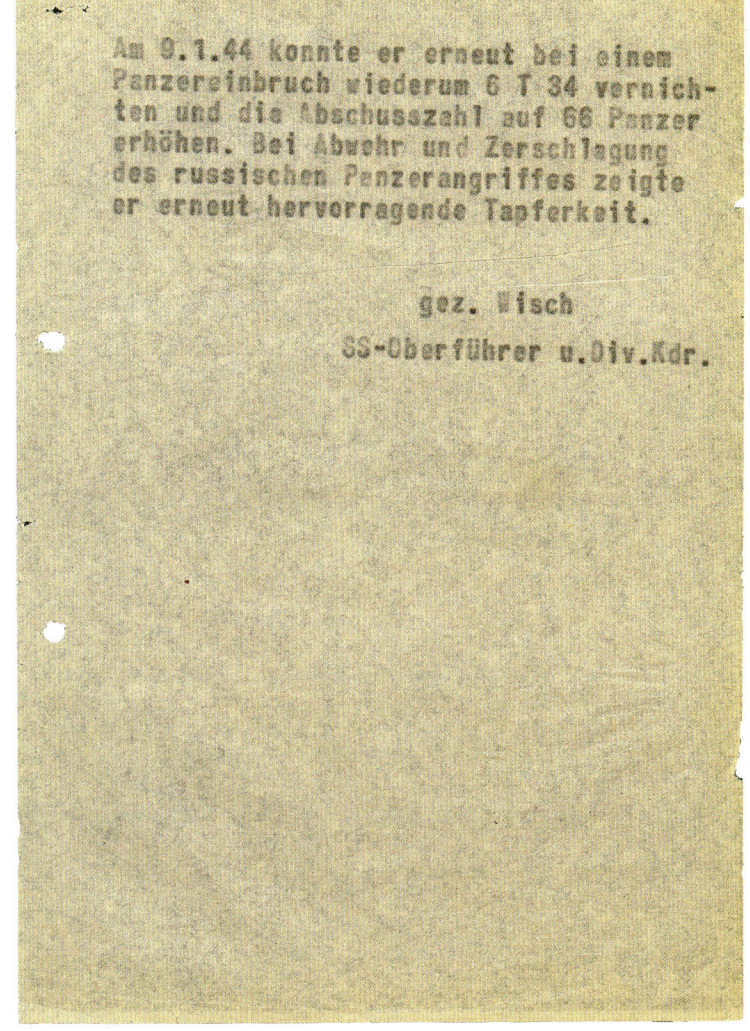

plètement de Tigres et il a désormais entre les mains une des armes allemandes des plus efficaces. Le premier jour de la bataille de chars de Kursk, il détruit 8 chars ennemis et 7 pièces d'artillerie. A la fin de cette offensive, il a ajouté 30 chars et 28 canons à son actif.

Un jour d'automne 1943, il détruit 10 chars dans un combat singulier ce qui porte son score à 56 et les 7 et 8 janvier 1944 avec sa section de Tigres, il empêche la percée d'une brigade blindée soviétique. Le 14 janvier, il reçoit la distinction de chevalier de la Croix de fer. Si, au dire de certains de ses camarades, cette distinction a eu du mal à venir, il n'en a pas été de même de la suivante. En effet, le 31 janvier il reçoit un télégramme l'informant que les « feuilles de chêne » lui avaient été décernées. Le télégramme était ainsi rédigé : *« Je vous remercie de votre action héroïque pour l'avenir de notre peuple et vous êtes le 320ᵉ soldat de l'armée allemande à recevoir les feuilles de chêne de la Croix de fer, signé, Adolf Hitler »*. En même temps, il était nommé *Obersturmführer*, lieutenant SS, et en avril il devenait commandant de la 2ᵉ compagnie du 101ᵉ bataillon de chars lourds SS.

C'est au cours de la bataille de Villers-Bocage que, le 13 juin 1944, Michael Wittmann va prendre place dans l'histoire de la guerre mondiale et, à partir de ce jour, on ne pourra plus parler de l'arme formidable qu'était le Tigre à cette époque, sans y associer son nom. A la suite de ce fait d'armes, l'*Obergruppenführer Dietrich*, général de corps d'armée, Chef du Iᵉʳ corps blindé SS, le propose pour recevoir la Croix de fer avec Glaives. Quelques jours après cette distinction, il est nommé *Hauptsturmführer*, capitaine, et il lui est proposé un poste à l'arrière dans une école d'entraînement où sa grande expérience aurait pu être utilisée pour l'instruction des futurs tankistes. Il refuse, préférant rester avec ses camarades.

Le 28 juin, le poste de commandement du 101ᵉ bataillon s'est installé à Clinchamps-sur-Orne et début juillet, Wittmann se trouve au centre de l'action de ses blindés entre Avenay et Vieux.

Le 7 août, le bataillon qu'il commande depuis le 10 juillet est rattaché à ce qu'il reste de la *12. SS-Panzer-Division « Hitlerjugend »*, et reçoit l'ordre de Kurt Meyer de capturer le village de Cintheaux, à une quinzaine de kilomètres au sud de Caen sur la route de Falaise pour protéger le flanc droit de la division. Il soutient dans cette contre-attaque des grenadiers, des *Panzer IV*, des chasseurs de chars d'une *« Jäger Abt. »*, des régiments de la *12. SS-Panzer-Division* qui font face à des unités anglaises, canadiennes et polonaises engagées dans l'opération « Totalize ». La bataille durera plusieurs heures et sera meurtrière. Plusieurs dizaines de chars resteront immobilisés dans la plaine de Caen, *Sherman, Cromwell, Panzer IV* et aussi 4 Tigres du bataillon 101 dont celui de Michael Wittmann, le *007*.

Le Tigre de Wittmann aurait été attaqué à l'est de Cintheaux par 5 *Sherman* de l'escadron A du *1st Northamptonshire Yeomanry* de la 33ᵉ Brigade Blindée anglaise et aurait explosé en flammes. Mais en fait, selon des témoins, le Tigre dont la tourelle était éjectée aurait été touché par des roquettes d'un Typhoon attaquant le char par l'arrière.

Le corps de Wittmann a été enterré sommairement avec ses coéquipiers dans un fossé sur le bord de la route et retrouvé en mars 1983. Ses restes et ceux de ses hommes ont été déposés dans une même tombe au cimetière allemand de La Cambe, bloc 47, rangée 3, tombe 120.

Il était le chef de char le plus réputé de l'histoire. Audacieux, jamais emporté, c'était un homme pensif, très admiré et estimé de ses camarades et très

of heavy Panzer VI and Tiger tanks, and a troop of light Panzer III tanks. Wittmann was assigned to this last troop. After the battle of Kharkov, the company was fully fitted out with Tigers and now had one of the most effective of German weapons in its hands. The first day of the Kursk tank battle, he destroyed 8 enemy tanks and 7 artillery guns. By the end of this offensive, he had added another 30 tanks and 28 guns to his tally.

One day in the fall of 1943, he destroyed 10 tanks in single combat, pushing his score up to 56, and on January 7th and 8th 1944 with his troop of Tigers, he prevented a Soviet tank brigade from breaking through. On January 14th, he received the distinction of Knight of the Iron Cross. If, as some of his comrades said, this award was a long time coming, the next one followed on almost immediately as on January 31st, he received a telegram informing him that he was being decorated with «Oak Leaves». The telegram was drafted as follows: «*I thank you for your heroic action for the future of our people and you are the 320th soldier of the German Army to receive Oak Leaves to the Iron Cross, signed, Adolf Hitler*». At the same time, he was promoted to Obersturmführer, SS lieutenant, and in April he was put in command of the 2nd Company of the 101st SS Heavy Tank Battalion.

It was during the Battle of Villers-Bocage, on June 13th, 1944, that Michael Wittmann earned a place in the history of the world war and, from that day on, it became impossible to speak of the formidable weapon that was the Tiger at that time without mentioning his name. Following this feat of arms, Obergruppenführer (lieutenant-general) Dietrich, commander of I SS Tank Corps, nominated him for an Iron Cross with Swords. A few days after this distinction, he was promoted to Hauptsturmführer (captain), and was offered a post in the rear at a training school where he might have put his considerable experiment to use in instructing trainee tank crews. He refused, preferring to remain with his comrades.

On June 28th, the 101st Battalion set up its headquarters at Clinchamps-sur-Orne and in early July, Wittmann was in the thick of the tank action in the area between Avenay and Vieux.

On August 7th, the battalion under his command since July 10th was attached to what was left of the 12. SS-Panzer-Division «Hitlerjugend», and was ordered by Kurt Meyer to capture the village of Cintheaux, some fifteen kilometers south of Caen on the Falaise road to cover the division's right flank. It lent its support to this counter-attack by grenadiers, Panzer IVs, the tank destroyers of a «Jäger Abt.», regiments of the 12. SS-Panzer-Division confronting the British, Canadian and Polish units committed in Operation Totalize. This fierce battle raged on for several hours. The plain of Caen was littered with dozens of knocked out Sherman, Cromwell and Panzer IV tanks, and also 4 of the 101st Battalion 'sTigers, including Michael Wittmann's, no. 007.

Wittmann's Tiger is thought to have been attacked east of Cintheaux by 5 Shermans of "A" Squadron 1st Northamptonshire Yeomanry, 33rd British Armored Brigade, and to have burst into flames. What actually happened however, according to eyewitnesses, was that the Tiger was hit by rockets from a Typhoon attacking the tank from the rear and the turret was ejected.

Wittmann's body was hastily buried with the rest of his crew in a ditch along the roadside and was found in March 1983. His own and his men's remains were placed in a single grave in the German cemetery at La Cambe, section 47, row 3, grave 120.

hautement considéré de ses supérieurs. Son incontestable habilité et sa vaillance personnelle lui ont assuré une place de tout premier ordre dans les annales de l'histoire militaire.

Villers-Bocage après le 13 juin

Le **14 juin,** les Allemands occupent Villers de nouveau. Ils pénètrent dans les immeubles pour rechercher les Britanniques attardés qui s'y sont réfugiés ou qui sont blessés. Sans pratiquer de pillage systématique, ils font quelques dégâts un peu pour se venger de l'accueil que les Français avaient réservé aux Anglais et, craignant une contre-attaque britannique, incendient l'Hôtel de Ville et quelques maisons et boutiques qu'ils ont visitées, pour cacher leurs méfaits. Dans les jours qui vont suivre, les alliés vont accentuer leur pression et, pratiquement tous les jours, vont laisser tomber quelques bombes et envoyer quelques obus pour rendre la ville difficile à traverser aux convois militaires qui montent au front, empêcher les rassemblements de troupes et essayer de déloger les batteries d'artillerie et de DCA.

Les Villérois doivent s'organiser pour continuer à vivre dans cette atmosphère particulière à laquelle ils ne sont pas préparés. La plupart des familles quittent la commune pour se réfugier dans la campagne voisine et abandonnent maisons et commerces sans s'inquiéter des pillages éventuels de soldats ou de civils. Ils se risqueront à revenir dans la journée pour récupérer dans leur foyer quelques objets dont ils ont besoin journellement dans leur nouvelle demeure, pour se loger, s'habiller, se nourrir et se protéger. Ceux qui ne peuvent quitter le bourg, qui sont indispensables à la vie de la commune, qui l'administrent tant bien que mal, ceux qui accueillent, renseignent, soignent, ravitaillent ou qui ne veulent pas quitter leur maison au risque de leur vie consolident leurs abris, leurs caves, leurs tranchées et réparent hâtivement les dégâts causés par les combats pour essayer de survivre.

Dans les villages voisins, à la campagne, la vie semble moins exposée bien que souvent des obus tombent à l'aveuglette, des avions en difficulté se débarrassent de leurs bombes un peu n'importe où, des civils se font mitrailler sur les routes et dans les champs par des aviateurs méfiants ou des Allemands soupçonneux. Toutefois, le ravitaillement se fait mieux là où il n'y a pas eu de combats, il y a des vaches dans les près, des porcs dans les « burets », des poules dans les cours de ferme. On peut trouver du lait, de la crème, du beurre, des œufs, de la viande. Il y a des légumes dans les jardins, de vieilles pommes de terre dégermées dans les caves si les nouvelles ne sont pas assez précoces. Il doit rester la farine que l'occupant n'a pas réquisitionnée en attendant que les blés mûrissent et il existe encore dans beaucoup de fermes des fours à pain.

Un endroit semble être une véritable oasis, le château, que des croix rouges installées par les Allemands sur le toit le 11 juin semblent protéger, bien que des obus soient tombés sur les communs. Il est entouré de bois et de bosquets où sont installées de batteries de DCA. Il a accueilli beaucoup de monde et 200 personnes y résident régulièrement. Le 16, l'hospice situé en plein centre de la ville, qui a souffert de la bataille et dont les vitres ont volé en éclats, y est évacué. Religieuses, vieillards, enfants orphelins y sont installés, les femmes dans l'aile droite, les enfants au premier étage, les hommes dans un pavillon proche et vont occuper les lits laissés par les Allemands le 13 juin lors de leur départ.

A partir de ce moment, le château va devenir presque un hôpital. Beaucoup de civils, fuyant les zones de

He was the most famous tank commander in history. Bold, without ever getting carried away, he was a thoughtful man, greatly admired and liked by his comrades and very highly thought of by his superiors. His undeniable ability and personal bravery earned him a very high place in the annals of military history.

Villers-Bocage after June 13th

On June 14th, the Germans reoccupied Villers. They entered buildings to flush out any British stragglers sheltering there or lying wounded. Without going in for systematic looting, they did perpetrate some damage in revenge for the reception which the French had reserved for the British and, fearing a British counter-attack, set fire to the Town Hall and some houses and shops they had visited, to hide their misdeeds. In the days that followed, the Allies increased their pressure, and practically every day dropped a few bombs and fired a few shells to make the town awkward for the military convoys to pass through on their way up to the front line, to prevent troops gathering and to try and dislodge the artillery and AA batteries.

The people of Villers had to organize themselves in order to continue living in this peculiar atmosphere for which they were unprepared. Most families moved out of town to shelter in the neighboring countryside, abandoning their houses and stores without worrying over possible looting by soldiers or civilians. They took the risk of returning home during the daytime to pick up whatever they might need each day where they were now staying, by way of board and lodging, clothing and protection. Those who were unable to leave town, who were needed to keep the place running as best they could, those who received, informed, looked after, supplied, or simply preferred to risk their lives than leave their homes, consolidated their shelters, cellars and trenches, and hastily repaired the damage sustained in the fighting in order to try and survive.

In the nearby villages, in the open country, life seemed less exposed although there would often be a stray shell falling, aircraft in distress getting rid of their bombs wherever they could, civilians machine-gunned on the roads or in the fields by overcautious airmen or suspicious Germans. However, supplies did get through better to places where there had been no fighting; there were cows in the meadows, pigs in the pigsties and hens in the farmyards. There was milk, cream, butter, eggs and meat. There were vegetables in the gardens, old potatoes with the sprouts removed in the cellars if the early new potatoes were not ready. And until the wheat ripened there was still some flour over that the occupying forces had not requisitioned, and many farms still had bread ovens.

One place seemed like a real oasis, the château, which red crosses placed on the roof by the Germans on June 11th seemed to protect, although some shells did fall on the outbuildings. It was surrounded by woods and thickets in which anti-aircraft batteries had been set up. The château had taken in a lot of people, with 200 people staying there on a regular basis. On the 16th, the hospice in the town center was evacuated there, after suffering damage in the battle with the window panes being blown out. Nuns, old people and orphaned children were brought there, the women in the right wing, the children on the first floor, the men in a nearby house, and they occupied the beds left by the departing Germans on June 13th.

From then on, the château was almost turned into a hospital. Many civilians fleeing the battlefield stopped there to be comforted, looked after and fed, before heading off for places as yet untouched by the war.

Page ci-contre : Un char Tiger a été détruit au hameau de Greland, sur la commune de Cahagnes, probablement celui du lieutenant Philipsen. Le marquage montre qu'il s'agit effectivement d'un char de la 1re compagnie. Henri Marie montre l'endroit exact (marqué par un piquet) où ce char a été détruit. (Coll. H. Marie et Heimdal.)

combat vont s'y arrêter pour y être réconfortés, soignés, nourris avant de prendre la direction d'endroits pas encore touchés par la guerre. Des équipes d'urgence de la Croix Rouge vont être constituées par Madame de Clermont-Tonnerre parmi les personnes réfugiées. Ces équipes auront le rôle de parcourir les communes voisines touchées par la bataille, en carrioles ou en voitures pour recueillir les isolés, les infirmes, les vieillards, soigner les blessés, rassembler les plus atteints et les transporter vers l'hôpital du Bon Sauveur à Caen tant qu'il ne sera pas libéré, c'est-à-dire jusqu'au 7 juillet. Les soins sont assurés par les sœurs de l'hospice et par le personnel médical et paramédical de Villers qui vient chaque jour soigner malades et blessés, après avoir répondu aux appels qui leur viennent du bourg et de la campagne voisine. Toute connaissance médicale est utilisée. C'est ainsi qu'un vétérinaire, le docteur Droulers a retiré avec succès la balle qu'un jeune homme de Tracy avait reçue dans le pied. Une maternité va être installée et quelques enfants vont y naître pendant cette période.

200 personnes vont vivre dans les appartements du château et dans les bâtiments de la ferme attenante, ils constituent un véritable village qu'il va falloir entretenir et alimenter. Le 20 juin, des jeunes iront chercher au château de Villy, malgré des duels d'artillerie, quatre tonnes de farine abandonnées par les Allemands. Des animaux affolés par les explosions errent dans les herbages, des vaches ont besoin d'être traites, elles fourniront du lait, d'autres sont blessées, elles seront abattues et dépecées. Par moment, il faudra parlementer avec les Allemandes à la recherche de nourriture fraîche et partager avec eux.

Vers la libération

Après le 15 juin, le front se stabilise sur la ligne Caumont-Bucéels. La *2. Pz.-Div.* va se fixer entre la *Panzer-Lehr-Division* qui est devant Tilly et la 326e division allemande qui se trouve devant Caumont. Elle est arrivée le 13 et son régiment blindé, le *Panzer-Regiment 3*, composé de chars *Panzer IV* et de *Panzer V, Panther*, n'arrivera que le 18. Elle s'est déployée devant Caumont au nord-ouest, Livry au nord, Villers-Bocage au nord-est et se trouve face à la *7th Armoured Division* qui va adopter un rôle défensif jusqu'au 30 juin. Pendant ce temps, l'activité des Britanniques et des Allemands va se limiter à une action intense de patrouilles. La 22e Brigade Blindée est en réserve à Sainte-Honorine-de-Ducy, la 131e Brigade d'Infanterie, les *Queen's*, et le 1er Bataillon de la *Rifle Brigade* sont en ligne, au sud-est de la Paumerie, à Livry, au Pont Mulot. Le *8th Hussars*, régiment très mobile, a le rôle de couvrir sur le flanc est, la Butte, au nord de Saint-Paul-du-Vernay avec le 50e Régiment d'Infanterie.

Le **15 juin**, les Anglais prennent Lingèvres et la Belle Epine. La *2. Pz.-Div.* reprend sur son aile droite le hameau de Launay près d'Anctoville et de Saint-Germain-d'Ectot. A l'aile gauche, 4 Tigres de la 1re compagnie de la *SS-Pz.-Abt. 101,* sous le commandement du lieutenant Philipsen, poursuivent les Britanniques qui se sont repliés vers le nord et qui vont perdre 5 blindés et plusieurs canons antichars. Le sergent Wendt, commandant du Tigre n° 132 a fait partie de ce groupe qui devait traverser un terrain en pente entrecoupé de nombreuses haies. Arrêté près d'une de ces haies, le long de la route Cahagnes Briquessard, entre le Quesnay et la Tringale, il a reçu, a-t-il raconté, probablement une charge creuse tirée de très près qui a pénétré son char et empli sa tourelle de flammes. Tout son équipage

Werner Wendt.

Hannes Philipsen.

Emergency Red Cross teams were set up from among the refugees by Madame de Clermont-Tonnerre. These teams were assigned to going the rounds of neighboring villages affected by the battle, in carts or by car, picking up isolated or disabled persons, old people, treating the injured, gathering up the worst cases and taking them to the Bon Sauveur hospital in Caen until it was finally liberated, on July 7th. They were looked after by the sisters of the hospice and the medical personnel and paramedics of Villers who came each day to tend the sick and wounded, after answering calls received from the town and neighboring countryside. Any medical knowledge was put to use. Thus Dr Droulers, a veterinary surgeon, successfully removed a bullet which a young man from Tracy had received in the foot. A maternity ward was set up and a few children were born there during this period.

200 people lived in the apartments of the château and in the adjoining farm buildings, making up a proper village that needing maintaining and feeding. On June 20th, some young people set off for the château de Villy, despite duelling artillery, to fetch four tons of flour abandoned by the Germans. Shell-shocked animals were wandering in the fields, cows needed milking and so provided milk; others were injured and so were slaughtered and cut up for their meat. Occasionally it was necessary to haggle with Germans also looking for fresh food and share it with them.

Towards liberation

After June 15th, the front was stabilized on the Caumont-Bucéels line. The 2. Pz.-Div. was to fix itself between the Panzer-Lehr-Division which was before Tilly, and the German 326th Division which was before Caumont. It arrived on the 13th and its tank regiment, Panzer-Regiment 3, made up of Panzer IVs and Panzer Vs (Panthers), did not arrive until the 18th. It was deployed before Caumont to the north-west, Livry to the north, Villers-Bocage to the north-east and facing the 7th Armoured Division, which took up a defensive role until June 30th. Meanwhile, the British and Germans reduced their activity to intensive patrolling action. The 22nd Armored Brigade was held in reserve at Sainte-Honorine-de-Ducy, the 131st Infantry Brigade, the Queen's, and the 1st Battalion the Rifle Brigade were in line, south-east of La Paumerie, at Livry and Pont Mulot. The 8th Hussars, a very mobile regiment, was assigned the task of covering La Butte, north of Saint-Paul-du-Vernay on the eastern flank, along with the 50th Infantry Regiment.

On June 15th, the British took Lingèvres and La Belle Epine. The 2. Pz.-Div. recaptured on its right flank the hamlet of Launay near Anctoville and Saint-Germain-d'Ectot. On the left flank, 4 Tigers of the 1st Company of SS-Pz.-Abt. 101, under the command of Lieutenant Philipsen, pursued the British who fell back northwards, losing 5 tanks and several anti-tank guns in the process. Sergeant Wendt, commander of Tiger "132", belonged to this group that had to cross some sloping terrain full of hedgerows. On stopping near one of the hedges, along the Cahagnes-Briquessard road between Le Quesnay and La Tringale, he recalls receiving probably a concussion shell fired from close range and which pierced his tank, filling the turret with flames. All the crew managed to bale out, but the gun-layer's hands were rather badly burned.

In the morning, Madame Béatrice Desmond, a farmer at Gournay, saw a Tiger entering her yard along with a large number of carriers full of Germans. She then went to Monsieur Raymond Dujardin's at Le

a pu être évacué, son pointeur toutefois a été assez grièvement brûlé aux mains.

Le matin, Madame Béatrice Desmond, cultivatrice à Gournay avait vu arriver dans sa cour un Tigre et un nombre important de chenillettes chargées d'Allemands. Elle s'est rendue alors chez Monsieur Raymond Dujardin, au Quesnay, pour qu'il prévienne les Anglais. L'après-midi, elle a voulu retourner à son domicile distant d'environ 500 mètres à travers champs mais s'est heurtée à des Allemands qui lui ont interdit le passage. Après avoir proféré quelques menaces acerbes aux soldats, elle a décidé de revenir chez elle en passant par la Tringale et la route de Villers-Bocage. Le soir, elle s'est réfugiée dans une tranchée recouverte avec son mari et sa mère pour y passer la nuit à l'abri. Les Allemands constatant son retour au cours de la nuit, ont lancé une grenade dans la tranchée. Les trois corps seront retrouvés plus de deux mois plus tard par des voisins. Le même jour, quelques Tigres de la 2ᵉ compagnie sont en surveillance cote 213. Le sergent Stubenrauch et le caporal pointeur Baube sont tués, 1 homme est blessé.

Le **16 juin**, les Britanniques se sont repliés vers Livry après avoir abandonné aux Allemands toute une zone mal délimitée comprise entre Amayé-sur-Seulles, Cahagnes et Caumont, tenue par la 2ᵉ Panzer-Division soutenue par deux sections de Tigres de la compagnie 1 de la *SS-Pz.-Abt. 101* incomplètes et réduites à 3 ou 4 chars qui couvrent le secteur. Deux de ces chars étaient aux environs du hameau de Benneville où la nuit du 15 au 16 l'un d'eux aurait été victime d'un vol de jerricans d'essence par un civil. Naturellement, ces Tigres ont été toute la journée sous le feu de l'artillerie britannique dès qu'ils étaient signalés dans la zone de tir anglaise.

Ce jour-là, un Tigre a été touché et incendié sur la route de Cahagnes à Villers, au hameau de Greland en pleine campagne, sur un terrain entièrement dégagé, assez loin de la zone du front et des combats. On peut se demander ce que faisait ce char arrêté sur la route, canon dirigé vers Villers sans être camouflé.

Monsieur Marcel Lebas l'a vu dans la nuit du 15 au 16 à Benneville et le soir, incendié à Greland. Selon ce témoin, le char avait reçu l'ordre de regagner le secteur de Noyers où le Quartier Général du bataillon quelque part vers Baron-sur-Odon. A ce moment, la 3ᵉ compagnie du bataillon 101 devait se trouver entre Noyers et Verson, derrière l'infanterie allemande prête à intervenir sur le front proche au premier signal. On peut supposer que la *2. Pz-Div.* bien implantée dans le secteur de Cahagnes sur le point de recevoir le renfort des Panzers IV et Panthers de son régiment blindé le *Panzer Regiment 3* pouvait libérer les Tigres venus en renfort.

Selon les témoins, le Tigre de Greland en panne, peut-être d'essence, aurait été repéré par un avion passé en rase-mottes peu de temps avant et aurait été touché par des roquettes lancées par des avions ou par des tirs d'artillerie. Deux hommes, le conducteur et le radio, n'ont pas su se dégager du char et sont morts brûlés vifs. A peu de distance, Madame Boiroux, d'Amayé a entendu leurs cris épouvantables après le passage d'avions. Le chef de char a pu s'extirper avec son pointeur et son pourvoyeur mais 50 mètres plus loin, il a été tué par des éclats.

Ce chef de char serait le lieutenant Hannes Philipsen de la 1ʳᵉ compagnie de la *SS-Pz.-Abt. 101*, tué officiellement le 16 juin 1944;

Le récit du sous-lieutenant Hahn, chef de char de la même compagnie, qui a rédigé un rapport sur la perte du Tigre de Philipsen, est un peu différent. Il écrit qu'il se trouvait une vingtaine de mètres derrière lui après avoir repoussé à la sortie de Cahagnes, aidé

Quesnay for him to warn the British. That afternoon she tried to make her way back home 500 meters away across the fields, but came up against some Germans who would not let her through. After making loud threatening noises at the soldiers, she decided to go back home via La Tringale and the Villers-Bocage road. That evening, she took shelter in a covered trench along with her husband and mother, to spend the night under cover. But the Germans on finding out that she had come back during the night, lobbed a grenade into the trench. The three bodies were recovered by neighbors more than two months later.

That same day, a few of the 2nd Company's Tigers were observing Point 213. Sergeant Stubenrauch and

A Tiger tank, probably Lieutenant Philipsen's, was destroyed at the hamlet of Greland at Cahagnes. The markings show at least that the tank belonged to the 1st Company. Henri Marie shows the exact spot (marked by a stake) where this tank was destroyed. (Coll. H. Marie and Heimdal.)

Dans la plaine, près de Villers-Bocage, des chars Shermans sont en position. (Coll. H. Marie.)

In the plain near Villers-Bocage, Sherman tanks are in position. (Coll. H. Marie.)

de grenadiers, des Britanniques d'un village. Ils avaient été pris sous le feu intense d'artillerie qui aurait touché le Tigre. Hahn serait alors descendu de son char pour aller constater que Philipsen gravement blessé au ventre était mort sur le coup.

Le **17 juin**, à l'aile gauche, le 304ᵉ régiment de panzergrenadiers soutenu par les canons du 74ᵉ régiment d'artillerie allemand attaquent les *Queen's* et les repoussent vers le hameau du Quesnay près de Briquessard. Cette attaque est contenue à Briquessard par les chars du *8th Hussars* et les fusiliers du *4th Northumberland*. Le *8th Hussars* perd 5 chars et subit 20 pertes. Les tirs des régiments d'artillerie, *RHA*, arrêtent l'attaque des Allemands qui abandonnent le Quesnay. Ce jour-là les Tigres n'interviennent pas. Les chars rescapés ont dû rejoindre leur bataillon.

Le **18**, les Allemands reprennent le Quesnay, retrouvant une partie du matériel qu'ils avaient abandonné la veille et se dirigent vers Briquessard. Ils ont ainsi repris un peu de terrain mais, à quel prix ? Les Anglais eux aussi ont subi des pertes importantes. L'escadron A du *4th CLY* est reconstitué sur la base de 2 *troops* et le capitaine Hiscock nommé Major en prend le commandement.

Le **19**, Tilly-sur-Seulles est pris par la *50th ID*.

Le **22**, le lieutenant-colonel Raukin prend le commandement du *4th CLY* et le 1ᵉʳ bataillon de la *Rifle Brigade* qui a perdu 10 officiers et 163 hommes de troupe en 15 jours reçoit au Pont Mulot un renfort du 8ᵉ Bataillon de la *60th Rifle*. Le *6th Queen's* qui, le 13 juin était resté à Briquessard en réserve, supporte maintenant toutes les tentatives allemandes.

Le **26**, le Lieutenant-colonel Forrester qui commande le bataillon est blessé et le *Major* Russel Elliot qui commande la compagnie A est tué.

his gun-layer Corporal Baube were killed, with 1 man wounded.

On June 16th, the British withdrew towards Livry after leaving in German hands a whole area roughly between Amayé-sur-Seulles, Cahagnes and Caumont, held by the 2nd Panzer Division with support from two understrength troops of Tigers of the 1st Company, sSSPzAbt101, down to 3 or 4 tanks to cover the sector. Two of these tanks were in the vicinity of Benneville where during the night of 15-16th one of them is thought to have had its jerrycans of fuel stolen by a civilian. Naturally, these Tigers came under British artillery fire all day long whenever they were sighted within firing range.

That day, one Tiger was hit and burst into flames on the Cahagnes road at Villers, at the hamlet of Greland in the open country on a perfectly clear piece of ground away from the combat zone and the front. One may wonder what this tank was doing stationary by the roadside, with its gun trained uncamouflaged on Villers.

Monsieur Marcel Lebas saw it during the night of 15-16th at Benneville and again that evening, burnt out at Greland. According to this eye-witness, the tank had been ordered to return to the Noyers sector or battalion HQ somewhere near Baron-sur-Odon. At that moment, the 3rd Company, 101st Battalion must have been somewhere between Noyers and Verson, behind the German infantry ready to intervene promptly in the front line close by. One may suppose that being well established in the Cahagnes sector and about to receive in reinforcement Panzer IVs and Panthers from its tank regiment, the 3rd Panzer Regiment, the 2. Pz Div. was able to release the Tigers that had arrived in reinforcement.

According to eye-witnesses, the broken-down Tiger at Greland (maybe out of gas) was spotted just before by a hedge-hopping aircraft and was hit by rockets fired from aircraft or by artillery fire. Two men, the driver and the wireless operator, were unable to bale out and were burned to death. Not far off, Madame Boiroux from Amayé heard their dreadful screams after the planes flew past. The tank commander did manage to bale out with his gun-layer and ammunition-server, but he was killed by shrapnel 50 meters down the road.

This tank commander seems to have been Lieutenant Hannes Philipsen of the 1st Company, sSSPz Abt101, officially reported killed on June 16th 1944. Second-Lieutenant Hahn, a tank commander with the same company who wrote a report on the loss of Philipsen's Tiger, gave a somewhat different account. He wrote that he was about twenty yards behind him after pushing some British troops out of a village to the exit from Cahagnes with the help of some grenadiers. They had been caught in heavy artillery fire and the Tiger had been hit. Hahn then got out of his tank to discover that Philipsen had died instantly from a severe wound to the stomach.

On June 17th, on the left flank, the 30th Panzergrenadier Regiment, supported by the guns of the German 7th Artillery Regiment, attacked the Queen's and pushed them back towards the hamlet of Le Quesnay near Briquessard. This attack was contained in Briquessard by the tanks of the 8th Hussars and the 4th Northumberland rifles. The 8th Hussars lost 5 tanks and sustained 20 casualties. The RHA artillery regiments' fire stopped the attack and the Germans abandoned Le Quesnay. The Tigers took no part in the day's fighting. The tanks that escaped had to return to their battalion.

On the 18th, the Germans retook Le Quesnay, regaining some of the equipment they had abandoned the day before, then moved on to Briquessard. They thus

A Villers, le **27 juin**, une escadrille de chasseurs-bombardiers P 47, les *Thunderbolt*, du *362nd Fighter Group* de la *9th US Air Force* en maraude derrière la ligne de front est prise à partie par la *DCA* allemande installée dans le parc du château. L'avion du chef de l'escadrille, le capitaine George Rarey, est touché. Le pilote est éjecté de l'appareil, parachute en flammes, s'écrase au sol et trouve la mort. Le P 47 continue sa chute et s'écrase sur un camion allemand camouflé sous une haie haute dans le chemin rural qui mène à l'équarrissage. Les Allemands inhumed a little ground but, at what cost? The British too suffered considerable losses. "A" Squadron 4th CLY was reorganized on the basis of two troops, and Captain Hiscock promoted to Major took over in command.

On the 19th, Tilly-sur-Seulles was captured by the 50th ID.

On the 22nd, Lieutenant-Colonel Raukin took over command of the 4th CLY and the 1st Battalion the Rifle Brigade, which had lost 10 officers and 163 other ranks inside a fortnight received reinforcements at Pont Mulot in the form of the 8th Battalion 60th Rifles. 6th Queen's, held in reserve at Briquessard on June 13th, now supported all the British efforts.

On the 26th, the battalion commander, Lieutenant-Colonel Forrester, was wounded and "A" Company commander Major Russel Elliot was killed.

At Villers, on June 27th, a squadron of P4 47 fighter-bombers, the Thunderbolts of the 362nd Fighter Group of the 9th US Air Force patrolling behind the front line came under fire from German AA in the grounds of the château. The flight commander Captain George Rarey's leading aircraft was hit. The pilot baled out, his parachute in flames, and crashed to his death on the ground. The P 47 in turn came down and crashed on a German truck camouflaged under a tall hedge in the country lane leading to the knacker's yard. The Germans hastily buried the airman in a ditch, in the bend on the «Vierge Noire» hill, at the foot of a concrete fence post. A macabre detail of this story tells of how the body, with only a few clods of earth to cover it, soon had a foot protruding from the soil. Civilians covered it over and put a fence round the grave. Some time afterwards, the body was exhumed and taken to the American cemetery at Colleville-sur-Mer. A commemorative plaque now recalls this episode of the war at the «Black Virgin».

On June 29th, the Panzer-Lehr-Division was ordered to prepare to leave the Tilly sector for the Cherbourg peninsula on July 5th. It was to be relieved by the 276th Infantry Division and had to leave behind a third of its tanks. The German command's intention was to reinforce the 17. SS-Pz.-Gren.-Division in order to launch an attack towards Carentan and the Baie des Veys so as to drive a wedge through the American Army. This attempted breakout failed and the ensuing American counter-attack ended in the capture of Saint-Lô on July 16th.

By the end of the month, the British were slowly approaching, with the artillery fire never letting up. An offensive seemed to be brewing and some of the casualties were civilians, including Monsieur Clément Busnel on June 27th and Monsieur Denis Hébert on the 28th.

On June 29th, the church bell-tower collapsed.

The bombardment of June 30th, 1944

During the afternoon, the shell fire intensified. The people of Villers abandoned their homes and took to their shelters and trenches away from the town center. The worst hit area seemed to be the northern end of the Rue Pasteur, and the inhabitants moved towards the trenches they had dug in their gardens between the Rue Saint Martin and the open country nearby.

At around 17.00, the first planes arrived in the west. These were mostly Lancasters which began to drop their bombs, escorted by fighter planes, British Spitfires and American Thunderbolts. There were 250 of them, and they dropped 1,100 tons of explosive bombs, first from west to east, then from north to south.

Le *Captain* George W. Rarey **(1)** pilote son P47 **(2)** du *362nd Fighter Group* le 27 juin lorsqu'il est abattu par la Flak. Il s'écrase au sol à cet endroit **(3)**. Un monument **(4)** rappelle la mémoire de ce pilote qui était aussi un artiste. (Coll. H. Marie et E.G./ Heimdal.)

Captain George W. Rarey **(1)** was flying his P47 **(2)** of the 362nd Fighter Group on June 27th when he was shot down by Flak. He crashed to the ground on this spot **(3)**. A monument **(4)** was raised to the memory of this pilot who was also an artist. (Coll. H. Marie and E.G./ Heimdal.)

L'église de Villers-Bocage, dont on voit ici l'intérieur dans la première moitié du XXe siècle, a été totalement détruite par le bombardement du 30 juin 1944. (Coll. H. Marie.)

2. La porte du tabernacle de cette église a été sauvegardée. Nous la voyons cadrée en couleur sur la photo précédente. (EG/Heimdal.)

3. Le bombardement du 30 juin 1944 sur Villers-Bocage. Le centre du bourg est particulièrement touché.

4. Le résultat de ce bombardement.

5. Depuis le nord-ouest, on remarque que le centre et le bas du bourg ont été écrasés sous les bombes. En haut à droite : la route menant à Aunay, en bas celle menant à Vire. (IWM.)

The church at Villers-Bocage (here we see the inside in the first half of the 20th c.) was completely destroyed in the bombardment of June 30th 1944. (Coll. H. Marie.)

2. *The church's tabernacle door was saved. We see it framed in color in the previous picture. (EG/Heimdal.)*

3. *The bombardment of Villers-Bocage on June 30th 1944. The town center was particularly badly hit.*

4. *The result of this bombardment.*

5. *From the north-west, notice how the center and bottom end of the town have been razed by bombs. Top right: the road to Aunay, below, the road to Vire. (IWM.)*

ment l'aviateur sommairement dans un fossé, le long de la courbe de la côte de la « Vierge Noire », au pied d'un poteau de barrière en maçonnerie. Détail macabre, la dépouille recouverte seulement de quelques mottes de terre laissera bientôt apparaître un pied sortant de terre. Des civils le recouvriront et entoureront la tombe d'une clôture Le corps bientôt relevé sera transporté au cimetière américain de Colleville-sur-Mer. Une plaque commémorative rappelle maintenant cet épisode de la guerre à la « Vierge Noire ».

Le 29 juin, la *Panzer-Lehr-Division* reçoit l'ordre de se préparer à quitter le secteur de Tilly le 5 juillet pour rejoindre le Cotentin. Elle sera relevée par la 276e Division d'Infanterie et devra laisser sur place un tiers de ses chars. Le but du commandement allemand est de renforcer la *17. SS-Pz.-Gren.-Division* pour lancer une attaque vers Carentan et la baie des Veys destinée à couper l'armée américaine en deux. La tentative de percée échouera et la contre-attaque américaine qui va suivre va se terminer par la prise de Saint-Lô le 16 juillet.

A la fin du mois, les Anglais approchent lentement, les tirs d'artillerie succèdent aux tirs d'artillerie. Une offensive semble se préparer et des civils sont parmi les victimes, le 27 juin, Monsieur Clément Busnel, le 28, Monsieur Denis Hébert...

Le **29 juin**, le clocher de l'église s'écroule.

Le bombardement du 30 juin 1944

Dans l'après-midi, les tirs d'obus s'intensifient. Les Villérois abandonnent leurs maisons, regagnent leurs abris et tranchées et quittent le centre de la commune. La zone la plus visée semble être la partie nord de la rue Pasteur et les habitants se dirigent vers leurs tranchées creusées dans leurs jardins entre la rue Saint-Martin et la campagne proche.

Vers 17 heures, les premiers avions arrivent à l'ouest. Il s'agit de Lancaster surtout qui commencent à lâcher leurs bombes, protégés par des chasseurs, *Spitfire* anglais et *Thunderbolt* américains. Ils seront 250, qui laisseront tomber 1 100 tonnes de bombes explosives, d'ouest en est d'abord, du nord au sud ensuite.

D'ouest en est, c'est la rue Pasteur qui est visée, les immeubles de chaque côté de la rue s'écroulent, bouleversant la rue et la recouvrant de pierres et gravats divers, l'interdisant complètement à la circulation. 2 chars allemands, un *Panzer IV* et un Tigre l'obstruaient depuis le 13 juin au niveau de la place de l'Hôtel de Ville. Les chars sont maintenant désarticulés, le *Panzer IV* mis en pièces, le Tigre réduit à sa carcasse.

Du nord vers le sud, le but est de couper les routes de Vire, d'Aunay, d'Evrecy et le nœud de communication important de la place Jeanne d'Arc. Les bombes tombent dans les herbages, au sud des écoles, au sud de l'église, le long du boulevard Joffre, dans le champ au sud du marché aux bestiaux et plus au sud après les établissements Rivière jusqu'à l'allée qui mène au château. Partout elles forment de profonds cratères donnant aux terrains un aspect lunaire.

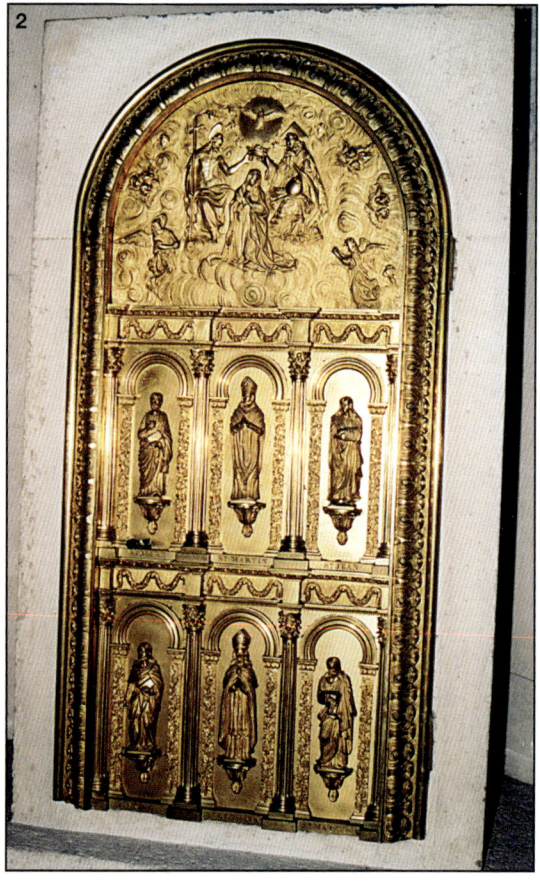

Le capitaine R.H. Boyers, de la 51e escadrille de la *R.A.F.* a été le premier pilote « *à laisser tomber ses bombes sur Villers, d'ouest en est, le long de la grand-rue* ». Selon lui, la résistance aurait demandé au « *Bomber-Command* » de bombarder Villers pour atteindre une division blindée allemande qui s'y serait abritée et il lui avait été dit que les habitants de la ville réfugiés au château étaient à l'abri. Parti à 14 heures d'une base aérienne du Yorkshire, il est arri-

vé 3 heures plus tard sur son objectif où, à cause des nuages, il a dû descendre de 12 000 à 3 500 pieds avant de commencer à lâcher ses bombes. Le bombardement terminé, les avions ont regagné leur base et l'artillerie a repris ses tirs.

Bois de l'Ecanet, les Allemands avaient camouflé deux camions de munitions dont, le contenu, touché, a éclaté en tous sens.

Le but de ce bombardement était de rendre impraticables les carrefours du bourg, d'obstruer les routes qui y menaient par des amoncellements de matériaux et d'y former des excavations difficiles à franchir. Comme dans la plupart des villes victimes de ce genre d'intervention aérienne, les Allemands ont vite trouvé la parade en contournant la ville en utilisant le moindre passage praticable et en passant à travers champs. Les routes qui passaient par Villers étaient importantes pour eux, pour assurer les transports de troupes et de ravitaillement vers le front de Tilly et vers Noyers où se déroulait la bataille de l'Odon.

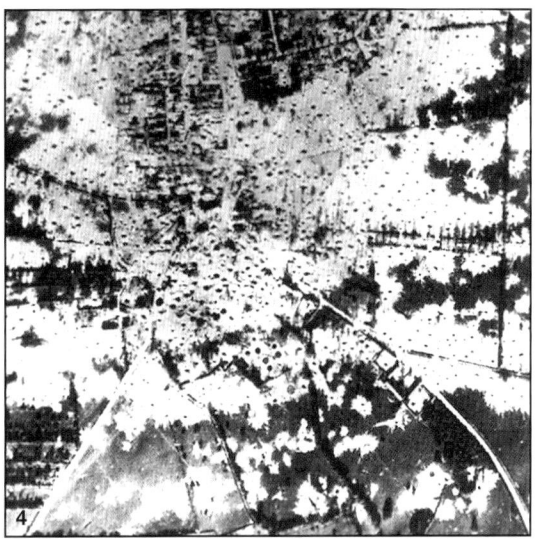

Heureusement que ce jour-là, le bourg avait été déserté par la plupart de ses habitants. Etaient restés du côté du boulevard Joffre et du quartier de la gare, quelques-uns dont un certain nombre de maisons avaient été épargnées par la bataille du 13 juin et qui s'étaient installés ou construits des abris suffisants, estimaient-ils, pour leur assurer une certaine sécurité. Cinq d'entre eux ont été tués, Mme Florentine Denis, Mme Annie Cervelle, M. Alphonse Lebel, Mme Marguerite Wilmann, M. Adjutor Lefèvre. Dès la fin du bombardement, les équipes d'urgence du château s'étaient précipitées munies de matériel de secours pour porter main forte aux sauveteurs, dégager les victimes des décombres, soigner sur place les blessés légers, transporter vers le château les blessés les plus atteints sur des brancards improvisés, réconforter les rescapés indemnes mais plus ou moins choqués et ceux dont un proche avait été tué. Les opérations de sauvetage ont duré toute la

nuit, à la lueur des incendies qui s'étaient déclarés, tandis que des obus continuaient de tomber sur Villers ou sifflaient en passant au-dessus de leurs têtes et allaient exploser plus loin. Le soir, l'Hospice était en flammes et le Dispensaire boulevard Joffre s'était écroulé sur la cave où se trouvaient les sœurs restées sur place près de la population et quelques voisins qui s'étaient joints à elles. Tous ont pu être retirés à temps de leur position inconfortable.

A partir de ce jour-là, le château va devenir un véritable hôpital que Mme de Clermont-Tonnerre va diriger. Elle sera secondée par le Docteur Dary et son épouse, infirmière, venus s'installer sur place le 28 juin, des sœurs de l'Hospice, auxquelles s'étaient jointes celles du Dispensaire, de Mlle Andrée Marie, préparatrice en pharmacie et les secouristes de la Croix Rouge. En deux jours furent installées deux salles de pansements au sous-sol où les blessés étaient soignés. Les plus légers restaient sur place mais les plus gravement atteints seront dirigés vers Caen tant que les communications le permettront.

C'est le 30 juin également que Thury-Harcourt a été bombardé pour le même objectif, empêcher les renforts allemands de gagner le sud de Caen. Il y a eu une quarantaine de victimes civiles.

Derniers jours de l'occupation

Entre la zone du front et la route Caen-Vire, les civils commencent à gêner les Allemands. Le 6 juillet, le commandement décide d'évacuer une zone qui va de Dampierre à Saint-Jean-des-Essartiers à l'ouest jusqu'à Mondrainville à l'est. L'évacuation doit commencer le vendredi 7 juillet à 0 heure et être terminée le 9 juillet à 24 heures dernier délai.

Les évacués en provenance des communes concernées doivent atteindre le 9, la ligne Saint-Sever-Vire-Falaise. Les ordres sont stricts et menaçants, toute personne trouvée dans la zone après la date fixée devait être fusillée. L'évacuation prescrite par les Allemands doit s'effectuer en bon ordre et les maires qui préviennent individuellement leurs administrés peuvent se faire aider par la *Feldgendarmerie* s'il le faut. Les évacués peuvent emporter ce qu'ils veulent, ravitaillement, linge, couvertures, bétail et se regrouper en colonnes formées d'habitants de la même commune encadrés par des responsables locaux et se signaler par des étoffes de couleur blanche. L'itinéraire à suivre fixé par les Allemands conduit les groupes ainsi formés jusqu'à une ligne Argentan-Sourdeval-Flers.

Après le bombardement, des soldats allemands passent à côté du char Sherman d'observation d'artillerie au milieu de la localité ruinée. (Heimdal.)

After the bombardment, German soldiers pass by the artillery observation Sherman in the middle of the ruined town. (Heimdal.)

From west to east, the target here was the Rue Pasteur; the buildings on either side of the street collapsed, damaging the street and covering it with stones and all kinds of rubble, making it completely unusable for traffic. 2 German tanks, a Panzer IV and a Tiger, had been blocking the way on a level with the Town Hall square since June 13th. These tanks were now dismantled, the Panzer IV torn to pieces, and the Tiger reduced to a carcass.

From north to south, the aim was to cut off the roads to Vire, Aunay and Evrecy and the major crossroads of the Place Jeanne d'Arc. The bombs fell into the fields, to the south of the schools, south of the church, along the Boulevard Joffre, in the field south of the cattle market and further south after the Rivière factory as far as the drive up to the château. Everywhere they formed deep craters like some moonscape.

Captain R.H. Boyers of the R.A.F.'s 51st Squadron was the first pilot to drop his bombs on Villers, from west to east, along the main street. According to him, the Resistance had asked Bomber Command to bombard Villers in order to flush out a German armored division thought to be sheltering there, and he had been told that the townspeople had taken refuge out of harm's way at the château. After taking off from an airfield in Yorkshire at 14.00, he arrived 3 hours later over his target where, owing to cloud cover, he had to come down from 12,000 to 3,500 feet before starting to drop his bombs. Once the bombing raid was over, the planes returned to base and the artillery resumed its barrage of fire.

In Ecanet Wood the Germans had concealed two ammunition trucks which were hit and the contents exploded in all directions.

The purpose of this bombing raid was to make several crossroads in the town impracticable, to block the roads to it with piles of material, and form excavations that would be hard to get across. As in most places that came in for this kind of air raid, the Germans soon found a way round the town, taking every possible route including across the fields. The roads that passed through Villers were important to them for transporting troops and supplies up to the front at Tilly and Noyers where the battle of the Odon was being fought.

Fortunately, that day the town had been deserted by most of its inhabitants. A few had stayed on in the Boulevard Joffre and the station quarter, some of whose houses had been spared in the battle of June 13th and others who had settled in or built shelters that they considered to provide them with an adequate degree of safety. Five of them were killed: Madame Florentine Denis, Madame Annie Cervelle, Monsieur Alphonse Lebel, Madame Marguerite Wilmann and Monsieur Adjutor Lefèvre. As soon as the air raid was over, emergency crews from the château had rushed up with first-aid equipment to assist rescuers pull victims out of the debris, tend the slightly injured on the spot, carry the badly injured to the château on makeshift stretchers, and comfort those who had survived uninjured but more or less in a state of shock, and those who had had a close relation killed. Rescue operations went on throughout the night, by the light of fires that had been started, as shells continued to fall on Villers or whistled overhead before exploding further on. That evening, the hospice was in flames and the dispensary in the Boulevard Joffre collapsed onto the cellar where the sisters had remained on the spot close to the population with some neighbors who had joined them. They were all rescued in time from their predicament.

From that day on, the château became a veritable hospital run by Madame de Clermont-Tonnerre. She

Les autorités administratives françaises donnent également des instructions et des conseils aux maires :
- terminer les opérations d'évacuation le 8 juillet à 12 heures pour que la zone soit complètement vidée de ses civils le 9 à 24 heures et éviter aux retardataires et aux réfractaires d'être appréhendés par l'armée allemande,
- délivrer à chaque habitant une pièce frappée du cachet de la mairie et portant le nom de la famille de l'intéressé et la signature du maire pour justifier de son identité d'évacué dans les jours à venir,
- constituer des équipes d'avant-garde à bicyclette pour aller préparer les cantonnements dans les communes traversées tous les 20 kilomètres environ,
- prévoir un service sanitaire avec du personnel sanitaire en fin de colonne pour récupérer les retardataires et les malades.

Heureusement, ces ordres impératifs et difficiles à mettre en application dans beaucoup de cas ne sont pas toujours respectés. C'est ainsi que le 13, une colonne d'évacués arrive au château pour faire soigner ses malades, se ravitailler et essayer de se débarrasser de la vermine qui a envahi leurs vêtements. Il s'agit d'habitants de Hottot-les-Bagues qui ont vécu depuis plusieurs semaines tout près des lignes anglaises, parfois à moins de 200 mètres, sans pratiquement pouvoir sortir de leurs maisons.

Aucune exception n'a été prévue pour le château, comme partout ailleurs, l'ordre arrive le 6 d'évacuer les 200 personnes qui y habitent avec comme destination pour les vieillards, les enfants, les infirmes, les blessés, soit une centaine de personnes, le château de Ville-Gontier en Ille-et-Vilaine. Mme de Clermont-Tonnerre se trouve devant une difficulté matérielle d'exécuter cet ordre. Elle décide alors d'aller demander au général allemand qui commande la région depuis le Mesnil-au-Grain de l'autoriser à laisser ouvert l'espace sanitaire que constitue le château. Le 9, elle reçoit l'avis que sa demande est acceptée mais elle doit tout de même évacuer les vieillards, les petits enfants et les blessés légers. Ainsi, la zone de paix protégée par les croix rouges du toit peut encore remplir son rôle. La limite d'évacuation étant la route Caen-Vire, les équipes mobiles à bord des véhicules facilement identifiables par les soldats allemands qui les laissent passer ont un peu plus de facilité pour assurer, non sans risques, le ravitaillement dans la campagne au sud de Villers.

Jusqu'au 17 juillet, la *7th Armoured Division* va être mise au repos. Dans les derniers jours de juin, elle a quitté les premières lignes pour s'installer aux alentours du hameau de Jérusalem près du Douet-de-Chouain, et a été remplacée par la 2ᵉ Division Blindé américaine.

Pendant cette période, les bataillons des *Queen's* sont réorganisés. Des renforts venant du *2th et 5th Flight* vont compléter les compagnies pour combler en partie les lourdes pertes subies depuis le 6 juin. De nouveaux officiers vont être nommés à la tête des unités. Le général F.C. Pepper prend le commandement de la 131ᵉ Brigade d'Infanterie, le lieutenant-colonel J.B. Ashworth celui du *5th Queen's*, le lieutenant-colonel J.H. Mason celui du *6th Queen's*, le lieutenant-colonel M.F.P. Lloyd celui du *7th Queen's*.

Le **17 juillet**, la brigade blindée est à Saint-Gabriel, prête à partir pour la plaine de Caen. A partir du 18 juillet, la *7th Armoured Division* va changer de terrain d'opérations. Elle va abandonner le bocage normand et son horizon souvent limité à de courtes distances pour la plaine de Caen plus propice aux combats de chars et où vont se dérouler du 18 au 20 juillet, l'opération *Goodwood* et, du 24 au 31 juillet, l'opération *Spring*. Elle va aussi quitter le *XXXᵉ corps*

was assisted by Dr Dary and his wife, a nurse, who came to stay on the spot on June 28th, by the sisters of the hospice, joined by those from the Dispensary, by Mademoiselle Andrée Marie, a chemist's assistant, and by Red Cross first-aid workers. Within two days two bandaging rooms had been set up in the basement where casualties were treated. The slightly injured were kept there, but whenever communications allowed, others who were more seriously wounded were taken to Caen.

Also on June 30th, Thury-Harcourt was bombarded for the same purpose, to prevent German reinforcements from arriving south of Caen. There were forty civilian casualties.

Final days of the occupation

Between the front sector and the Caen-Vire road, civilians were beginning to get in the Germans' way. On July 6th, the German command decided to evacuate a zone from Dampierre to Saint-Jean-des-Essartiers in the west to Mondrainville in the east. The evacuation was to begin on Friday July 7th at 00.00 and be completed by 24.00 on July 9th at the latest.

Evacuees from the communities concerned had to reach the Saint-Sever-Vire-Falaise line by the 9th. The orders were strict and threatening, and anyone found in the area after the deadline was to be shot. The evacuation ordered by the Germans was to be carried out in an orderly fashion and if necessary, the mayors, who informed their citizens individually, could get help from the Feldgendarmerie. The evacuees could take whatever they liked by way of supplies, clothing, blankets and livestock, and had to gather in columns formed by inhabitants of the same community supervised by local officials, and carry pieces of white cloth as identification. The route to be followed as set by the Germans took the groups thus formed to the Argentan-Sourdeval-Flers line.

The French administrative authorities also issued instructions and advice to the mayors:

- to complete evacuation operations by 12.00 hours on July 8th so that the area was completely cleared of civilians by 24.00 hours on the 9th and thus avoid anyone lagging behind deliberately or otherwise from being arrested by the German Army,

- to issue each inhabitant with a document bearing the Town Hall's stamp and the person's surname together with the mayor's signature in proof of the evacuee's identity in the days to come,

- to organize teams to go on ahead on bicycles and prepare quarters in the communities to be passed through, every 20 kilometers or so,

- to arrange for a medical service with medical personnel at the rear of the column to recover stragglers and the sick.

Une partie de l'équipe médicale installée au château de Villers-Bocage, photo prise le 10 août 1944 : - traction 15 CV : Mme de Clermont-Tonnerre, Dr Davy, Abbé Vigou ; Hotchkiss : Paulette Orenne, Stanislas de Clermont-Tonnerre, - Citroën 11 CV : Mme Lemoal (infirmière), Mlle Marie, Jacques de Clermont-Tonnerre ; - moto : Philippe de Vendeuvre.

Part of the medical team based at the Château de Villers-Bocage, photo taken on August 10th 1944: - 15 HP front-wheel drive: Madame de Clermont-Tonnerre, Dr Davy, Abbé Vigou; Hotchkiss: Paulette Orenne, Stanislas de Clermont-Tonnerre, - 11 CV Citroën: Madame Lemoal (nurse), Mademoiselle Marie, Jacques de Clermont-Tonnerre; - motor-cycle: Philippe de Vendeuvre.

140

that day on became a proper hospital run by Madame de Clermont-Tonnerre, helped by Dr Dary and his wife, a nurse, and by the sisters of the hospice and the dispensary, by Mademoiselle Marie (a chemist's assistant) and by Red Cross first-aid workers.

We have several views of the château here, in 1944 and today, and a small house in the outbuildings where some old people lived. (Coll. Mme Dary and EG/Heimdal photographs.)

L'équipe médicale (médecins, infirmières et équipes d'urgence) est installée au château de Villers-Bocage pendant la bataille. Le château avait été épargné et, dès la fin du bombardement du 30 juin 1944, les équipes d'urgence étaient parties, munies de brancards improvisés pour participer au déblaiement et au sauvetage des victimes. Tous les blessés étaient dirigés vers le château qui, à partir de ce jour-là, devint un véritable hôpital organisé par Madame de Clermont-Tonnerre, aidée du Docteur Dary et de son épouse, infirmière, des sœurs de l'Hospice et du dispensaire, de Mlle Marie (préparatrice en pharmacie) et des secouristes de la Croix Rouge.

Nous voyons ici plusieurs aspects du château, en 1944 et actuellement, ainsi que d'une petite maison des communs où résidaient des personnes âgées. (Coll. de Mme Dary et photos EG/Heimdal.)

During the battle the medical staff (doctors, nurses and emergency crews) were based at Villers-Bocage in the château. The château had been saved, and at the end of the bombardment of June 30th, 1944, the emergency crews had gone out with makeshift stretchers to take part in clearance work and rescuing casualties. All the wounded were taken to the château, which from

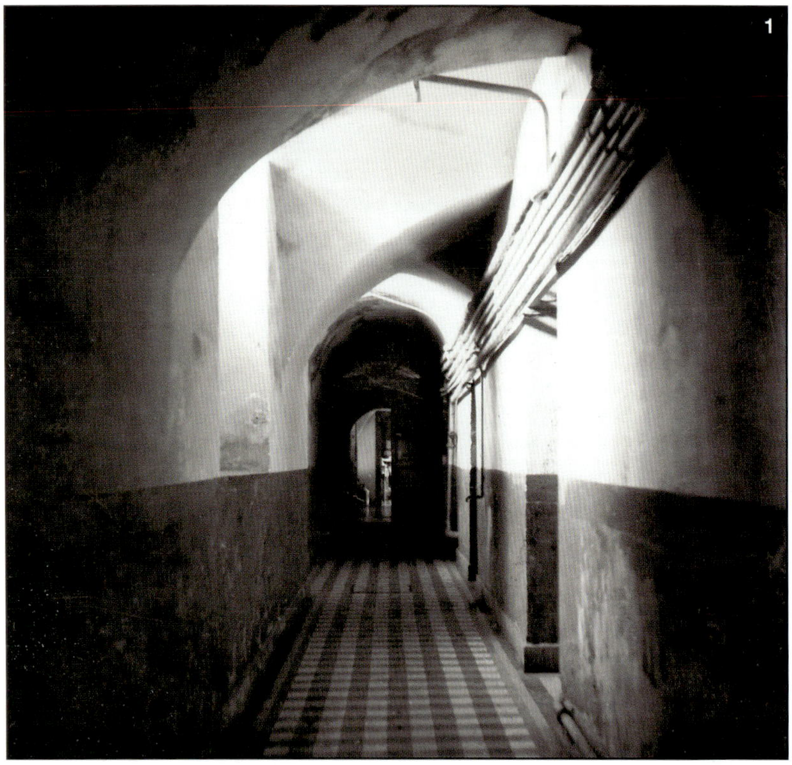
1

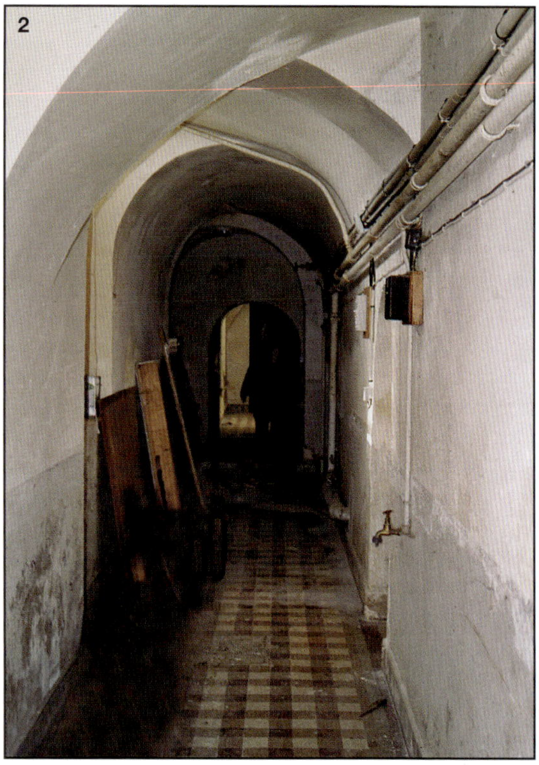
2

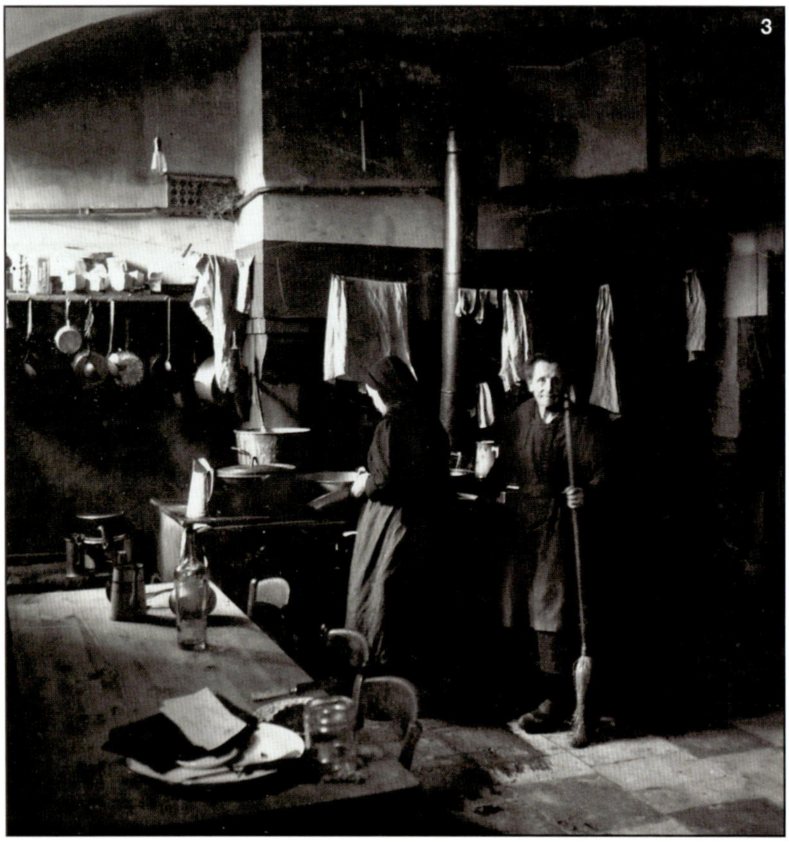
3

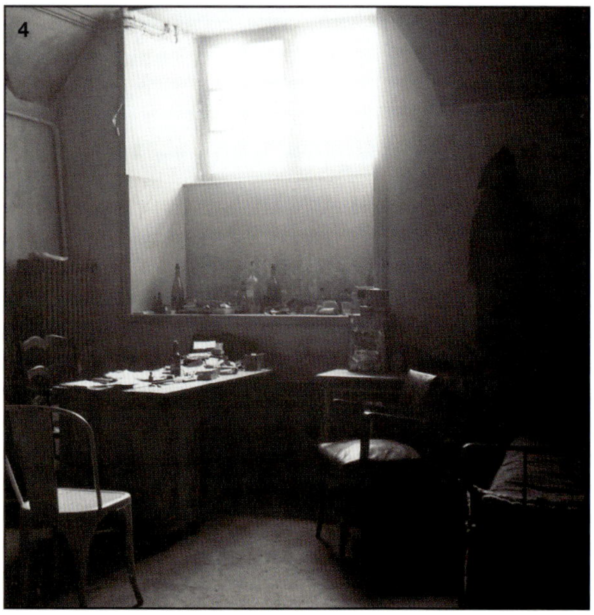
4

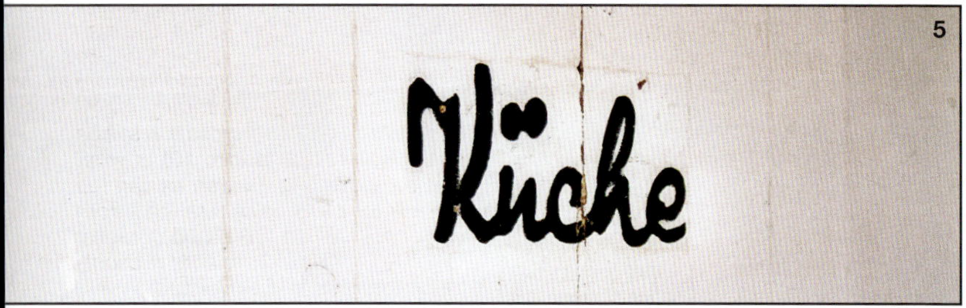

5

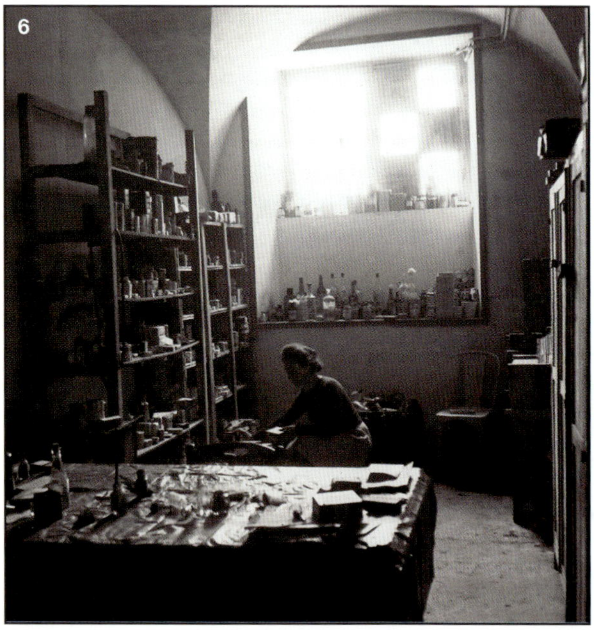
6

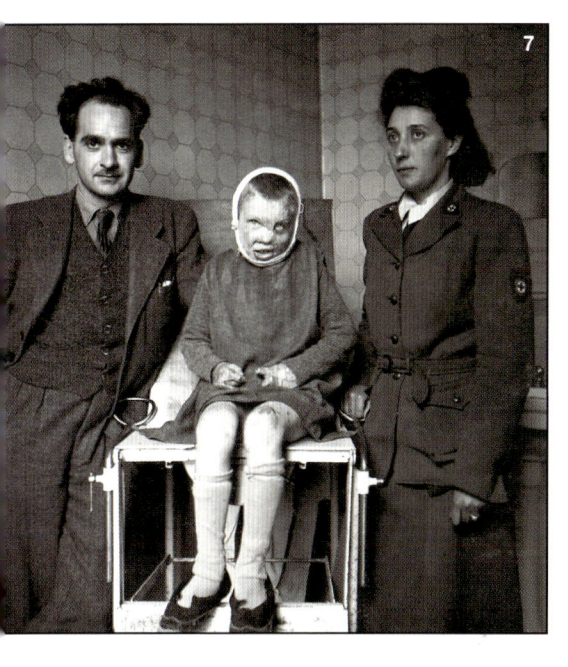

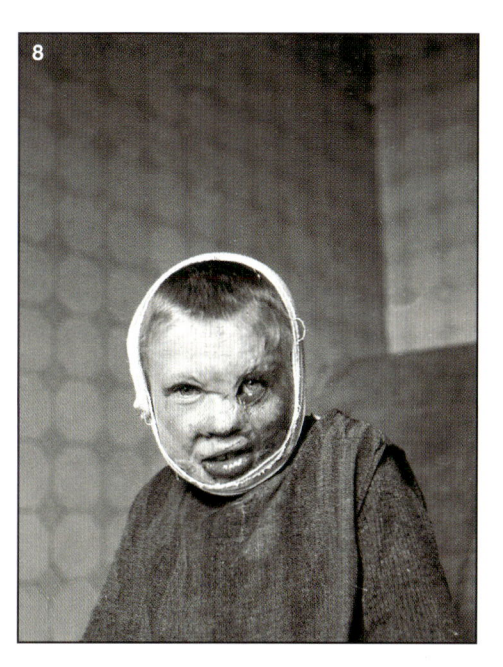

Le couloir du sous-sol du château (**photo 1**, en 1944, et **photo 2**, actuellement) servit de refuge à toute la population de Villers à partir du 13 juin et pendant deux mois pendant les bombardements. Au même niveau, la cuisine (**photo 3**) servit 3 000 repas aux réfugiés. A proximité se trouvent la salle de consultation du Docteur Dary (**photo 4**) et la pharmacie (**photo 6**). Actuellement, on trouve encore une inscription en allemand sur la porte de la cuisine (**photo 5**) dans ces sous-sols. Un petit sinistré, sérieusement blessé, a été adopté par le château (**photos 7 et 8**). Mais les pertes humaines ont été lourdes. Dans le parc s'alignent les tombes du cimetière français (**photo 9**) et celles du cimetière allemand (**photo 11**). Pendant cette période, il y a aussi un nouveau né (**photo 12**) dans la maternité du château. Actuellement, on peut encore voir la croix (ancienne chapelle) sur le sol de ce qui fut la salle d'opération (**11**) et la trace de la cloison la séparant de la salle post-opératoire. (Photos de 1944 : coll. Mme Dary ; photos actuelles : EG/Heimdal.)

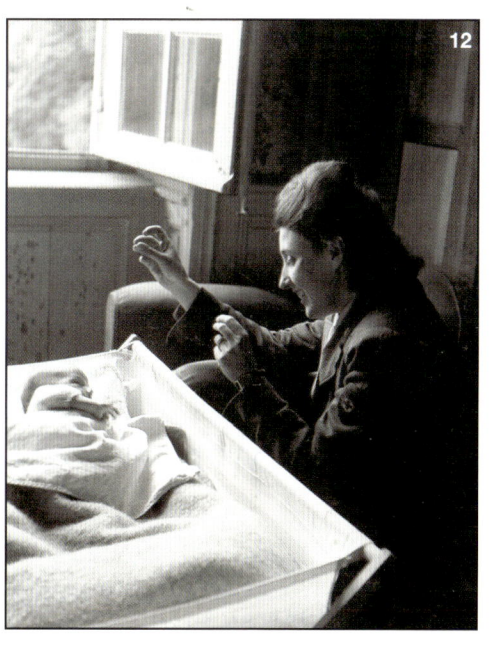

*The corridor of the château basement (**photo 1**, in 1944, and **photo 2**, today) was used to shelter the entire population of Villers during the bombardments for two months from June 13th. On the same level, the kitchen (**photo 3**) served 3,000 meals to the refugees. Close by are Dr Dary's surgery (**photo 4**) and pharmacy (**photo 6**). Today, there is still a German inscription on the basement kitchen door (**photo 5**). One young victim who was badly wounded was adopted by the château (**photos 7 and 8**). But human losses were heavy. In the grounds are rows of graves from the French cemetery (**photo 9**) and those of the German cemetery (**photo 11**). There was also a baby born during this period (**photo 12**) in the château maternity ward. Today, the old chapel cross can still be seen on the floor (**photo 10**) of what was the operating theater, as can the mark of the partition separating it from the recovery room. (1944 photographs: Coll. de Clermont-Tonnerre; present-day photographs: EG/Heimdal.)*

143

pour le *VIIIe*, combattre aux côtés de la 11e Division Blindée, la Division Blindée de la Garde, et le *IIe corps canadien*, l'objectif final de ces opérations étant Falaise. Elle va combattre au centre du dispositif de l'offensive mais n'atteindra pas l'objectif qui lui était fixé et ne dépassera pas Bourguébus.

Pendant ce temps, **à Villers**, l'artillerie continue ses tirs, le **20 juillet**, des éclats d'obus criblent la façade sud du château et tuent trois personnes, M. Geffroy, de Hottot, M. J. Lebosquain, de Vendes et un Monsieur âgé, imprudent, resté près d'une fenêtre. Le **25**, les Allemands sentant la bataille approcher veulent récupérer trois salles au sous-sol pour y installer un poste de secours d'un bataillon en première ligne. Cette fois-ci, il faut évacuer dans des cars munis de croix rouges avec un médecin, le docteur François, des vieillards, quelques religieuses et quelques personnes qui peuvent se déplacer. Resteront au château, 148 personnes dont un tiers de malades, le personnel soignant, et tous ceux qui participent au ravitaillement de tous.

Côté allemand, les événements évoluent au sein de la 2. Pz.-Div. qui était en position dans le secteur Caumont-Tilly et qui doit faire face à toutes les escarmouches. Son régiment de chars, le *Pz. Reg. 3*, a été souvent appelé pour prêter main forte aux unités allemandes voisines, vers le 26 juin à la division *Hitlerjugend* dans la bataille de l'Odon ou le 18 juillet à la *Panzer-Lehr-Division* au sud de Saint-Lô. Aussi la division est-elle relevée par une division d'Infanterie entre le 21 et le 24 juillet. Elle se rassemble à l'est de Thury-Harcourt mais laisse dans son ancien secteur comme réserve mobile son IIe bataillon de chars, la II./Pz. Reg. 3.

Le **25 juillet**, la division des « rats du désert », la *7th Armoured Division*, passe sous le commandement du *IIe corps canadien* et elle est maintenue à Ifs où elle se tient prête à intervenir dans l'opération *Spring*. En fait, cette opération n'est qu'une manœuvre de diversion pour maintenir le plus possible de blindés allemands au sud de Caen et permettre à l'armée américaine sa percée sur Avranches.

Le 20 juillet 1944, des éclats d'obus criblent la façade sud du château de Villers-Bocage et tuent trois personnes. Les impacts sont encore visibles. (EG/Heimdal.)

The artillery fire went on at Villers-Bocage on July 20th, and the southern frontage of the château was riddled with shrapnel, killing three. (EG/Heimdal.)

Fortunately, in many cases, being difficult to apply, these imperative orders were not always followed. Thus on the 13th, a column of evacuees arrived at the château to tend their sick, pick up supplies and try to get rid of the vermin infesting their clothing. They were inhabitants of Hottot-les-Bagues who had been living for several weeks very close to the British lines, sometimes less than 200 meters away, practically without being able to venture outdoors.

No exception was made for the château; like everywhere else, the order arrived on the 6th to evacuate the 200 people living there, with the old people, children, the disabled and the injured – a hundred people altogether – to be sent off to the château at Ville-Gontier in the Ille-et-Vilaine department. Madame de Clermont-Tonnerre in complying with this order was confronted with a very practical difficulty. So she decided to go and ask the German general commanding the area from Le Mesnil-au-Grain for permission to keep the château open as a field hospital. On the 9th, she was notified that her request had been accepted but that all the same the elderly, small children and slightly injured had to be evacuated. So this haven of peace protected by the red crosses on the roof could continue to serve its purpose. With the evacuation perimeter extending as far as the Caen-Vire road, it became a little easier for mobile teams aboard vehicles easily identifiable by the German soldiers, who let them pass, to ensure, albeit at some risk to themselves, that supplies got into the country area south of Villers.

*The 7th Armoured Division was rested **until July 17th**. In the last days of June, it left the front lines to settle in the vicinity of the hamlet of Jerusalem, near Douet-de-Chouain, to be replaced by the 2nd US Armored Division.*

During this period, the Queen's battalions were reorganized. The companies received reinforcements from the 2nd and 5th Flights to partly make up for the heavy losses sustained since June 6th. New officers were appointed to command the units. General F.C. Pepper took over command of the 131st Infantry Brigade, Lieutenant-Colonel J.B. Ashworth that of 5th Queen's, Lieutenant-Colonel J.H. Mason that of 6th Queen's, and Lieutenant-Colonel M.F.P. Lloyd that of 7th Queen's.

On July 17th, the armored brigade was at Saint-Gabriel, ready to leave for the plain of Caen. From July 18th, the 7th Armored Division was to transfer its operations and leave the hedgerow country and its horizon often limited to short distances for the plain of Caen more suitable for tank battles, and where Operation Goodwood was fought from July 18th to 20th, and Operation Spring from July 24th to 31st. It also left XXX Corps to join VIII Corps, to fight alongside the 11th Armored Division, the Guards Armoured Division, and Canadian II Corps, Falaise being the final objective of these operations. It fought in the center of the offensive disposition but failed to achieve its assigned target or advance beyond Bourguébus.

*Meanwhile, the artillery fire went on **at Villers** on **July 20th**, and the southern frontage of the château was riddled with shrapnel, killing three: Monsieur Geffroy from Hottot, Monsieur J. Lebosquain from Vendes, and an elderly gentleman who had unwisely been standing at a window. On the 25th, feeling the battle approaching, the Germans wanted to take over three basement rooms to use as a first-aid post for a front line battalion. This time, the old people, a few nuns and some others who could be moved had to be evacuated in a bus marked with red crosses, accompanied by a physician, Dr. François. This left 148 people at the château, a third of whom were*

La *7th Armoured Division* a compté 400 pertes durant cette période. Le **28 juillet**, la Brigade des *Queen's* est relevée par la 9ᵉ Brigade canadienne et la division va rejoindre le *XXXᵉ Corps* dans le secteur de Caumont avant de se lancer dans l'opération *Bluecoat* qui va permettre la libération de Villers-Bocage.

Le **29 juillet** marque la journée d'adieu du *4th CLY* à la *7th Armoured Division*. Le régiment est remplacé par un régiment frais qui vient d'Angleterre, le *5th Royal Inniskilling Dragoons Guards*, les « *Skins* », commandé par le lieutenant-colonel J.E. Swetham. Tristes, les *Sharpshooters* abandonnent leurs chars sur l'aérodrome de Carpiquet, pris le 4 juillet, à un autre régiment. Ils regrettent surtout le *Firefly* qui, pour eux, était le meilleur char britannique capable de mettre le Tigre hors de combat. Le *4th CLY* rejoint le *3th CLY* au sein de la 4ᵉ Brigade Blindée indépendante pour former le *3/4 CLY*. Les vaillants *Sharpshooters* ne vont pas participer à la libération de Villers-Bocage.

Le 25 juillet, dans la Manche, les Américains après une formidable attaque aérienne avaient lancé l'opération « Cobra » et fonçaient vers le sud. Le 28, ils prenaient Coutances, le 30, Granville et Avranches et le 31 franchissaient la Sélune à Pontaubault.

Dans la plaine de Caen, les opérations *Goodwood* et *Spring* lancées par Anglais et Canadiens ont fait progresser leur avance vers Falaise, mais les Allemands solidement installés de chaque côté de l'Orne doivent continuer à se maintenir sur le défensive pour stopper la progression alliée et maintenir au sud de Caen et à l'est de Noyers une grande partie de leurs blindés.

Il se trouve entre ces deux zones, Sud Manche et Plaine de Caen, toute une zone de bocage plus facile à défendre qu'à conquérir qui va de Noyers à Vire et même Saint-Lô, limitée vers le sud du département par une ligne de collines bien défendues dominées par la cote 309, près de Saint-Martin-des-Besaces, la cote 301 près de Jurques et le Mont Pinçon qui culmine à 365 mètres.

Le général Montgomery envisage de lancer dans cette zone la presque totalité de la 2ᵉ armée du général Dempsey dans cette opération massive qui débutera le 30 juillet au petit matin. Ce sera **l'opération Bluecoat**.

Vont participer à cette opération 3 Corps d'armée. Un qui partira de l'ouest de Caumont, le *VIIIᵉ Corps*, composé de la *11th Armoured Division*, de la *15th Infantry Division* écossaise et de la Division Blindée de la Garde. L'objectif de ce Corps d'armée sera d'occuper la cote 309 et le terrain jusqu'à Vire. Un autre, à l'est du front, depuis le secteur de Noyers, le *XIIᵉ Corps* qui devra renforcer la tête de pont établie à l'est de l'Orne, atteindre Thury-Harcourt et libérer Villers-Bocage. Il est composé de deux divisions d'Infanterie, la 53ᵉ *(Welsh)* et la 59ᵉ *(Staffordshire)*.

Au centre, sera le *XXXᵉ Corps*, composé de la *43th ID. Essex*, de la *50th ID Northumbrian* qui tient le front de Tilly à Caumont depuis six semaines et la *7th Armoured Division*, les « rats du déserts », qui n'entrera en action que le 31. Ces trois divisions ont pour objectif Aunay-sur-Odon et le Mont Pinçon. Leur point de départ sera la ligne Tilly-Caumont.

Face aux divisions d'Infanterie et aux divisions de chars alliées dont sensiblement le nombre est égal à celui engagés dans l'opération *Goodwood*, se trouvent deux divisions d'Infanterie allemande, à l'est la 276ᵉ qui a remplacé la *Panzer-Lehr-Division* et a conservé en renfort des chars du *Panzer-Lehr Reg. 130* et, au centre du front, la 326ᵉ qui a remplacé la

patients, auxiliary nursing staff, and all those involved in providing supplies for everyone.

On the German side, things were moving in the 2. Pz.-Div. which was in position in the Caumont-Tilly sector and had all the skirmishing to deal with. Its tank regiment, Pz. Reg. 3, was frequently called upon to support the neighboring German units, the Hitlerjugend Division in the battle of the Odon on around June 26th, and the Panzer-Lehr-Division south of Saint-Lô on July 18th. The division was then relieved by an infantry division sometime between July 21st and 24th. It mustered east of Thury-Harcourt, leaving its 2nd tank battalion, II./Pz. Reg. 3, behind as a mobile reserve in its old sector.

On July **25th**, the «Desert Rats» division, 7th Armoured, was placed under the command of Canadian II Corps and was held at the ready at Ifs to intervene in Operation Spring. This was in fact no more than a diversionary operation to pin down as much as possible of the German armor south of Caen so as to allow the American Army to break out at Avranches.

The 7th Armoured Division counted 400 losses during this period. On July 28th, the Queen's Brigade was relieved by the 9th Canadian Brigade and the division joined XXX Corps in the Caumont sector before taking part in Operation Bluecoat, which saw the liberation of Villers-Bocage.

July 29th was the 4th CLY's last day with the 7th Armoured Division. The regiment was replaced by a fresh regiment brought over from England, the Royal Inniskilling Dragoon Guards, the "Skins", commanded by Lieutenant-Colonel J.E. Swetham. The Sharpshooters sadly handed over their tanks to another regiment at Carpiquet airfield, captured on July 4th. They especially hated parting with the Firefly which, for them, was the best British tank for dealing with the Tiger. The 4th CLY joined the 3rd CLY to form 3/4 CLY with the 4th independent Armoured Brigade. The valiant Sharpshooters took no part in the liberation of Villers-Bocage.

On July 25th, after a tremendous aerial attack in the Manche department, the Americans launched Operation Cobra and raced southwards. They took Coutances on the 28th, Granville and Avranches on the 30th, and crossed Sélune at Pontaubault on the 31st.

In the plain of Caen, Operations Goodwood and Spring launched by the British and Canadians boosted their advance on Falaise, but the Germans were firmly entrenched on either side of the Orne river and stayed on the defensive to stop the Allied advance and pin down much of their armor south of Caen and east of Noyers.

Between these two sectors, the southern Manche and the plain of Caen, was a whole area of mixed wood and pasture more suited to defense than to attack, stretching from Noyers to Vire and even Saint-Lô, and bordered to the south of the department by a line of well-defended hills, the highest being Hill 309 near Saint-Martin-des-Besaces, Hill 301 near Jurques, and Mont Pinçon, rising to 365 meters.

In this sector General Montgomery planned to launch almost the whole of General Dempsey's 2nd Army in this massive operation codenamed Operation **Bluecoat,** set to begin early on the morning of July 30th.

The lineup for this operation included three corps. One starting from west of Caumont, VIII Corps comprising the 11th Armoured Division, the 15th Scottish Infantry Division and the Guards Armoured Division. This corps' assigned objective was to occupy Hill 309 and the area as far as Vire. Another, XII Corps, along the eastern section of the front, starting from the Noyers sector, was to reinforce the bridgehead established east of the Orne, advance to Thury-Har-

2 Pz.-Div. et a conservé les chars de la *II./Pz Rgt. 3* comme réserve mobile.

A l'ouest se trouvent des blindés, ce qu'il reste de la *21. Panzer-Division*, soit la valeur d'un bataillon, des Tigres I et II de la *Pz.-Abt. 503*, quelques formations autonomes de chasseurs de chars armés de canons de 88 mm tel le 654ᵉ bataillon et des formations de canons autotractés de 105 mm, les canons d'assaut *Sturmgeschütz*.

Au début de l'offensive, les résultats alliés sont prometteurs. L'objectif du *VIIIᵉ Corps* à l'Ouest, dont les éléments avancés ont atteint Vire et qui a abandonné la prise de la ville à la *1ʳᵉ Division U.S.*, et maintenant les secteurs de Condé-sur-Noireau et de Flers. Celui du *XXXᵉ Corps* est étendu vers l'est, la rive gauche de l'Orne et Thury-Harcourt.

Le commandement allemand réagit et décide alors de ramener des divisions blindées du *II. SS-Pz.-Korps* vers le Bocage virois et de créer une zone infranchissable de blindés afin de stopper le *VIIIᵉ Corps anglais* dans sa progression vers le sud et confie son secteur dans la plaine de Caen à la *1. SS-Pz.-Div.*, la *Leibstandarte (LSSAH)*.

C'est ainsi que la *9. SS-Pz.-Div.*, la *Hohenstaufen*, en position à l'est de l'Orne va se placer face au *VIIIᵉ Corps*, au sud de Bény-Bocage, devant la ligne Vire-Condé, et que la *10. SS-Pz.-Div. Frundsberg*, dont les premiers éléments arrivent à Aunay le 1ᵉʳ août, prend place face à la *43th I.D.*, la *9th Armoured Division* et la *50th I.D.* de Jurques à Aunay pour défendre le Mont Pinçon. Ces deux divisions allemandes pourront disposer des Tigres de la *(s.) SS-Pz.-Abt. 102* qui ont abandonné la cote 112 près d'Evrecy et vont se fixer plus au sud vers Saint-Jean-le-Blanc.

L'offensive est générale. Le 30 juillet, les tirs d'artillerie reprennent de plus belle sur Tracy-Bocage et Amayé-sur-Seulles. Le 31, une colonne d'infanterie allemande s'installe dans le parc du château de Villers et occupe les salles du sous-sol que 10 enfants devront libérer. Ils sont évacués sur Carouges. Ce mouvement d'Allemands déclenche des tirs d'artillerie qui ne blessent heureusement aucun des 148 civils réfugiés au château.

L'objectif du *XXXᵉ Corps* était de prendre Aunay, Villers restant dans celui du *XIIᵉ Corps*.

Le **1ᵉʳ août**, le *5th Queen's* et le *5th RTR* atteignent le hameau de Breuil et le plateau autour de la Jatte du Val.

Les « Skins » qui ont remplacé le *4th CLY* dans la *7th A.D.*, sont à Coulvain, au carrefour de Quéry. Ils vont recevoir le 3 le renfort du *65th Antitank Regiment, the Norfolk Yeomanry* qui, avec ses canons autotractés munis d'obus de 17 livres, sera le bienvenu car des Tigres rodent dans les parages. Il doit s'agir d'éléments de la *(s.) SS-Pz.-Abt. 503*.

Le **3 août**, après trois jours de l'opération *Bluecoat*, « les rats du désert » sont encore à 10 kilomètres de leur objectif, Aunay. Ce jour est une journée noire pour la division qui est mise en échec et déplore au moins 200 pertes autour d'Aunay. Le *5th Queen's* nettoie les bois autour de Breuil. Le *7th Queen's* en fait de même entre Villers et Aunay, appuyé par le *8th Hussars*. Il subit des pertes sensibles. Le *6th Queen's* qui a conquis du terrain autour de Saulques, subit une contre-attaque de la 326ᵉ Division allemande au sud-ouest d'Aunay soutenue par le 10ᵉ Rgt d'Artillerie SS. et les blindés du *SS-Pz.-Rgt. 10 (10. SS-Pz.-Div.)*. Le *5th RTR* lance une attaque vers les hameaux de Courcelles et la Lande. Il sera le plus touché de la division. Son attaque est repoussée autour de la cote 188 par les grenadiers armés de *panzerfaust du SS-Pz.-Gren.-Rgt. 21* soutenus par

court and liberate Villers-Bocage. It was made up of two infantry divisions, the 5th (Welsh) and the 59th (Staffordshire).

In the center, XXX Corps comprised the 43rd ID Essex, the 50th ID Northumbrian, holding the front from Tilly to Caumont for the past six weeks, and the 7th Armoured Division, the "Desert Rats", which did not join the fray until the 31st. These three divisions' objectives were Aunay-sur-Odon and Mont Pinçon, starting from the Tilly-Caumont line.

Facing the Allied infantry divisions and tank divisions, in numbers roughly equal to those committed in Operation Goodwood, were two German infantry divisions, in the east the 276th, replacing the Panzer-Lehr-Division and keeping the tanks of Panzer-Lehr Reg. 130 in reinforcement, and in the center of the front, the 326th, which replaced the 2. Pz.-Div. and held the tanks of II./Pz Rgt. 3 as a mobile reserve.

In the west were what was left of the 21. Panzer-Division's tanks, i.e. a battalion's worth, the Tigers I and II of SS-Pz.-Abt. 503, a few independent formations of tank destroyers armed with 88 mm guns, like the 654th Battalion, and formations of 105 mm self-propelled guns, the Sturmgeschütz assault guns.

At the start of the offensive, the Allies achieved promising results. The objective of VIII Corps in the west, its forward elements having reached Vire and left the U.S. 1st Division to take the town, was now the Condé-sur-Noireau and Flers sectors. The objective of XXX Corps extended eastwards, to the left bank of the Orne and Thury-Harcourt.

The German command reacted by deciding to bring the armored divisions of II. SS-Pz.- Korps back into the hedgerow country around Vire and create an impregnable buildup of armor to halt the southward advance of British VIII Corps, and by placing its sector in the plain of Caen in the hands of the 1. SS-Pz.-Div., Leibstandarte (LSSAH).

Thus 9. SS-Pz.-Div. Hohenstaufen, positioned east of the Orne, stood facing VIII Corps south of Bény-Bocage, in front of the Vire-Condé line, and 10. SS-Pz.-Div. Frundsberg whose leading elements arrived at Aunay on August 1st took up position opposite the 43rd I.D., the 9th Armored Division and the 50th I.D. from Jurques to Aunay in defense of Mont Pinçon. These two German divisions had at their disposal the Tigers of (s.) SS-Pz.-Abt. 102 which abandoned Point 112 near Evrecy and dug in further south near Saint-Jean-le-Blanc.

This was an all-out offensive. *On July 30th, the artillery fire was stepped up on Tracy-Bocage and Amayé-sur-Seulles. On the 31st, a column of German infantry parked itself in the grounds of the château de Villers and occupied the basement rooms, evicting 10 children who were evacuated to Carrouges. The Germans' move drew artillery fire which fortunately spared all of the 148 civilian refugees at the château.*

XXX Corps' objective was to take Aunay, while Villers remained a target of XII Corps.

*On **August 1st**, the 5th Queen's and 5th RTR reached the hamlet of Breuil and the flat land around La Jatte du Val.*

The «Skins», who had taken over from the 4th CLY in the 7th A.D., were at Coulvain, at the Quéry crossroads. On the 3rd they received the 65th Antitank Regiment, the Norfolk Yeomanry which, with its self-propelled guns firing 17-pdr shells, was a welcome reinforcement against the Tigers lurking in the area. These would have been elements of (s.) SS-Pz.-Abt. 503.

les chars du *SS-Pz.-Rgt. 10* et une section de *Panther*. Les chars du *5th RTR* doivent se replier à Breuil d'où ils étaient partis, en laissant de nombreux chars sur le terrain.

Ce manque de résultats dans le secteur de Saint-Georges-d'Aunay va provoquer une réaction du général Montgomery. Le 3 août, il va limoger les généraux Bucknall et Erskine qui seront remplacés le 4 par le général B.G. Horrocks à la tête du *XXXe Corps* et par le général Verney, *Brigadier* de la 6e brigade de la Garde, à la tête de la *7th Armoured Division*.

Bien que les actions de la *7th Armoured Division* au sud de Villers, de la *50 I.D.* au nord et à l'ouest, des *53th* et *59th I.D.* à l'est ne donnent pas de résultats rapides, elles risquent de créer autour de la ville une poche où des unités allemandes pourraient être encerclées.

C'est alors que le **3 août, à une heure du matin, Hitler en personne décide de raccourcir le front** et de libérer des unités de panzers pour les lancer dans l'offensive vers Mortain. Ainsi, la *10. SS-Pz.-Div.* se replie vers Vassy tout en combattant. Les Allemands vont se retirer, ce qui va permettre aux Anglais de libérer rapidement toute une région malgré quelques combats d'arrière-garde et malgré les nombreuses mines disséminées dans les champs, les habitations, les carrefours, qui feront des victimes parmi les alliés et parmi les civils dès qu'ils reviendront.

La bataille approche de Villers, les Allemands songent à évacuer la ville et à battre en retraite protégés par des Waffen-SS qui ont pris position alentour et dans la campagne. Avant de partir, ils voudraient enlever les croix rouges fixées sur le toit du château puis ils changent d'avis et même envoient un message aux Anglais qui approchent pour les prévenir que 150 civils y séjournent. Le message est entendu et le château sera épargné.

On **August 3rd**, after three days of Operation Bluecoat, the "Desert Rats" were still 10 kilometers short of their objective, Aunay. This was a black day for the division, which had failed to break through and had suffered at least 200 losses in the Aunay area. The 5th Queen's mopped up in the woodlands around Breuil. The 7th Queen's did likewise between Villers and Aunay, with the 8th Hussars in support. It too sustained heavy casualties. The 6th Queen's which had gained ground around Saulques, came under counter-attack south-west of Aunay by the German 326th Division supported by the 10th SS Rgt Artillery and the tanks of SS-Pz.-Rgt. 10 (10. SS-Pz.-Div.). The 5th RTR launched an attack on the hamlets of Courcelles and La Lande. It was the division most badly hit. Its attack was repulsed around Hill 188 by grenadiers of SS-Pz.-Gren.-Rgt. 21 armed with Panzerfausts supported by the tanks of SS-Pz.-Rgt. 10 and a troop of Panthers. The 5th RTR's tanks had to fall back to their start line at Breuil, leaving many tanks behind in the field.

This lack of results in the Saint-Georges-d'Aunay sector brought a reaction from General Montgomery. On August 3rd, he dismissed Generals Bucknall and Erskine, whom he replaced on the 4th with General B.G. Horrocks in command of XXX Corps and Brigadier-General Verney of the 6th Guards Brigade, in command of the 7th Armored Division.

Although the actions by the 7th Armoured Division south of Villers, by the 50th I.D. to the north and west, and by the 53rd and 59th I.D. to the east, brought no immediate results, they were likely to create a pocket around the town in which the German units might be encircled.

It was then, at one in the morning on August 3rd, **that Hitler himself decided to shorten the front** and release his panzer units to launch them in the Mortain offensive. Thus, 10. SS-Pz.-Div. fell back to Vassy, fighting as it went. The Germans withdrew, enabling

Cahagnes, 2 août 1944. Un char Sherman traverse le village dévasté par les combats en soulevant un nuage de poussière. (IWM.)

2 August 1944. A Sherman tank throwing up clouds of dust as it speeds through Cahagnes. (IWM.)

147

2 août 1944.

1. L'église et le cimetière de Cahagnes ont été dévastés par les combats.

2. Près de Benneville, des fantassins et des engins de reconnaissance avancent sur une route menant à Villers-Bocage entourée d'arbres hachés par les éclats d'obus.

3. Deux chars Sherman et un camion traversent Cahagnes.

(IWM.)

Le 3 août, des éléments avancés de la *50th I.D.* et du *11th Hussars* sont à Tracy. Ils sont dirigés vers « la poste », à la limite de leur secteur d'intervention.

Le 4 août, à une heure du matin, les préparatifs du départ commencent, d'une façon un peu désordonnée. Les Allemands manquent de véhicules et réquisitionnent ceux qu'ils trouvent, même les bicyclettes leur sont bonnes pour partir et quitter le plus rapidement possible une zone qu'ils ne peuvent plus contrôler. Dans les premières heures de la matinée, tous partent. Un dernier barrage d'artillerie précipite leur départ et, en début d'après-midi, les premiers Anglais arrivent. Ce sont des hommes de la 59ᵉ Division d'Infanterie Britannique, partie du *XIIᵉ Corps*.

Villers-Bocage était libérée pratiquement sans combat.

Des combats, il y en a eu encore dans les environs. C'est ainsi qu'au village de la « poste », à Maisoncelles-Pelvey, les habitants à leur retour d'exode ont trouvé entre autres, un char anglais détruit et une tombe surmontée d'une croix portant cette inscription, Tpr Foxs Tan XII, R.I.P. Un *trooper*, un soldat de 2ᵉ classe du *XIIᵉ Corps* (R.I.P. signifie, *Rest In Peace, Requiescat In Pace, Repose en Paix*).

Les soldats de la *59th I.D.* ont trouvé une ville en ruines, vidée de ses habitants et de ses occupants. Seuls les 148 réfugiés au château pouvaient accueillir leurs libérateurs et faire revivre la cité.

Le lendemain 5 août, Aunay et Evrecy étaient libérés de la même façon que Villers, Aunay par le *11th Hussars* qui entrait dans la ville en ruines par l'ouest et Evrecy par la *53rd I.D.* qui prenait possession aussitôt de la cote 112 abandonnée par les Allemands.

the British to liberate a whole area in quick time despite a few rearguard actions and numerous mines laid in fields, in houses and at crossroads, causing casualties among the Allies and among the civilians upon their return.

As the battle approached Villers, the Germans began to think of evacuating the town and of beating a retreat protected by the Waffen-SS who had taken up position in the vicinity and in the countryside. Before leaving, they wanted to remove the red crosses on the roof of the château, but they changed their minds and even sent a message to the approaching British to warn them that there were still 150 civilians there. The message was received and the château was spared.

On August 3rd, forward elements of the 50th I.D. and the 11th Hussars were in Tracy. They were directed to "La Poste», which was as far as they were to go.

On August 4th, at one in the morning, preparations for departure started, albeit in somewhat disorderly fashion. The Germans were short of vehicles and requisitioned any they could find, even grabbing bicycles to get out as quickly as possible from a zone that was no longer in their control. They all left during the small hours. One final artillery barrage hurried them on their way and the first British arrived early that afternoon. These were men of the British 59th Infantry Division, XII Corps.

Villers-Bocage was liberated almost without a shot being fired.

But there was more fighting in the surrounding area. Thus at Maisoncelles-Pelvey, the village where "La Poste» was, among other things the inhabitants found

2 August 1944.

1. The devastated church and churchyard at Cahagnes.

2. A scene near Benneville, showing infantry and vehicles along a road which divides a shell torn wood on the road to Villers-Bocage.

3. Sherman tanks passing through the ruined streets of Cahagnes.

(IWM.)

Les Britanniques libéraient des villes meurtries, endeuillées et écrasées par des bombardements massifs que beaucoup jugeront avoir été inutiles. En effet, pour retarder l'arrivée des renforts allemands, les alliés avaient écrasé sous les bombes des localités parce qu'elles se trouvaient aux carrefours de routes importantes et essayé de créer des obstacles qui avaient rapidement été contournés. Aunay et Evrecy bombardés le 12 juin et dans la nuit du 14 au 15 faisaient partie de ces localités. On peut supposer que ces bombardements avaient pour but de préparer d'abord et de protéger ensuite la tentative de percée vers Villers et la cote 213. **Il y a eu 175 morts à Aunay et 130 à Evrecy.**

La bataille de Villers-Bocage a eu lieu le 13 juin 1944, vidant pratiquement la ville de ses habitants. On peut s'imaginer ce qu'aurait été le bombardement du 30

on returning home were a destroyed British tank and a grave surmounted by a cross bearing the inscription Tpr Foxs Tan XII, R.I.P.

The soldiers of the 59th I.D. found a town in ruins, deserted by inhabitants and occupying forces alike. Only the 148 château refugees were there to welcome their liberators and bring the town back to life.

*On the following day, **August 5th**, Aunay and Evrecy were liberated as Villers had, Aunay by the 11th Hussars, entering the ruined town from the west, and Evrecy by the 53rd I.D. which took possession of Point 112 as soon as it was abandoned by the Germans.*

The towns the British liberated were battered, in mourning, and crushed by massive bombardments that many would consider to have been pointless.

2 août 1944.

1. Des prisonniers allemands sont rassemblés au centre de Cahagnes détruit par les combats tandis que des véhicules britanniques montent vers le front.

2. La progression continue au sud de Villers-Bocage. Ici, un soldat britannique a mis un fusil-mitrailleur Bren en position dans Jurques, sur la route Caen-Vire. (IWM.)

*Indeed, to hamper the arrival of German reinforcements, the Allies had bombed to pieces towns and villages located at major crossroads and tried to create obstacles that had been easily skirted. Among these places were Aunay and Evrecy, bombarded on June 12th and during the night of the 14-15th. We may surmise the purpose of these bombardments to have been first to soften things up and then to protect the attempted breakout towards Villers and Point 213. **There were 175 killed at Aunay and 130 at Evrecy.***

The battle of Villers-Bocage was fought on June 13th 1944, practically emptying the town of its inhabitants. One can imagine what the bombardment of June 30th would have been like had the town not been in

juin, si la ville n'avait pas été dès les premiers jours du débarquement, au cœur de la bataille et si ses habitants n'avaient pas pris conscience d'une telle situation.

Un autre événement important pour notre région a été la prise du Mont Pinçon où étaient installés observatoire et artillerie lourde.

La *7th Armoured Division* lançait une attaque à l'est, depuis la vallée de Hamars tandis que le *43rd Essex* partait de l'ouest avec l'appui du *13/18th Hussars*, régiment constitué le 29 juillet avant le début de l'opération *Bluecoat* et incorporé à la 8e Brigade Blindée indépendante. Des combats meurtriers avaient raison des troupes allemandes dont certaines se défendaient avec acharnement. C'est en définitive le *13/18th Hussars* qui, le premier, s'emparait du sommet du Mont le 7 août.

Le 4 août, Villers-Bocage était libérée mais la guerre ne se terminera que le 8 mai 1945. Au fur et à mesure que les régions de France seront libérées, les habitants regagneront leur ville. Ils n'y trouveront que des ruines. La vie reprendra immédiatement et commencera alors la période de la reconstruction qui ne se terminera peut-on dire qu'en 1959, avec la construction de l'Hôtel de Ville qui sera inauguré officiellement par le général de Gaulle lors de son passage à Villers le 8 juillet 1960.

the heart of the battle from the very first after D-Day, and if its inhabitants had not become aware of such a situation.

Another significant event for our area was the capture of Mont Pinçon, where an observatory and heavy artillery had been set up.

The 7th Armoured Division launched an attack from the Hamars valley to the east, while the 43rd Essexes started out from the west with the 13/18th Hussars in support, a regiment raised on July 29th, before the launch of Operation Bluecoat, and attached to the 8th Independent Armoured Brigade. A fierce battle overcame the German troops, some of which put up some stout resistance. In the end, the 13/18th Hussars were first to take the top of Mont Pinçon, on August 7th.

Villers-Bocage was liberated on August 4th, but the war did not end until May 8th 1945. As the regions of France were liberated, the inhabitants returned to their home town, to find nothing but ruins. Life started again immediately, followed by the period of reconstruction which really went on until 1959 and the building of the Town Hall, which was officially opened by General de Gaulle on a visit to Villers on July 8th 1960.

2 August 1944.
1. German prisoners assembled in the centre of battered Cahagnes watching British armor passing through the village on its way into battle.

2. A Bren gunner takes up position near corected houses on the roadway leading to the village of Jurques on the main Caen-Vire road. (IWM.)

Villers-Bocage, le 5 août 1944. Les troupes britanniques ont repris ce bourg tant disputé et maintenant inutilement anéanti par les bombes. Le *Sergeant* Mapham a réalisé alors plusieurs clichés, six que nous présentons dans cette double page et d'autres montrant des épaves de char, celle du Cromwell du *Captain* Victory et celles situées à la hauteur de la mairie.

Des camions arrivent par la route de Caen et entrent dans le bourg encore signalé par une pancarte allemande **(photo 1)**. Ce secteur est relativement épargné, on reconnaît au fond le virage situé en haut de la rue Clemenceau. Une photo prise avant la guerre **(photo 2)** permet une bonne comparaison ainsi qu'un détail de la maquette **(photo 3)** montrant le même secteur. Place Richard-Lenoir **(photo 4)**, le *Lieutenant* F. Garstang, journaliste bien connu du *News Chronicle* de Manchester avant la guerre et maintenant *Press Officer* de la *59th Division*, parle avec les deux gendarmes restants de la brigade de Villers, M. Delafontaine et Alphonse Noblet. Une photo du début du XXᵉ siècle **(5)** montre cette place avant guerre. On retrouve le même endroit **(photo 7)** traversé par un officier britannique sur une moto. Sur ces deux clichés on notera l'emblème de la *50th Division* passée par là, les deux « T ». Un peu plus loin **(photo 6)**, derrière la mairie, place du monument aux morts, deux bulldozers dégagent la rue Saint-Germain, côté ouest, direction place de l'église. Au fond à gauche, la carcasse de salle paroissiale et les ruines de l'église. Dans le fond et au centre, l'immeuble Broussieu, route d'Aunay. L'un des deux bulldozers, « Ann Newcastle » **(photo 8)**, comble un trou de bombe. On peut apercevoir, au nord de l'excavation, deux obus de gros calibre de la guerre 14-18 qui servaient de supports aux chaînes entourant le monument aux morts. La rue dégagée est la rue Saint-Germain, partie est, qui mène vers la place du marché. Un peu plus loin encore **(photo 9)**, un convoi allié traverse Villers, venant du haut du bourg. La rue Pasteur est coupée, les véhicules doivent traverser la place de la mairie et rejoindre le boulevard Joffre (on comparera avec la photo 1 page 8 prise au même endroit). C'est également ce chemin que doivent prendre les convois pour rejoindre Epinay-sur-Odon et Evrecy, la place du marché n'étant pas accessible depuis la rue principale. On reconnaît ici deux White Scout Cars. (IWM : **1, 4, 6, 7, 8, 9**. Coll. H. Marie : **2, 5**, photo Vandevorde : **3**.)

Villers-Bocage, August 5th, 1944. British troops recaptured this hotly disputed town now needlessly destroyed by bombs. Sergeant Mapham then took a number of pictures, six that we present in this two-page spread, and others showing the wreckage of tanks, Captain Victory's Cromwell and those on a level with the Town Hall.

*Trucks arrive by the Caen road and enter the town still announced by a German sign **(photo 1)**. This sector was comparatively spared, and the bend at the top of the Rue Clemenceau is recognizable in the background. A photo taken before the war **(photo 2)** offers a good comparison, and also a detail of the scale model **(photo 3)** showing the same sector. In the Place Richard-Lenoir **(photo 4)**, Lieutenant F Garstang, a well-known journalist working for the Manchester News Chronicle before the war and now Press Officer of the 59th Division, chats with the two remaining gendarmes of the Villers brigade, M. Delafontaine and Alphonse Noblet. A photograph dating from the early 20th century **(5)** shows the square prewar. The same spot is shown **(photo 7)** as a British officer passes by on a motorcycle. Notice in these two pictures the insignia of the 50th Division which passed*

through - two "T"s. A little further on *(photo 6)*, behind the Town Hall, in the square where the war memorial stands, two bulldozers clear the Rue Saint-Germain on the western side towards the church square. In the background on the left, what is left of the parish hall and the ruins of the church. In the background and the center, the Broussieu building on the Aunay road. One of the two bulldozers, "Ann Newcastle" *(photo 8)*, fills in a bomb crater. To the north of the excavation, two large shells from the 1914-18 war can be seen which were used to hold up chains around the war memorial. The cleared street is the eastern section of the Rue Saint-Germain, leading to the marketplace. A little further on again *(photo 9)*, an Allied convoy passes through Villers, down from the top of the town. The Rue Pasteur is blocked, and the vehicles have to cross the Town Hall square into the Boulevard Joffre (compare with photograph 1 on page 8 taken at the same spot). Convoys heading for Epinay-sur-Odon and Evrecy also had to come this way, the place of the marketplace being inaccessible from the high street. Two White Scout Cars are recognizable here. (IWM: 1, 4, 6, 7, 8, 9. Coll. H. Marie: 2, 5, photo Vandevorde: 3.)

Evaluation de l'engagement de la *schwere SS-Panzerabteilung 101* en Normandie du point de vue tactique

Assessment of the commitment of schwere SS-Panzerabteilung 101 in Normandy from the tactical viewpoint

par Wolfgang Schneider

Ce texte est une mise au point nécessaire. Elle est proposée par le lieutenant-colonel Wolfgang Schneider, officier instructeur à l'école des blindés de Munster, de l'actuelle Bundeswehr. C'est un praticien qui connaît particulièrement l'utilisation des blindés, qui a rencontré des centaines de vétérans ayant participé au combat des chars Tiger. Auteur de plusieurs ouvrages sur ce blindé, il a vérifié en détail le combat de Villers-Bocage ; son examen critique est du plus haut intérêt.

This text offers a much-needed clarification. It is by Lieutenant-Colonel Wolfgang Schneider, an officer and instructor at the tank school at Munster of what is now the Bundeswehr. He is someone with considerable practical experience of the use of tanks and who has met hundred of veterans who took part in Tiger tank combat. He has written several books on this tank, and has checked out the details of the battle of Villers-Bocage; his critical examination is of the utmost interest.

Lors du débarquement allié en Normandie, aucune unité de chars Tiger n'a pu participer aux combats. Les deux premières compagnies de Tiger de la *schwere SS-Panzerabteilung 101* ne pourront parvenir à l'est de Villers-Bocage que six jours plus tard. Des 14 panzers de l'effectif théorique de la 1re compagnie, seuls huit parviendrons au but et six seulement pour la 2e compagnie. Le reste du bataillon et les panzers de l'autre bataillon, le 102, parviendront en ordre dispersé dans les semaines suivantes. Il n'est donc tout d'abord pas question de penser à un engagement massif de tout le bataillon. Bien plus, sur place, la situation était beaucoup plus sérieuse que ce qui était prévu. C'est pourquoi les panzers vont être engagés par petits groupes, sans connaître la situation au départ, et en ordre dispersé. Il n'est donc pas étonnant que le premier engagement des Tiger en Normandie ait eu lieu par hasard.

Le mystère de Villers-Bocage

Presque tous ceux qui s'intéressent à l'engagement des chars Tiger connaissent les nombreux récits concernant l'engagement du « chef de char au score le plus élevé », l'*Obersturmführer* Wittmann. Malheureusement, la plupart d'entre eux ne savent pas que presque tous les récits sont complètement faux ! Tous les témoignages qui ont suivi doivent être passés au crible de l'étude critique. Enfin, depuis un certain temps, il est devenu possible de vérifier sur place ces faits jusque dans le moindre détail.

Les événements du 12 juin 1944

La *schwere SS-Panzerabteilung 101* devait être tenue disponible sur l'aile gauche du *I. SS-Panzerkorps* et, dans cette optique, un secteur de concentration lui avait été attribué juste à l'est de Villers-Bocage. Ces dispositions vont rencontrer les préparatifs des Britanniques et des Canadiens pour l'Opération

At the time of the Allied landings in Normandy, there was no Tiger tank unit at hand to take part in the battle. The first two Tiger companies of schwere SS-Panzerabteilung 101 managed to arrive east of Villers-Bocage only six days later. Of the 1st Company's theoretical strength of 14 panzers, only eight reached their destination and only six of the 2nd Company's did so. The remainder of the battalion and the panzers of the other battalion, the 102nd, arrived in extended order during the following weeks. So first of all, there is no question of thinking that the whole battalion was committed en masse. Much more importantly, the situation when they got there proved much more serious than expected. This is why the panzers were engaged in small groups, without knowing the situation at the outset, and in extended order. It therefore comes as no surprise that the first engagement with the Tigers in Normandy should have been a chance occurrence.

The Villers-Bocage mystery

Just about anyone interested in Tiger tank battles will know of many accounts concerning the action of the "tank commander with the highest number of kills», Obersturmführer Wittmann. Unfortunately, most of them do not know that almost all these accounts are completely untrue! All subsequent eye-witness accounts are subject to critical scrutiny. Lastly, for a while now, it has become possible to check these facts on the spot down to the last detail.

The events of June 12th 1944

The schwere SS-Panzerabteilung 101 was to be held available on the left flank of I SS-Panzerkorps and, accordingly, it had been allocated a mustering sector just east of Villers-Bocage. These dispositions clashed head-on with the British and Canadian preparations for Operation Perch. The purpose of this operation was to close round Caen from the north-

154

« Perch ». Celle-ci avait pour but d'envelopper Caen par le nord-ouest et de prendre pied sur le cours de l'Odon. Dès la nuit, un violent tir d'artillerie britannique s'était abattu sur le secteur. Les panzers, qui étaient en train d'arriver, durent changer trois fois de position. La 2e compagnie rejoint alors, avec six panzers, un chemin creux situé à environ 100 mètres au sud de la Route Nationale 175, en dessous de celle-ci, sur les versants méridionaux de la hauteur de Montbroq. Ce chemin est situé à deux kilomètres à l'est de Villers-Bocage et à environ 500 mètres au sud-ouest de la cote 213. Les huit panzers de la 1re compagnie se trouvent encore un peu plus loin au nord-est, dans un secteur d'attente de l'autre côté de la Route Nationale.

Le chemin creux déjà cité procure une bonne protection mais il ne permet pas de faire dégager sur le côté tous les panzers ensemble. En outre, l'un des panzers (celui de l'*Oberscharführer* Lötzsch) est victime d'un incident à une chenille et n'est pas apte au combat. Devant se trouve le panzer de l'*Unterscharführer* Stief qui subit un incident de moteur !

Lors de cette mise à l'abri, le commandant de compagnie, l'*Obersturmführer* Wittmann est encore en route à cause de missions de liaison et pour rameuter d'autres panzers, il rejoindra dans la nuit. Après les progressions nocturnes qui ont duré plusieurs jours, les hommes sont totalement épuisés. Pour chaque panzer, à tour de rôle, un homme est en observation, les autres se reposent. Un contrôle technique est hautement nécessaire, il est prévu pour le lendemain.

Depuis 5 heures, parmi d'autres unités, la *7th Armoured Division (« Desert Rats »)* est engagée dans le cadre de l'Opération « Perch ».

Rappelons maintenant les événements :

Pas déployé (!) mais avançant en file sur la route Caen/Villers-Bocage, le groupement de la 22e Brigade blindée arrive jusqu'à proximité de la cote 213 et il s'arrête là. Sans s'en rendre compte, ce groupement s'est faufilé dans une brèche existant entre la *352. Infanterie-Division* et la *Panzer-Lehr-Division*.

Soudain, dans le chemin creux déjà évoqué, le chef de *2./SS-101* (Wittmann) est alerté : des blindés, probablement britanniques, avancent sur la route en direction de l'est. Il saute dans le premier panzer, envoie le commandant (Stief) vers les autres panzers pour les mettre en état d'alerte. Le conducteur reçoit de Wittmann l'ordre de démarrer. Au bout de 20 à 30 mètres, l'équipage signale à son chef que le moteur ne marche pas bien. Wittmann sort et court vers le panzer suivant qui est juste en train de sortir du chemin. C'est le « 222 » de l'*Unterscharführer* Sowa. Il le fait sortir de son panzer et le prend pour lui.

L'*Obersturmführer* Wittmann, dont la compagnie n'est pas encore en ordre de combat, attaque l'élément de pointe des Britanniques *(A Squadron, 4th City of London Yeomanry* et éléments du *1st Battalion Rifle Brigade)*. Tout d'abord, en direction de Caen, il détruit un **Cromwell** puis un **Firefly** du *A Squadron* qui se trouvait déjà sur la cote 213. Ensuite, il roule parallèlement à la route menant à Villers-Bocage. A bout portant, il tire dans la masse de la *1st Rifle Brigade* (13 véhicules semi-chenillés M3, trois chars de reconnaissance Stuart, deux chars Sherman d'observation d'artillerie, le Scout Car Daimler de l'*Intelligence Officer* et le M3 du médecin militaire (!), ainsi qu'une douzaine de Bren et Lloyd Carriers (dont la batterie antichar), en roulant tout au long de ces engins. A la limite de la localité, il détruit **trois** des quatre **Cromwells** du groupe de commandement du régiment, le *4th CLY*. A l'issue, il entre tout seul (!) dans Villers-Bocage, localité occupée par l'ennemi,

west and to gain a foothold along the Odon River. As soon as night fell, fierce British artillery fire rained down on the sector. The panzers, as they arrived, had to change position three times. The 2nd Company, with six panzers, then reached a sunken lane some 100 meters south of Highway 175 below them, on the southern slopes of the high ground at Montbroq. This road lay two kilometers east of Villers-Bocage and about 500 meters south-west of Hill 213. The eight 1st Company panzers were a little further again to the north-east, in a standby position on the other side of the highway.

The aforementioned sunken lane offered good protection but did not allow the panzers to pull clear sideways all together. Moreover, one of the panzers (Oberscharführer Lötzsch's) had problems with one of its tracks and was not battleworthy. And in front was Unterscharführer Stief's panzer, which encountered engine trouble!

At the time these tanks took shelter, the company commander, Obersturmführer Wittmann, was still on his way liaising and gathering up other panzers, and he caught up during the night. The men were completely exhausted after advancing under cover of night for several days. For each panzer, one man kept lookout in turns while the others rested. A vehicle inspection was overdue, and was scheduled to take place the following day.

For 5 hours, the 7th Armored Division («Desert Rats»), among other units, had been engaged as part of Operation Perch.

Let us now recall what happened:

Not deployed (!) but advancing in line along the Caen/Villers-Bocage road, the 22nd Armored Brigade group came up to Hill 213 where it halted. Without realizing, this group had slipped through a breach between the 352. Infantry-division and the Panzer-Lehr-Division.

Suddenly, in the sunken lane already mentioned, the commander of 2./ss-101 (Wittmann) was alerted: tanks, probably British, were advancing eastwards along the road. He jumped into the first panzer, sending the commander (Stief) off to alert the other panzers. The driver was ordered to get going by Wittmann. After 20 or 30 meters, the crew announced to its commander that the engine was not working properly. Wittmann climbed out and ran up to the next panzer as it came out of the lane. This was Unterscharführer Sowa's tank «222». Wittmann got him to get out of his panzer and took over himself.

Obersturmführer Wittmann, whose company was not yet in battle order, attacked the leading British element (A Squadron, 4th City of London Yeomanry and elements of the 1st Battalion Rifle Brigade). Facing in the Caen direction, he started by destroying a **Cromwell** then a **Firefly** of A Squadron which was already on Hill 213. He then drove down alongside the road leading to Villers-Bocage. At point-blank range, he fired at the massed 1st Rifle Brigade (13 M3 half-tracks, three Stuart reconnaissance tanks, two artillery observation Shermans, the Intelligence Officer's Daimler Scout Car and the medical officer's M3 (!), also a dozen Bren and Lloyd Carriers (including the anti-tank battery), while driving alongside these vehicles. At the town exit, he destroyed **three** out of four **Cromwells** of the regiment's command group, 4th CLY. At the end, he entered alone (!) Villers-Bocage, a locality occupied by the enemy, pursued by the fourth Cromwell, two of whose shells (fired from 50 meters away!) skidded off his armor. A few hundred meters further on, the panzer was knocked out by an anti-tank shell in the suspension and tracks. The crew baled out. He reported back to Panzer-Lehr-Division headquarters at Orbois-Sermentot.

poursuivi par le quatrième Cromwell dont deux obus (tirés à 50 mètres de distance !) ripent sur le blindage. Quelques centaines de mètres plus loin, le panzer est immobilisé par un obus antichar dans le train de roulement. L'équipage abandonne le panzer. Il rejoint le poste de commandement de la *Panzer-Lehr-Division* à Orbois-Sermentot.

Les trois autres chars opérationnels de la 2ᵉ compagnie, après en avoir reçu l'ordre de leur chef, vont se mettre en position à l'est de Villers-Bocage (au sud de la route) et détruisent **deux** autres **Cromwell** (Sowa) et **trois** chars **Sherman** (Oberscharführer Brandt), environ 230 prisonniers se rendent.

A partir de **8 heures**, la 1ʳᵉ compagnie (*Hauptsturmführer* Möbius) attaque avec huit *Tiger*, également le long de la N 175 en direction de Villers-Bocage. **Cinq** chars **Cromwell** se trouvant encore là sont abandonnés intacts par leurs équipages qui s'enfuient. Plusieurs Panzer IV de la *Panzer-Lehr-Division*, qui se trouvaient à Parfouru-sur-Odon, prennent aussi part à l'attaque. Deux chars *Tiger* et un Panzer IV avancent le long de la rue principale (rue

Pasteur). Au travers de deux fenêtres d'angle, le Tiger « 112 » (*Oberscharfürer* Ernst) est détruit par un Firefly du *B Squadron*, par l'arrière alors qu'il avançait. Lors d'une autre avance, après un changement de position, le Panzer IV est détruit par une pièce anti-char. Le Tiger « 121 » (*Untersturmführer* Lukasius), qui roulait devant, est détruit par l'arrière par un Firefly. Plus tard, les Anglais incendient volontairement ces panzers. Cinq autres Tiger avancent vers le sud par les rues. Un Tiger est détruit par un canon anti-

After receiving orders from their commander, three more of the 2nd Company's operational tanks were set in position east of Villers-Bocage (south of the road) and destroyed a further **two Cromwells** *(Sowa) and* **three Shermans** *(Oberscharführer Brandt), with some 230 prisoners surrendering.*

From **08.00 hours***, the 1st Company (Hauptsturmführer Möbius) attacked with eight Tigers, also along Highway 175 towards Villers-Bocage.* **Five Cromwells** *were still there after being abandoned intact by their fleeing crews. Several of the Panzer-Lehr-Division's Panzer IVs at Parfouru-sur-Odon also took part in the attack. Two Tigers and a Panzer IV advanced along the main street (Rue Pasteur). Through two corner windows, Tiger «112» (Oberscharführer Ernst) was destroyed by a B Squadron Firefly from the rear as it moved forward. During another advance, after a change of position, the Panzer IV was destroyed by an anti-tank gun. Tiger «121» (Untersturmführer Lukasius), which was in front, was destroyed from the rear by a Firefly. Later, the British deiberately set fire to these panzers. Five other Tigers advanced southwards through the streets. One Tiger was destroyed by an anti-tank gun in the Rue Emile Samson. Two other Tigers were knocked out.*

Tiger «132» (Unterscharführer Wendt) stayed right on the edge of the town. During the night, «132» was on Hill 213, and four of the 1st Company's Tigers were in position south of Villers-Bocage. The 2nd Company settled back in the sunken lane running alongside Highway 175.

The battalion's losses for the day came to three NCOs and seven troopers killed. The 1st Company lost three Tigers. Overall, the British lost 26 tanks, 14 M3s, eight Bren Carriers and eight Lloyd Carriers.

Meanwhile, the 3rd Company reached Falaise.

The following day, the general commanding I SS-Panzerkorps, SS-Obergruppenführer Dietrich, nominated Wittmann for a Knight's Cross of the Iron Cross with Oakleaves and Swords. This was awarded on June 22nd 1944, along with promotion to the rank of Hauptsturmführer.

Some of the terms of the nomination for this award are noteworthy:

- «the Wittmann company... was... near Hill 213... ready for combat»

- «Wittmann could no longer issue orders to his men who were some distance away...»

- (at Villers-Bocage, after having to evacuate, being immobilized): «he further destroyed... all the vehicles within range...»

- (after reaching Panzer-Lehr-Division headquarters): «he set off again for Villers-Bocage... engaged (the 1./SS 101)».

- the number of tanks destroyed is noted as being 25.

Let us review what actually happened.

In examining these events, they should not be taken lightly! We have to think of the following:

Wittmann was well-liked by his subordinates and was appreciated by his superiors. In the Balkans campaign and particularly on the Eastern front, he had fought bravely and destroyed many enemy tanks with his Sturmgeschütz and later his panzer.

On the morning of June 12th 1944, the situation was not at all clear.

The decision to attack an enemy about to carry out a decisive breakout had to be taken.

Wittmann's own action was energetic and courageous.

Cette photo a été prise probablement le 14 juin, après les combats, elle a été prise en même temps que la photo de la page 14. Nous voyons le même char de la 2ᵉ compagnie. Il s'agit du « 212 » ou « 232 » (de la compagnie commandée par Wittmann) du lieutenant Hantusch (en manteau à droite sur le char) qui a pris en remorque le « 231 » qui est en panne. Ce dernier Tiger a donné sa tourelle vers l'arrière pour éviter que son canon ne vienne heurter par accident la tourelle du char tracteur. Comme nous le voyons sur ce reportage, ce cliché a été pris sur la route N 176 entre la cote 213 et Villers. (BA/738/275/20a.)

This photo was probably taken on June 14th, after the battle, it was taken at the same time as the photo on page 14. We see the same 2nd Company tank. It is tank "212" or "232" (belonging to the company commanded by Wittmann) of Lieutenant Hantusch (in the coat on the right on the tank) which is towing tank "231" which has broken down. The latter Tiger has turned its turret to the rear so as to prevent the gun accidentally hitting the turret of the towing tank. As we see in this reportage, this picture was taken on Highway N 176 between Point 213 and Villers. (BA/738/275/ 20a.)

char dans la rue Emile Samson. Deux autres Tiger sont immobilisés.

Le Tiger « 132 » (*Unterscharführer* Wendt) est resté à la limite de la localité. Dans la nuit, le « 132 » est sur la cote 213, quatre Tiger de la 1re compagnie sont en position au sud de Villers-Bocage. La 2e compagnie s'installe de nouveau dans le chemin creux parallèle à la N 175.

Pour cette journée, les pertes du bataillon s'élèvent à trois sous-officiers et sept hommes de troupe tués. Trois Tiger de la 1re compagnie ont été perdus. Au total, les Britanniques ont perdu 26 chars, 14 M3, huit Bren Carriers et huit Lloyd Carriers.

Entre-temps, la 3e compagnie atteint Falaise.

Le lendemain, le général commandant le *I. SS-Panzerkorps*, le *SS-Obergruppenführer* Dietrich, propose Wittmann pour la Croix de Chevalier de Croix de fer avec feuilles de chêne et épées. L'attribution, ainsi que la nomination au grade de *Hauptsturmführer* aura lieu le 22 juin 1944.

Quelques termes de la proposition d'attribution méritent d'être notés :

- « La compagnie Wittmann… se trouvait … près de la cote 213… prête au combat. »
- « Wittmann ne pouvait plus donner d'ordre à ses hommes qui étaient éloignés… »
- (à Villers-Bocage, après qu'il ait dû évacuer, ayant été immobilisé) : « il détruisit… encore tous les véhicules se trouvant à portée… »
- (après avoir atteint le poste de commandement de la *Panzer-Lehr-Division*) : « il repartit sur Villers-Bocage… l'engagea (la 1./SS 101) ».
- Le nombre des chars détruits est noté pour 25.

Voyons les faits.

En examinant ces événements, on ne doit pas les prendre à la légère ! Il faut réfléchir à ce qui suit :

Wittmann était aimé de ses subalternes et apprécié de ses supérieurs. Dans la campagne des Balkans et particulièrement sur le Front de l'Est, il avait combattu courageusement et avait détruit de nombreux blindés ennemis avec son Sturmgeschütz puis avec son panzer.

Au matin du 12 juin 1944, la situation est bien loin d'être claire.

La décision d'attaquer un ennemi sur le point de mener une percée décisive était exigée.

L'engagement personnel de Wittmann était énergique et courageux.

Cependant, s'élèvent un bon nombre de questions critiques. Il est facile de juger du contenu de la proposition d'attribution rédigée par Sepp Dietrich. Toutes les affirmations citées plus haut sont carrément fausses. La compagnie de Wittmann n'était pas prête au combat (voir plus loin) ; les panzers ne se trouvaient pas « éloignés » ; il y avait assez de temps pour donner les ordres ; après l'impact dans le train de roulement, l'équipage s'est enfui, il ne pouvait pas s'attaquer à d'autres cibles ; pendant l'attaque de la compagnie de Möbius, Wittmann n'avait pas encore atteint le poste de commandement de la *Panzer-Lehr-Division* ; il n'a pas pris part à l'attaque sur Villers-Bocage. Le lecteur peut assez facilement calculer le nombre de coups au but ; il y en a **dix**. Si on veut aussi prendre en compte le char d'observation d'artillerie « armé » d'un canon en bois et les légers chars Stuart, on n'arrive quand même pas à 25. Il reste l'allusion au fait que « la menace d'un grave danger » a été écartée par l'engagement déterminé de Wittmann. Normalement, les différents degrés, de la Croix de chevalier sont attribués pour des exploits individuels « déterminants pour le combat »,

However, this raises a whole string of critical questions. It is easy to judge the contents of the award nomination drafted by Sepp Dietrich. All the aforementioned assertions are downright untrue. Wittmann's company was not ready for battle (see below); the panzers were not «some distance away»; there was enough time to issue orders; after being hit in the tracks, the crew ran away, it could not attack any other targets; during the attack by Möbius's company, Wittmann had not yet reached Panzer-Lehr-Division headquarters; he took no part in the attack on Villers-Bocage. The reader can quite easily work out how many shots were on target: **ten**. And even if we take into account the artillery observation tank «armed» with a wooden gun and the light Stuart tanks, the number still does not come to 25. This leaves the reference to the fact that «the threat of serious danger» was removed by Wittmann's determined action. Normally speaking, the various grades of the Knight's Cross were awarded for individual feats that were «decisive for the battle», and not specially for a high score. I will come back to this point later.

After the event, it is difficult to know whether there was enough time to wait until the company's other (three) panzers were in battleworthy condition. The fact that Sowa's panzer was able to follow quickly indicates that it would have taken just a few more minutes.

If the British had not put themselves in such a position as not to be ready for battle and had not been so lax, Wittmann could certainly have destroyed several enemy vehicles, but obviously, he would have been stopped, at least by a shell in his tracks. The combat distances were fairly close, so that the Cromwell tanks, usually deprived of success, might well have hit the Tiger's flanks. This judgement presupposes that Wittmann could have fallen back into an observation position and waited for his panzers to join him, and then attacked with a much larger fighting force with the means of covering each other. Given that the British column was at a halt, obviously there was still some little time to issue orders (by radio, for example).

Even if several enemy tanks had advanced in the Caen direction from Hill 213, they would have passed the 1st Company by.

The single-handed attack of an enemy-occupied town is not debatable. But that does not explain the point of it all to the reader.

Neither does the attack on Villers-Bocage which followed, by panzers, elements of Panzer-Lehr-Division and of 1./SS-101 almost at full strength, with no supporting infantry, follow the fundamental principle whereby tanks are supposed as far as possible to bypass any towns or face quick losses.

In short, it can be established that the action by the two Tiger companies (partly with elements of the Panzer-Lehr-Division) averted the threat of a decisive British breakout. However, the critical remarks regarding the conduct of the operation still stand, as does the question of whether Wittmann's personal action was decisive in this operation.

In the days that followed, 101 took part in further battles for Cahagnes and with the 3rd Company, which arrived at Evrecy later on, before the British could first launch their attack towards the Odon River.

General analysis

The action of the 1st and 2nd Companies of the schwere SS-Panzerabteilung 101 was everything but awe-inspiring. SS-Panzerkorps propaganda then gave

pas particulièrement pour un score important. J'y reviendrai plus tard.

Après coup, il est difficile de savoir s'il y avait assez de temps pour attendre que les autres (trois) panzers de la compagnie soient aptes à combattre. Le fait que le panzer de Sowa a pu suivre rapidement indique qu'il aurait fallu quelques minutes de plus.

Si les Britanniques ne s'étaient pas mis en position de ne pas être aptes au combat et ne s'étaient pas comportés avec insouciance, Wittmann aurait pu certainement détruire plusieurs engins adverses mais, de toute évidence, il aurait été stoppé, au moins par un obus dans son train de roulement. Les distances de combat étaient limitées si bien que les chars Cromwell, habituellement dépourvus de succès, auraient pu atteindre les flancs du Tiger. Un pareil jugement supposerait que Wittmann aurait pu se permettre d'aller en position d'observation, de se faire rejoindre par ses panzers et, ensuite, de s'engager avec une force combattante bien plus importante et avec les moyens de se couvrir mutuellement. Etant donné que la colonne britannique était arrêtée,il y avait, de toute évidence, encore un peu de temps pour transmettre les ordres (par radio, par exemple).

Même si plusieurs blindés ennemis s'étaient avancés en direction de Caen, depuis la cote 213, ils seraient passés à côté de la 1re compagnie.

L'attaque solitaire d'une localité occupée par l'ennemi n'est pas discutable. Elle n'explique pas au lecteur quel était son but.

L'attaque qui a suivi, de panzers, d'éléments de la *Panzer-Lehr-Division* et de la *1./SS-101* presque complète sur Villers-Bocage, sans infanterie de soutien, ne correspond pas non plus aux principes de base selon lesquels les localités doivent être le plus possible contournées par les panzers. Sinon les pertes surviennent promptement.

En résumé, on peut établir que l'engagement des deux compagnies de Tiger (en partie avec des éléments de la *Panzer-Lehr-Division*) a éliminé la menace d'une percée britannique décisive. Restent les remarques critiques sur la conduite de l'opération, reste aussi la question de savoir si l'action personnelle de Wittmann a été décisive dans cette opération.

Les jours suivants, la *101* participera à d'autres combats pour Cahagnes et avec la 3e compagnie qui arrivera à Evrecy, avant que les Britanniques puissent d'abord lancer leur attaque en direction de l'Odon.

Analyse générale

L'engagement des 1re et 2e compagnies de la *schwere SS-Panzerabteilung 101* fut tout sauf grandiose. La propagande du *SS-Panzerkorps* l'a ensuite falsifiée de façon déterminante. Comment peut-on l'expliquer ?

D'abord, il faut savoir que la Waffen-SS - contrairement à la Wehrmacht - ne disposait d'une arme blindée expérimentée. A la lumière des brillants exploits des « vieilles » divisions de panzers (de la Wehrmacht), la Waffen-SS ne pouvait espérer des succès analogues. A la rigueur, l'engagement du *II. SS-Panzerkorps* en Russie, dans le secteur Sud, lors de l'Opération « Zitadelle » en juillet 1943, impose le respect. Par conséquent, Sepp Dietrich va tout faire pour fabriquer un héros avec l'*Obersturmführer* Wittmann.

Sur le Front de l'Est, l'attribution de la Croix de chevalier, ainsi que celle des Feuilles de chêne, était liée à des « victoires ». Ceci est d'autant plus étonnant que, dans l'arme blindée de la Wehrmacht,comme dans les chasseurs de chars, de nombreux soldats ont eu des scores bien plus élevés !

a decisively misleading account of it. How can this be explained?

First, we have to remember that – unlike the Wehrmacht – the Waffen-SS did not have a experienced tank arm. Compared with the brilliant exploits of the «old» (Wehrmacht) panzer divisions, the Waffen-SS could not hope for similar successes. At a pinch, the II SS-Panzerkorps's action in Russia, in the southern sector, during Operation Zitadelle in July 1943, commands respect. So with Obersturmführer Wittmann, Sepp Dietrich tried his utmost to manufacture a hero.

On the Eastern front, the Knight's Cross was awarded for «kills», as were Oakleaves. This is all the more astonishing as, both in the Wehrmacht tank arm and among the tank destroyers, many soldiers had much higher scores!

As we know, the legend of the "Second World War tank commander with the highest number of kills" has been kept up to this day. This judgement is completely wrong, in terms both of the actual score and the tactics employed!

A competent tank company commander does not accumulate so many serious mistakes as Wittmann made.

1. The company commander knows exactly the technical status of all his panzers. He does not place a vehicle which has engine trouble at the head of a stationary column; the risk of blocking all the other panzers is just too great.

2. A sunken lane can afford some protection, but it is not a suitable standby base when the enemy's position is unknown. The major concern is for the panzers to be disposed in such a way as not to hamper their freedom of movement.

3. In a concentration sector, all battleworthy panzers are placed in alert positions. These positions and the roads leading to them have to be carefully reconnoitered. When placed on the alert or when the concentration sector is attacked, upon orders, the crews join them individually. But when the morning alert came, none of the company's tank commanders knew what to do.

4. After the first intelligence of the enemy, the company should have been placed on the alert at once with orders to ready itself for battle as quickly as possible. Invaluable time would certainly have been saved and it could have regrouped before engaging the enemy. Such action would have been more effective.

5. As the enemy's position was not clear, it was all the more necessary to work out a well-conceived counter-attack. On the basis of observation relying on an overall view of the situation, valuable intelligence could have been obtained before engaging. Such overhastiness was uncalled for, as the next company (1./SS-101) was in a favorable position further north-east, and it could have attacked the enemy forces when they advanced.

6. The hasty, single-handed attack on the large and powerful British force may seem brave, but it goes against all the rules (no centre of gravity, no concentration of forces, importance of the moment of surprise). The action that followed by the bulk of the 2nd Company and by Möbius 1st Company came up against an enemy who had gone onto the defensive.

7. The carefree advance of a single panzer into a town occupied by the enemy is pure folly.

Thoughtlessness of this kind was to cost the "tank commander with the highest number of kills" his life on August 9th 1944, near Gaumesnil, during an attack casually launched in open country with an exposed flank.

On sait que la légende du « chef de char de la Seconde Guerre mondiale ayant le score le plus élevé » s'est maintenue jusqu'à aujourd'hui. Ce jugement est complément faux, sur le plan du score obtenu comme sur celui du comportement tactique !

Un chef compétent d'une compagnie de chars ne commet pas une telle accumulation de fautes graves comme celles commises par Wittmann.

1. Le commandant de compagnie connaît précisément l'état technique de tous ses panzers. Il ne place pas un véhicule qui a un problème de moteur au début d'une colonne qui est arrêtée ; le danger de bloquer tous les autres panzers est trop grand.

2. Un chemin creux peut procurer une protection mais il ne convient pas comme base d'attente quand on ne connaît pas la position ennemie. Le but essentiel est que les panzers soient disposés de façon à ce que leur liberté de mouvement ne puisse être entravée.

3. Dans un secteur de concentration tous les panzers (aptes au combat) sont attribués à des positions d'alerte. Celles-ci et les chemins qui y mènent doivent être soigneusement reconnus. Lors de la mise en état d'alerte ou lors d'une attaque contre le secteur de concentration, sur ordre, les équipages les rejoignent individuellement. Lors de la mise en alerte matinale, aucun des chefs de char de la compagnie ne savait quoi faire.

4. Après les premières informations concernant l'ennemi, la compagnie aurait dû aussitôt être mise en état d'alerte avec l'ordre de se mettre aussi vite que possible en ordre de combat. Un temps précieux aurait certainement été épargné et un engagement regroupé aurait été favorisé. Un tel engagement en bloc aurait été plus efficace.

5. Comme la situation de l'ennemi n'était pas claire, il fallait d'autant plus réfléchir à une contre-attaque bien conçue. Sur la base d'une observation s'appuyant sur une vue générale de la situation, des informations valables pour l'engagement auraient pu être obtenues. Une hâte excessive n'était pas de mise car la compagnie voisine (la *1./SS-101*) se trouvait en position favorable, plus loin au nord-est, et elle aurait pu s'attaquer à des forces ennemies qui se seraient avancées.

6. L'engagement hâtif et solitaire contre les Anglais nombreux et puissants peut paraître courageux mais il va à l'encontre de tous les principes de base (absence d'un centre de gravité, pas de concentration des forces, importance du moment de surprise). Les engagements suivants du gros de la 2e compagnie ainsi que celui de la compagnie Möbius (la 1re) vont se heurter à un ennemi passé à la défensive.

7. L'avance insouciante d'un seul panzer dans une localité occupée par l'ennemi est insensée.

Une légèreté semblable coûtera la vie, le 9 août 1944, au « commandant de char au score le plus élevé », près de Gaumesnil, lors d'une attaque menée de manière insouciante sur une surface dégagée avec un flanc exposé.

Bibliographie
Bibliography

- J.L. Cloudsley-Thompson, *The Royal Arm. Corps Journal*, avril 1959, *Arromanches to Bucéels*, octobre 1961, *Sharpshooters at Villers-Bocage, Escape to Villers-Bocage. Back to the beaches.*
- Patrick Delaforce, *Churchill's desert rats*, 1994, Alan Sutton Publishing Ltd Gloucestershire.
- Henry Maule, *Caen, The brutal battle and breakout for Normandy*, London, 1976. *7 Armoured Division History*, *14/15 June 44.*
- Lt General Forty (RTR), *Deserts rats at war*, II (After the battle magazine).
- Christopher Milner, *Villers-Bocage, 13/14th June 44.*
- Lt-Colonel Desmond Gordon, *The action by the 1/7 Bn, the Queen R.R. at Villers-Bocage.*
- Colonel Goulburn, *8th Hussar*, rapport privé, 14 juin 44.
- Max Hastings, *Overlord*, London, 1985.
- Alexander McKee, *La bataille de Caen*, Paris, 1965.
- Everslay Belfield & H. Essame, *The battle for Normandy*, London, 1982.
- John Keegan, *Six armies in Normandy.*
- W.D. Allen and R.F.H. Cawston, *Carpiquet Bound*, 1997.
- Gordon Williamson, *Aces of the Reich*, 1989.
- Daniel Taylor, *Villers-Bocage through the lens*, After the battle.
- Patrick Agte et Georges Bernage, *Tiger*, Editions Heimdal.
- Jean-Pierre Benamou, *Opération Perch*, 39/45 Magazine, n° 41, Editions Heimdal.
- Georges Bernage, *Album Mémorial Normandie*, Bayeux, 1983.
- Yves Buffetaux, *Le choc des blindés*, Paris, 1991.
- Eric Lefèvre, *Les Panzers*, Bayeux, 1978.

Témoignages

Allen Walter D. Anne Raymond. Bernouis Pierre. Bertou Pierre. Boyers R.H. Mme Boiroux. M. Brière. Clermont-Tonnerre Stan. Cloudsley-Thompson L.C. Chevalier Henriette. Currie H.J. M. Delafontaine. Dyas Patrick. Fisher Jack. Forrester Mickael. Mme Fromont. Grant L.C. Gell Harry. Gilmann Neville. Guilbert Joseph Hammacot Donald. Key George. Loisel André. Lepoil Paul. Lokwood Stan. Loir André. Loir Auguste. Lebas Marcel. Marie Jean. Marie Andr. Milner Christopher. Moore Bob. Mulot Roger. Ozenne Marcel. Pierce Charles. Pichard André. Queruel Pierre. Roger René. Robine Maurice. Mme Julien THorel. Watt Douglas. Weight K.K. Zekar François.

Achevé d'imprimer sur les presses d'Arti Grafiche à Pomezia (Italie)
pour le compte des Editions Heimdal à Damigny/Normandie (France),
Georges Bernage, éditeur
Seconde édition – décembre 2010